ENVIRONMENTAL MANAGEMENT OF MINERAL WASTES

NATO ADVANCED STUDY INSTITUTES SERIES

Proceedings of the Advanced Study Institute Programme, which aims at the dissemination of advanced knowledge and the formation of contacts among scientists from different countries.

The series is published by an international board of publishers in conjunction with NATO Scientific Affairs Division

A	Life Sciences	Plenum Publishing Corporation
B	Physics	London and New York
C	Mathematical and Physical Sciences	D. Reidel Publishing Company Dordrecht and Boston
D	Behavioural and Social Sciences	Sijthoff & Noordhoff International Publishers B.V. Alphen aan den Rijn, The Netherlands and Winchester, Mass., USA
E	Applied Science	

Series E: Applied Science — No. 7

ENVIRONMENTAL MANAGEMENT OF MINERAL WASTES

edited by

GORDON T. GOODMAN, D. Phil.

professor of applied biology
University of London, U.K.

and

MICHAEL J. CHADWICK, D. Phil.

lecturer in biology
University of York, U.K.

SIJTHOFF & NOORDHOFF   1978
Alphen aan den Rijn — The Netherlands

Proceedings of the NATO Advanced Study Institute
on Waste Disposal and the Renewal and Management
of Degraded Environments
Yorkshire, Cardiff, Cornwall (UK), and Ruhr (BRD)
July 13-28, 1973

ISBN-13: 978-94-009-9919-0   e-ISBN-13: 978-94-009-9917-6
DOI: 10.1007/978-94-009-9917-6

PREFACE

G.T. Goodman and M.J. Chadwick

With the increasing pressure worldwide on mineral and fossil fuel resources, there is a growing awareness of the need to utilize all available workable deposits as thoroughly and efficiently as possible. This inevitably means disturbance, often of high quality environments. Adverse environmental impacts on land, water and air, and frequently to human health and biological resources, have been taking place for many years in different parts of the world. In total cycle of mineral extraction, transportation use and disposal involves all compartments of the biosphere. The need to minimise these impacts by wise environmental management has never been more important than it is today. There is great interest, therefore, in pooling our understanding of the scientific processes involved and our technical capability of dealing with these problems.

Scientists and engineers who are expert in handling different types of mineral process have much to learn from one another particularly when codes of practice from different countries in the field of land reclamation are compared. In order to illustrate the scientific and technical expertise required in the necessary detail it was deemed essential in this Advanced Study Institute to concentrate on specific case studies utilizing a range of actual sites. The examples of Coal, China Clay and Nonferrous Metals were chosen as displayed in the United Kingdom and Germany. The availability of lecturers from Canada and the United States of America enabled comparisons to be made with North American experience.

A detailed discussion on the range of problems posed by each type of mineral was followed by site visits to compare attitudes and exchange views in the field as a preparation for the presentation of the final review lectures. The main review lectures are

reprinted here as papers. First, there are a group of three
contributions dealing with the general problems of management
for mineral extraction in North America, the United Kingdom and
the Ruhr District of the Federal Republic of Germany. This is
followed by a series of papers specifically reviewing the
individual problems of coal, china clay and non-ferrous metals.

Taken as a whole the papers represent an up to date approach
to the land reclamation aspects of modern waste management which
will be a continuing concern in our increasingly environmental
society.

ACKNOWLEDGMENTS

The result of holding an Advanced Study Institute in four different locations from north eastern England through to south western England, Wales and the Federal Republic of Germany is that many people are involved in organisation. In particular, thanks are due to officials of the National Coal Board in Yorkshire; the Welsh Office in Cardiff and the Welsh Water Authority; staff of English China Clays Ltd. in Cornwall; officials of the Ruhr Regional Planning Authority (Siedlungsverband Ruhrkohlenbezirk) for making local arrangements. These, together with reporters and chairmen of the discussion sessions, and the lecturers who gave the papers that are presented here made ten days in July, 1973 a most valuable educational event.

Thanks are also due to Mrs I. Fletcher, Mrs L. Wainhouse, Miss M. Conway and Miss J. Atkins for help in organisation and with typing. Mrs S. Sparrow drew most of the diagrams. Hospitality was provided by the University College Swansea, English China Clays Ltd. and Siedlungsverband Ruhrkohlenbezirk, for which we are most grateful.

TABLE OF CONTENTS

X

APPROACHES TO LAND RECLAMATION IN BRITAIN AND NORTH AMERICA

J.V. Thirgood

University of British Columbia,
Canada

INTRODUCTION

Most reclamation research is preoccupied with the basic techniques required to solve the problems posed by plant growth on
hostile media. This nearly always leaves unformulated a number
of basic assumptions concerning the landscape goal of much of the
reclamation activity. In Britain the implicit goal is the re-
production of the familiar well-managed landscape; in North
America the close historical association with wilderness leads
to the very different reclamation ideal of the high priest landscape. Most of the differences between the British and American
reclamation philosophies arise out of four causes - time, scale,
geography and the pattern of society.

Britain, with one thirtieth of the land surface of the
United States, but with one third the population, is in a very
different situation from that of the North American continent,
where Canada, with a population of 22 million, occupies an area
greater than that of the whole of Europe and where even the industrialised eastern United States has land not under husbandry.
The north American is greatly impressed when he sees that most
land in the United Kingdom is being managed and producing goods
and services. Britain, with 621 people per square mile, the U.S.
with 60 and Canada with 5.5 are obviously in very different
categories. These British urban or intensely farmed landscapes
saw the birth of the industrial revolution. North American reclamation concerns are often in pristine environments and even
untouched wilderness. Thus the biological and social differences
arose. In Britain there can be, or should be, no unused or uncared
for areas.

The structure and organisation of the societies also exhibit significant differences that are important in determining attitudes toward land reclamation. The British, despite industrialisation, are a nation of farmers – essentially settled rural people in outlook. Although once densely forested, the British are not a forest people. Only 8 percent of its surface area is under forest and it has less woodland than any other European country with the exception of Ireland. To the British living in a man-made environment, the countryside is represented by the estate and the farm.

The North American, on the other hand, occupies a country that, until relatively recently, was, and over great areas in all essentials still remains, a wilderness, in large part forested. Seventy-five percent of the land under forest at the time of European settlement of the present-day United States is still under trees, while the great Canadian wilderness, in relative terms, remains unchanged. The frontier tradition impinges largely on the American consciousness. The American solution to disturbance, if it is not to be left to take care of itself, is to return the land to a state approaching wilderness condition.

The very concept of wilderness, that plays such an important role in current American thought, is barely understood, much less sought after by the British. I suspect most people in Britain are happier when they can see the hand of man – a cultural land-scape – which of course by no means implies that is unsightly.

Further, as a consequence of the limited land base, people in Britain have been obliged to learn how to live together leading to cultural assimilation. The tendency to take up en-trenched positions – the politics of confrontation that have been such a strong element in the American environmental movement in recent years, is much less apparent. The owner's unrestricted use of his own land, a basic article of faith for many Americans, has, in the United Kingdom, long been subject to public control. British industry in general does not yet recognise that its use of the land is of an interim nature and that with this use lies the responsibility to return it to a reasonably productive state. The American industrialist at least gives lip-service to this social responsibility and legislation is increasingly giving substance to it.

The British industrial attitude probably results, not so much from the absence of eco-power, as from the fact that the British industrial regions have been ravaged for so long that it is still possible to see the workings of that derelict land mentality that accepts industrial mess as part of the natural order of things. Perhaps, subconsciously, the old adage "Where there's muck there's money", still influences the body of British industrial thought despite all the efforts of the enlightened regional planning of the past several decades.

Nevertheless and notwithstanding the foregoing general-isations, the pioneer nature of North American society has not been favourable to the development of a land ethic such as is found in Britain, where the concept of stewardship and good husbandry - land management as distinct from the extremes of un-considered exploitation or of preservation, is engrained. Whereas the American ideal, at least in theory, is the natural ecosystem, the British take it for granted that their ecosystems have been, and will be, manipulated.

Differences in government structure enter in. In the American system each state is sovereign and has determined the form of its own reclamation legislation. This may still take a quite rigid form with little provision for decision making at the local level. In Britain, with a strong tradition of local government, industrial land reclamation is the concern of Local Authorities that operate within the framework of the national Planning Acts. These authorities have a major role in determining the character of the urban and rural environment but may differ in their emphasis and level of the professional skills deployed. National government policies are felt through grants in support of approved projects. There has been an absence of concerted efforts either in research or practice by, for example, the Forestry Commission although notable success has been achieved locally where a local representative of this agency has played an active role, as for example, in the Forest of Dean and in parts of South Wales where mine dumps have been situated within areas of afforestation. Except for the Opencast Executive of the National Coal Board, industry has played little or no part.

In the United States, while public agencies such as the land grant universities and their associated experiment stations and especially the U.S. Forest Service have a research commitment, and have contributed much to the development of reclamation techniques, the major reclamation effort has been by the mining industry itself, either through company effort or through co-operative associations, albeit in response to public pressure and reclamation legislation. In this work foresters have played a major role.

SCALE OF THE PROBLEM

The most widespread form of disturbance in America has resulted from strip-mining of coal. The United States contains about 40 percent of the world's coal reserves in an area comprising about 15 percent of its surface area. If marble, limestone,

talc, phosphate, potash, oil shale, sand, gravel, commercial clays,
and other minerals are included, then the strippable area extends to at least a third. Surface-mining tools of almost un-
believable size, power and effectiveness have been developed.
"Big-Muskie", a walking dragline operating in Ohio, stands as
high as a twenty story building, lifts a boom 310 feet long
operated by a five-inch steel cable and scrapes away 325 tons of
overburden per scoop. The "Gem of Egypt", another giant earth-
moving dragline weighs 7,000 tons and has a 125 cubic yard bucket
capable of removing 200 tons of overburden at each bite. In
1965 there were 3,357 square miles of surface-mined land east of
the Mississippi. Seventy percent of these lands were disturbed
for coal, and about 2,000 square miles of these were in need of
rehabilitation. During the last ten years the area disturbed has
greatly increased. The working of other minerals, though of lesser
magnitude accounts for significant acreages. In Florida surface
mining of the vast phosphate beds has disturbed 300,000 acres.
It has been estimated that in the United States 650 acres are
strip-mined a week.

It was not until after the Second World War that the American
reclamation movement really got underway. Today, something of the
order of 100,000 acres of surface coal mined land a year is being
reclaimed and it is probably fair to say that almost all newly
mined land is now treated to at least minimal standards, but
there remains the problem of the so-called orphan lands - the
very considerable acreages that pre-date the reclamation laws -
and until recently there has been little real regard for returning
much of the mined land to a truly productive state.

Today we are seeing the extension of the surface mining
industry from the humid east to the more arid, mountainous and
sparsely populated west. Here the area disturbed by surface
mining has increased by thirty percent in the past seven years.

In North America western coal is being mined, in the case of
the United States, to meet burgeoning national energy needs and,
in the case of Canada, for export to Japan. The distance from
the eastern population centres has, in the past, prevented the
large-scale exploitation of these coals. Today, the situation
has changed. In the United States gasification and electric
generator plants are being linked in a national grid. In western
Canada modern technology, massive capital investment, container
trains and bulk loading has opened up an export market that has
raised production from about half a million tons to 14 million
within five years and is still increasing. Because these western
coals are of low sulphur content and possibly because it has been
thought that lesser populations equate with lesser public op-
position, these developments have been seized upon as a major contribution to the American national energy supply, without the

environmental problems of mining and processing coal in the
eastern metropolitan areas. However, environmental problems of
a different sort, and possibly greater magnitude, occur in the
new mining areas. Conditions are often considerably more adverse
than have been encountered in the past. Because of the great
diversity of climate, geology, soils, elevation, land form and
vegetation, the direct transfer of the proven established reclamation methods of the east is impossible and a new attitude
toward mining and reclamation must be developed.

Proven coal fields extend from the Yukon in Canada almost
to the southern border of the United States. Those in the
western United States comprise more than 100 million acres. Coal
underlies a considerable part of Texas, a third of Oklahoma, large
areas in New Mexico, Arizona, Colorado, Utah, Montana, Wyoming,
Washington and the Dakotas. The nearly untouched coal deposits
of Montana alone contain 13 percent of the known coal reserves of
the United States. Some 23 billion tons of Montana coal are known
to be strippable. In Canada, British Columbia and Alberta have
even greater reserves although, because of the mountainous topography, much of the coal cannot be won by surface methods and
underground mining will play a significant role. Operations are
large scale, as an example, one British Columbia company has
reserves of 120,000 acres of coal-bearing lands. In just one
valley in the Canadian Rockies, perhaps eighteen miles long, the
reserves are 18.5 billion tons. Wyoming has coal reserves amounting
to more than 500 billion tons, of which some 23 billion are
strippable. To date, only about three square miles have been
strip-mined in the lignite fields of North Dakota, but the beds
contain an estimated 351 billion tons and more than two million
acres are already under lease to mining companies with high gasification and electric power plants projected for the state.
Strippable coal in the western United States is regarded as that
occurring in seams at least six feet thick, with overburden not
exceeding 150 feet thick, however in British Columbia overburden
up to 400 feet thick has been removed from steeply pitching 50
feet thick coals under severe mountainous conditions, so it is
probable that technological developments will bring deeper stripping to the American West in the future.

Considerable oil shale deposits extend over about 16 million
acres in Utah, Wyoming and Colorado, while the vast Athabasca tar
sands of northern Alberta contain half as much crude oil as the
total of the world's present proven reserves, though in this
instance the field is just being developed and it is not known
yet how much will be recoverable. So far, recovery has been through
a strip-mining process which can only reach a small proportion
of the bitumen-bearing sands from which the liquid hydrocarbons
are refined out, leaving a residue of sand, water and bitumen.

Uranium, the demand for which will increase with the increasing acceptance of nuclear power, is found throughout the intermountain region in the United States and also in Canada.

The largest phosphate deposits in the United States, accounting for 42 percent of the known reserves, extend over 10 million acres of the western states and the full extent of the field is not yet known. Metallic ore bodies and numerous deposits of less valuable minerals occur. A similar story might be told of western Canada.

The development of new mining techniques in hard rock mining that permits the mining and processing of low grade ores has radically changed that industry. In British Columbia alone there are currently 37 large open pit mines, with mining rates of over 30,000 tons of ore per day. In one typical situation the removal of an ore body of 293 million tons over a 21 year period will require the removal of 80 million tons of glacial-till overburden and 254 million tons of waste rock. At the conclusion of this operation the following acreage will have been disturbed:

Open pit - 537 acres

Plant site - 227 acres

Waste disposal area - 1,535 acres

Tailings Pond - 1,370 acres

For a total of 3,669 acres, all at an elevation of between 3,500 and 6,500 feet, in a semi-arid climate and short growing season. It is instructive to compare these figures with the three million ton waste dump of the British coal fields.

Associated with metal mining are atmospheric pollution effects when the ores are smelted. The Trail, B.C. smelter was a famous case in the 1930's. Less well-known is the long abandoned and quite isolated Anyox copper smelter in north coastal British Columbia that also resulted in gross denudation and death of trees. Sudbury, Ontario produces about 50 percent of the world's nickel supply and a number of other metals in significant amounts. The ore used to be burned in open piles. Today there is a landscape of barren rocks where before there was northern coniferous forest. Forty square miles are classified as severely barren and another 140 square miles as having impoverished vegetation and poorly productive forest. There are similar situations at Wawa, in Ontario, in the Gaspe in Quebec, in the Tennessee Copper basin and in other places in Canada and in the United States.

By comparison, the total area of industrial dereliction
in Britain is small. It results not from the impact of the
mining industry but from the misuse of land over two hundred
years by various industries and urbanisation. The reclaimer's
concern here is to make good the damage of the past. The area in
need of reclamation varies according to the criteria adopted but
appears to be only about 300,000 acres, or 0.6 percent of the
total British land surface. However, in such a highly industrialised and densely populated country this wasted land has
significance far beyond that indicated by its extent. Unlike
much of the north American disturbance, the British wastes, which
may be in small parcels, are mostly situated in or near large conurbations. They have a major impact on the quality of life.

To consider only one region, the South Staffordshire Black
Country. The 18th and 19th century exploitation of the mineral
resources of this region had a disastrous impact on the original
agricultural landscape. Huge hollows resulted from former open
pits while many hundreds of small bell-pit and shaft mines scarred
the countryside through subsidence and spoil heaps. The waste
coal dumps were often on fire. Ironstone was mined to supply
open hearth furnaces that discharged sulphurous smoke. Fireclays, the seat earths of the coal seams, formed the basis for a
refractory-brick industry. Quarrying of Etruria Marl, raw
material for the famous Staffordshire blue brick industry, left
great open pits. A pottery industry left mounds of pot shards.
There were limestone workings in the surrounding hills, outcrops
of sandstone were quarried for building stone and grindstones,
huge dolerite quarries produced valuable road metals. In 1843 the
whole district was desribed as "a vast warren with quarries and
pits, mines and galleries honeycombing the land surface" - "...an
interminable village, composed of cottages and very ordinary
houses ... interspersed with blazing furnaces, heaps of burning
coal in process of coking, piles of calcining ironstone ... forges,
pit banks and engine chimneys".

Atmospheric pollution had hastened the decay and abandonment
of once prosperous farmland. "Vegetation shrivelled, faded slowly
and perished as of blight". By the 1880's "the earth had become
one vast unsightly heap of dead ashes and dingy refuse". Gradually
as the coal was worked out and the iron industry declined, the
atmosphere, in relative terms, cleared and rough grasses even
began to colonise the waste ground. This century the various
public health acts had an alleviating effect. In the 1920's the
area of derelict land was put at 30,000 acres. By 1945 only about
8,000 acres remained and by 1952 a further 3,000 acres had been
levelled or built upon. Nevertheless the urban areas still sat
in a matrix of abused land and industrial waste. Today further
progress has been made and land has been reclaimed for housing,
new industrial plant and as parks, but the old Black Country still

persists.

In the rural British coal-fields the total area affected by waste dumps and subsidence is less but coal tips may still dominate the landscape. There are no modern open pit metal mines as yet in Britain.

THE NATURE OF THE WASTE

There are significant differences between the materials that have to be reclaimed in Britain and in North America. In North America the matter is one of extensive areas disturbed by single industries when the waste comprises the overburden of surface and sub-surface materials that overlie the economically important minerals. Thus the American reclaimer most often works with a mixture of top-soil, sub-soil and weathered or fractured rock that may or may not be separated by differential placement during disposal.

The relatively new development of large-scale open pit metal mining has introduced a new element. Here the waste material comprises waste rock, usually only distinguished from the economic ore body by a lesser concentration of the sought-after metal. A further difference results from the nature of the mining operation. If the surface mining of sedimentary lowland deposits may be likened to the ploughing of a field, an open pit metal mine may be likened to the digging of a well. Whereas in ploughing the land is backfilled as the work progresses, a well cannot be filled while it is being excavated. Thus the waste rock from open pit mines has to be removed to outside the pit, forming dumps on mountainsides that may themselves be above the natural angle of repose. These dumps may remain active throughout the life of the mine. Only occasionally are the dumps built up as terraces. The slopes may be several thousand feet in length. Tailings ponds, formed behind dams often built of the tailings themselves, are an associated problem. Tailings vary from coarse sand to colloidal slimes. At best tailings are sterile but they may contain noxious residues. Tailings dams already exist that are planned to reach eight hundred feet in height and over a quarter of a mile across, with the pond extending back a distance of four miles. Mountains are being removed   The technology of mining low-grade ores is still developing.

Many mines are in undeveloped parts of the country. The Athabasca tar sands, previously mentioned, are a classic example of the Canadian dilemma. Exploitation could result in the formation of a huge expanse of sterile sands situated in a hostile climate. The question is how to exploit these and the other potentially rich mineral resources that are necessary for the development of the country with minimal environmental disturbance.

In the U.K. the situation is very different. There are the open-cast coal lands, the excellent reclamation of which is expedited by the fact that the entire operation is controlled by a nationalised industry. Sites seem to be considered almost entirely on their impact on the aesthetic environment. Although the British land base is finite, economic land values appear not to enter directly to any great extent except in the case of urban or industrial development on or adjacent to the restored land.

The most prominent features are the huge waste dumps resulting from the underground mining of coal that until recently were inseparable from the British industrial landscape. Comprised largely of shales, mudstones, waste rock and general mine refuse, these heaps were built up over many years and dominate the lowland landscape for miles around, often in areas of high population density.

TYPES AND PURPOSES OF RECLAMATION

The kind of reclamation and its purpose is as varied as the sites. In British Columbia the horticulturally famous Butchart Gardens near Victoria on Vancouver Island were developed before the First World War in extensive limestone workings by the wife of the quarry owner because she disliked the view from her house overlooking the quarries. More recently, in Vancouver, during the last few years, a most attractive park has been constructed in an old stone quarry that few now recognise as such.

The earliest British reclamation efforts have a similar history. They are to be found in the older coalfields where owners of landed estates wished to cover small coal banks on their properties. The banks were planted with trees by the estate forester in the normal routine of estate management and the dumps have long been lost from sight. This in itself is indicative of success. Thus, in fact, although often unrecorded, tree planting on spoil heaps has a quite long history in Britain, though the activity has been rather sporadic and few plantations exist that have had constant care since planting. Efforts were made by a voluntary association of public-spirited people in the Black Country where the first plantings date from 1885 and at least a dozen mounds were planted by the turn of the century with a total of 83 acres by 1916. By the mid-1950's there were perhaps some one hundred examples of tree planting on deep mined spoil in Britain, a respectable proportion dating from the first decades of this century and earlier. Apart from the original Black Country reclamation there is at Scremerston in northern Northumberland a reclaimed area on which the trees are now 80 years old. These early plantings

certainly did not sample all the possible conditions, and all possible species had not been tried, but taken together they
presented a most encouraging picture. Few failures could not be
ascribed to causes that were remediable, principally relating to
vandalism, animal damage or incorrect choice of species, usually
injured by smoke pollution.

During the past twenty years with greater interest in regional
planning, there has been continued activity, but except in South
Wales and North East England, trees have not played such a prominent role as in the earlier period.

In Britain today, as in the past, with the exception of the
surface coaling operations when the land is restored to its former
state of productivity, most reclamation appears to be undertaken
from the point of view of local landscaping or urban planning, the
objective being primarily the removal of unsightly vistas. The
planner and landscape architect plays a large role, the forester
much less.

THE RESTORATION OF SURFACE-MINED COAL LANDS IN BRITAIN

The British restoration of coal stripped lands is deserving
of special attention. It provides a case study for successful
reclamation. The most thorough and comprehensive efforts are
made to restore stripped lands to their original or even improved
productivity. This concern commences long before stripping begins.
There is provision for all concerned parties to make their views
or reservations known and all local and national government
agencies have to be satisfied with the proposals before mining is
permitted. The land is carefully surveyed, photographed and
topographically mapped. Powers of acquisition exist but remain
unused as farmers and land owners voluntarily make their land
available for open-cast mining. Existing buildings, woodlands and
hedges are removed, and the topsoil is scraped off in different
strata which are segregated and stored in grass covered heaps.
The overburden is removed and the coal is then taken out and the
layers replaced in their natural order. Each is compacted. The
surface is restored to its original configuration or to a shape
fitted to the environs. The Coal Board exercises close and continuing supervision. Tile field drains are laid and hedges or
fences restored. The land is fertilised and seeded to grass under
the direction of the government agricultural service and managed
by it for several years to bring it back to good heart before it
is returned to the farmer. There are examples of restored park-
land right up to the walls of great country houses that can only
be distinguished from the original landscape by the lesser age of
the park trees and plantations on the restored land. This highly
successful programme indicates what can be done through rational

operations by a specialised agency with reclamation completely integrated into the mining process and the final process of restoration to productive husbandry directed by specialised land
managers.

THE AMERICAN SCENE

In January 1, 1965, an estimated 3.2 million acres of land had
been disturbed by surface mining. Of this amount, around 2 million
acres was in need of conservation treatment and rehabilitation.
About 91 percent of the affected acreage is in private ownership,
and nearly half of that is in small scattered ownerships.

Table 1 is taken from the report of the national study,
Surface Mining and Our Environment, by the U.S. Department of the
Interior. All States have land that has been affected by surface
mining. More than 80 percent of the mined land surveyed was a
mile or more from towns of 200 or more people. About 40 percent
of the mined land was more than 5 miles from towns of this size.
About one-third of the land disturbed by surface mining has been
rehabilitated. Some of this was restored by natural seeding, but
more than half was treated through the efforts of private owners.

Information in the national study showed that 127,747 acres
of surface-mined land was successfully treated in 300 conservation
districts in 5 years, 1960-64. This work involved 5,255 district
co-operators in 31 States. During the same period, State forestry
agencies, in cooperation with the U.S. Forest Service, provided
assistance for restoring 36,710 acres of surface-mined land.

Mining firms have greatly increased their conservation efforts
in recent years. Surface-mine operators in 22 States have formed
the Mined Land Conservation Conference to bring about restoration
of mined land. In addition, associations have been formed for
this purpose in several States.

Associations in the Appalachian Region have been responsible
for reforestation and seeding of 74,000 acres.

The National Sand and Gravel Association reports that its
members rehabilitated more than half of the acreage they mined in
1965. Two years earlier, they treated only one-fourth of the
total acreage disturbed.

More than a fourth of the land disturbed by surface mining has
a potential for farm and forest recreation. Smaller percentages
of the affected areas have potential for cropland, residential,
institutional or industrial uses.

State	Land requiring reclamation[1]	Land not requiring reclamation[1]	Total land disturbed[2]
Alabama	83.0	50.9	133.9
Alaska	6.9	4.2	11.1
Arizona	4.7	27.7	32.4
Arkansas	16.6	5.8	22.4
California	107.9	66.1	174.0
Colorado	40.2	14.8	55.0
Connecticut	10.1	6.2	16.3
Delaware	3.5	2.2	5.7
Florida	143.5	45.3	188.8
Georgia	13.5	8.2	21.7
Hawaii	(3)	(3)	(3)
Idaho	30.7	10.3	41.0
Illinois'	88.7	54.4	143.1
Indiana	27.6	97.7	125.3
Iowa	35.5	8.9	44.4
Kansas	50.0	9.5	59.5
Kentucky	79.2	48.5	127.7
Louisiana	17.2	13.6	30.8
Maine	21.6	13.2	34.8
Maryland	18.1	7.1	25.2
Massachusetts	25.0	15.3	40.3
Michigan	26.6	10.3	36.9
Minnesota	71.5	43.9	115.4
Mississippi	23.7	5.9	29.6
Missouri	43.7	15.4	59.1
Montana	19.6	7.3	26.9
Nebraska	16.8	12.1	28.9
Nevada	20.4	12.5	32.9
New Hampshire	5.1	3.2	8.3
New Jersey	21.0	12.8	33.8
New Mexico	2.0	4.5	6.5
New York	50.2	7.5	57.7
North Carolina	22.8	14.0	36.8
North Dakota	22.9	14.0	36.9
Ohio	171.6	105.1	276.7

State	Land requiring reclamation[1]	Land not requiring reclamation[1]	Total land disturbed[2]
Oklahoma	22.2	5.2	27.4
Oregon	5.8	3.6	9.4
Pennsylvania	229.5	140.7	370.2
Rhode Island	2.2	1.4	3.6
South Carolina	19.3	13.4	32.7
South Dakota	25.3	8.9	34.2
Tennessee	62.5	38.4	100.9
Texas	136.4	29.9	166.3
Utah	3.4	2.1	5.5
Vermont	4.2	2.5	6.7
Virginia	37.7	23.1	60.8
Washington	5.5	3.3	8.8
West Virginia	111.4	84.1	195.5
Wisconsin	27.4	8.2	35.6
Wyoming	6.4	4.0	10.4
Total	2,040.6	1,147.2	3,187.8

[1] Compiled from data supplied by Soil Conservation Service, U.S. Department of Agriculture.

[2] Data compiled from reports submitted by the States on U.S. Department of the Interior form 6-1385X, from Soil Conservation Service, and estimates.

(3) Less than 100 acres.

Table 1. Status of land disturbed by strip and surface mining in the United States as of January 1, 1965, by State (Thousand acres).

Many of the restored sites have the potential for multiple uses. The sum of the acreages reported for the various uses in the survey exceeded the total area of the sample by 65 percent. They were:

Potential use	Percent of disturbed areas
Woodland	39.2
Wildlife habitat	38.6
Farm and forest recreation	27.2
Pastureland	17.8
Rangeland	
Ponds and reservoirs	9.9
Residential, institutional, and industrial	6.5
Cropland	3.4
Other (unspecified)	8.6
	164.5

In the eastern American States until recently, while the possibility of returning land to production is not ignored, reclamation has been something of an afterthought. Treatment has mostly consisted of the planting of trees to screen mined-out land and to provide a green cover for land left in its original rough post-mining configuration. There are notable areas where there has been differential placement of spoil, development of recreational facilities, establishment of potentially productive forests and prosperous farming enterprises, and undoubtedly some States and some individual companies have done excellent work. But in the majority of cases the objective has been to meet minimal reclamation requirements. Particularly in Appalachia surface mining under mountain conditions has led to problems. During the last two or three years there have been major revisions in most state reclamation laws. Requirements have been tightened up. Frequently major constraints have been placed on the actual mining techniques employed and on the reclamation standards required. For example, new Ohio legislation has resulted in a radical change in technique. Topsoil must now be conserved for respreading, high walls are not permitted and the land must be graded to a rolling configuration. If trees are to be established the maximum slope permitted is 25 percent; if the site is to be grassed then the slope cannot exceed 15 percent; and all sites must be grassed initially even if tree planting is intended. A very different landscape is being produced.

In the American and Canadian west very different attitudes from those of the east are appearing that bode well for the future.

Much of the western lands have remained in public ownership, as for example in British Columbia where 93 percent of the Provincia land area has remained with the Crown, and in Alberta where the eastern slope of the Rockies is either Provincial Forest Reserve or National Park. Similarly, in the western United States where a majority of the mineral bearing lands are in public ownership and administered by agencies such as the U.S. Forest Service, the Bureau of land Management and the Department of Indian Affairs, or are owned by the individual states. In total, in 1970 there were 94,402 mineral leases covering 58,905,110 acres of public lands in the interior American west, an area exceeding the combined areas of the seven states of New Jersey, New York, Rhode Island, Massachusetts, Vermont, Connecticut and New Hampshire. Potential mineral claims could occupy as much as one third of the western national forest lands. More than 800,000 claims have been registered on the national forests alone and new claims are being filed at a rate exceeding 10,000 per year. Hence, public land managers are necessarily becoming involved in mining and reclamation questions. The public interest is of major concern and the concept of the public good prevails. In the case of the U.S. National Forests, the Forest Service has a statutory requirement to manage the public estate for multiple use objectives. Further, the conservationist traditions of the Forest Service, that in its early days was concerned in protecting the forest and range lands against private interests, are still strong. Thus in both the Canadian and American west we find strong insistence that mining is to be considered as an interim use only and carried on with due regard to other values such as timber, watershed management, recreation, wild life and aesthetics. There is insistence that the land be returned to a state compatible with regional patterns. In British Columbia, for example, the Potential Land Productivity Classification of the recently completed Canada Land Inventory is increasingly being used as a guide to reclamation objectives and in both countries there is increasing emphasis on detailed reclamation pre-planning prior to mining, ecological base line studies that document the pre-mining environment, detailed environmental impact studies, integration of the mining and reclamation programmes, and long term monitoring, all with the purpose of developing sound procedural measures to ameliorate the environmental impact and to facilitate reclamation. Reclamation is ceasing to be considered a one-shot deal but a matter for continuing management. Watershed questions are a major concern. Emphasis is on individual site evaluation.

Is this concern necessary in the great expanses of the sparsely settled American west? In terms of present market land values most definitely not – but can one put a value on the scenic beauty of the high Rockies, on the U.S. National Forests or the Indian Reservations, upon the Bighorn sheep whose winter ranges are at 6000 or 7000 feet elevation in the coal zone and yet comprise no more than 70 slopes covering 200 square miles in Alberta and a much lesser area in British Columbia; or the water quality of the Saskatchewan and other rivers upon which the farmers of the Canadian prairies are dependent. In some alpine ranges 30 percent of the forage area has been lost through coal exploration roads alone. Few doubt that the minerals will be developed. A 'band-aid' approach after the damage is sustained would be fatal.

As in the eastern United States foresters are playing the major role in reclamation but here they are not merely operating as tree planters but in their full capacity as land managers and planners and an interdisciplinary approach is adopted. The way these matters are being approached may be seen in the regulatory requirements in the several jurisdictions.

All this is not far advanced. Administrative structures are still being developed along with technical expertise. The resources in danger are sometimes not even fully inventoried. However, concern has been generated and it will be more difficult for new mining developments to be designed around the maximum production – minimum cost goal. Too much is at stake. The impact of mining on other resources will have to be considered in the initial stages of mine development planning and not as an afterthought as in Britain and the eastern United States when rehabilitative measures can now be accomplished only with considerable time, effort and expense.

REFERENCES

A voluminous literature exists on many aspects of surface mining and mined-land reclamation, particularly in the United States. The subject matter varies widely, ranging from the results of individual research to nation-wide in-depth studies. Because of the sheer volume and diversity of this literature no attempt is made to present a comprehensive list of references. The following titles provide more detailed treatment of the subject matter discussed.

Canada

Hogg, J.L.E., Mined land reclamation in British Columbia: the importance of institutional and environmental factors in the formulation of policy, For. Chron. 47, 335, 1971.

Peterson, E.B. & Etter, H.M., A background for disturbed land reclamation and research in the Rocky Mountain region of Alberta, For. Res. Lab. Canad. For. Serv. Edmonton, Alberta, Info. Rep. A - X - 34, 1969.

Thirgood, J.V., Land disturbance and revegetation in Canada, Canad. Mining Jl., 90, 33, 1969.

Thirgood, J.V., The planned reclamation of mined lands, Western Miner, June 22, 1970.

Thirgood, J.V., The rehabilitation of the mining environment in British Columbia. Canad. Mining & Metall. Bull., August, 1971.

Thirgood, J.V., Planned reclamation. Proc. 1st Res. & Appl. Techn. Symp. on Mined Land Reclamation, Pittsburg, National Coal Association, Washington, D.C., 1973.

Watson, W.Y. & Richardson, D.H., Appreciating the potential of devastated land. For. Chron., 8, 312, 1972.

United Kingdom

Beaver, S.H., Report on derelict land in the Black Country. Minist. Town & Country Planning, London, 1946.

Civic Trust, Derelict Land, A Study in Industrial Dereliction and How it May be Redeemed, Civic Trust, London, 1964.

Ministry of Housing and Local Government, New Life for Dead Lands, Derelict Areas Reclaimed, H.M.S.O., London, 1963.

Wise, M.J., The Midland Reafforesting Association, 1903-1924, and the reclamation of derelict land in the Black Country. Inst. Landsc. Arch. J., 57, 13, 1962.

Wood, R.F. & Thirgood, J.V., Tree planting on colliery spoil heaps, For. Comm. Res. Branch, Pap., 17, 1955.

United States

Caudill, H.M., Farming and mining, there is no land to spare. Atlantic, 232, 85, 1973.

Conaway, J., The last of the west: Hell, strip it! Atlantic 232, 91, 1973.

18

Copeland, O.L. & Packer, P.E., Land use aspects of the energy crisis and western mining. J. For., 671, 1972.

Paller, W. & Schultz, D.A., Planning approaches to surface mining in the National Forests, Proc. 1st. Res. Appl. Techn. Symp. on Mined Land Reclamation, Pittsburg, National Coal Association, 1973.

ENVIRONMENTAL MANAGEMENT THROUGH PLANNING - THE WEST GERMAN
EXPERIENCE IN THE RUHR

Heinz Neufang

Siedlungsverband Ruhrkohlenbezirk, Essen, Federal
Republic of Germany

INTRODUCTION

The Siedlungsverband Ruhrkohlenbezirk (SVR) is a local
government association. Its territory situated between the Dutch
border in the West, Unna rural district in the East, the heights
south of the river Ruhr and the lowland north of the river Lippe,
embraces 18 autonomous cities, 6 rural districts and parts of
three others. Its area of about 4 600 km^2 covers some 14 per cent
of Northrhine-Westphalia, and its population of 5.5 million is
about 33 per cent of the Land total. The area produces more than
30 per cent of Northrhine-Westphalia's gross social product.

The existence of the organization is based upon the Ordinance
establishing the Ruhr Regional Planning Authority, a law adopted
in 1920 at Land level. The SVR being an instrument of cooperation
between the local authorities within its area is based on a constitution which corresponds to that of the local authorities with
one exception: its members (cities and rural districts) elect
three-fifths only of the members of the Assembly, which is the
supreme council of the SVR. The remaining two-fifths of the
current 88 members of the Assembly are chosen from lists put forward by local Chambers of Industry, Commerce and Agriculture,
Trade Unions and Employers' Federations, Public Housing Bodies
and Settlement Associations, Transport and Water Supply and Sewage
Disposal Authorities.

The tasks of the SVR are based on its Ordinance. In addition,
the SVR is the Land Planning Authority for its area, that is to
say one of the three Land Planning Authorities in the Land Northrhine-Westphalia.

PLANNING STRUCTURE

In order to get a clear understanding of the tasks and
activities of the Planning Board in its capacity of a Land
Planning Authority, it is necessary to explain the 4-tier
planning system adopted in the Federal Republic of Germany.

At the Federal level (first tier) there is no central
physical planning, but there is central specialized planning in
certain fields, such as National Railways, National Roads, National
Defence. There further exists central responsibility for outline legislation as far as general targets and principles in
physical planning and Land planning are concerned. Such power
is retained by the Federal Government and laid down in the Federal
law on physical planning (Bundesraumordnungsgesetz).

At Land level (second tier) is established Land planning as
institutionalized physical planning. Almost all the Länder have
a Land planning law covering responsibilities, instruments and
procedures. The Land Northrhine-Westphalia by its Land planning
law has laid down Land planning as a joint task of the State and
of the local authorities.

The government of Land Northrhine-Westphalia prepares a Land
development programme taking the form of a law to be passed by
the Landtag (Land Parliament). This programme is developed in
more detail in Land development plans. These latter are published
by the Land Government in conjunction with the Planning Committee
of the Land Parliament. They are guidelines which must be followed
by all public bodies. The individual citizen is not affected
legally neither by the Land development programme nor by the
Land development plans.

The regional level (third tier) of planning in Northrhine-
Westphalia is formed by the three Land Planning Authorities:
Ruhr Regional Planning Authority; Land Planning Authority Rhineland;
Land Planning Authority Westphalia.

The Land Planning Authorities have the task of preparing
the goals for the spatial development of their region within the
framework of the Land Development Programme, and to present
these goals in the Area Development Plan. The procedure is handled
in such a way that the opinions of a great variety of interests
are considered by the Land Planning Authority. The Development
Plan is subject to approval by the Land Government.

The second most important instrument at the disposal of the
Planning Board (SVR) is the procedure obliging local authorities
when preparing their urban plans to conform to the goals of
regional planning. At an early stage, local authorities preparing

their plans will determine what is compatible and what is not with the plans set up at regional level.

At the communal level (fourth tier) urban plans are prepared which take the form of a (preparatory) land-use plan and of communal plans. These plans are designed to enable everybody to become informed on community development and to make sure that the State and local authorities jointly enforce compliance with this concept - in agreement with the goals of Land Planning - in the procedure for granting building permits.

PLANNING PROCEDURE AND GOALS

The main tasks currently facing the Ruhr Regional Planning Authority will only be mentioned quite briefly here:

1. securing the open space and separating functions of the so-called regional green wedges in the congested core zone of the Authority's area

    a) statutorily by the Authority's own development plans on the same lines as the local authority building plans.

    b) privately through financial encouragement to local authority land purchases in these green spaces.

2. securing routes and land for regionally significant transport facilities through SVR building plans.

Planning responsibility of the SVR in individual cases is justified by the Authority's Designation Schedule. The right of the SVR to lay down conditions for a particular object in a building plan instead of the local authority is confirmed by the inclusion of the object in the Designation Schedule (Verbandsverzeichnis).

Regional landscape conservation, renewal of regional green spaces

The SVR promotes measures in the interest of woodland protection and landscape conservation, of planting waste tips and other derelict land. The start was made from the Authority's own resources and with limited Land funds to free green spaces in the central conurbation from other uses and to protect the "green belt" function.

Provision of regional leisure and recreation facilities

The SVR encourages and participates in the development of large

recreation areas (the Nature Park Hohe Mark as a member of the
Nature Park Union; the Haltern reservoir area as a member of the
Haltern reservoir planning association; green areas in connection
with the University of Bochum in the KemnadeRuhr recreation centre
project; in the Emscher-zone the development of five regional
leisure centres set up on the lines conceived by the SVR).

Refuse disposal in the Ruhr

The difficulties of the cities in disposing of solid refuse
from domestic, industrial and commercial sources have led the
SVR to establish an Information and Advisory Office for refuse
disposal. This Office is at the disposal of all local authorities
in the SVR area. The SVR is operating several controlled tipping
areas of considerable size and established and operates a system
of large-scale facilities in the field of refuse disposal.

CONCLUSION

From these activities it will be evident that the SVR embraces Regional Planning, Urban Planning and Investment Decision
Making in a way that is not commonly encountered in Local
Authorities. The efficiency of the Planning Board (SVR) depends
upon this combination of functions and the continuous flow of
information involving all agencies dealing with the development
of the region rather than on its financial strength.

THE RE-ESTABLISHMENT AND MAINTENANCE OF OPEN COUNTRYSIDE IN THE
RUHR REGION OF THE FEDERAL REPUBLIC OF GERMANY

Gerhard Petsch

Landscape Architecture and Forestry Section,
Siedlungsverband Ruhrkohlenbezirk, Essen, Federal
Republic of Germany

INTRODUCTION

There are two main functions of the Landscape Architecture
and Forests Section of the S.V.R. The first includes the
coordination of planning of green areas in schemes undertaken
by the member communes of the Ruhr Regional Planning Authority
(SVR). It tries to assure that in the decision making process
at regional level due consideration is given to special problems
of landscape architecture and forestry. Decisions are weighed
and consequences of actions evaluated against the demands and
requirements the forestry and landscape needs.

This activity is important since of the total area of the
Ruhr Regional Planning Authority (4, 593.3 km^2) 2, 356.3 km^2
(51.3 per cent) are agriculturally used, 730.6 km^2 (15.9 per cent)
are made up of woodlands, and 110.2 km^2 (2.4 per cent) consist
of water areas. If playground areas and inner town green spaces
are included these give 72.4 per cent of the whole territory of
the Ruhr Regional Planning Authority under some form of vegetational
cover.

Particular attention is being paid to what are called
"regional green wedges". Running north to south between the major
urban conurbations, these green wedges, located in close proximity
to the towns and cities, improve the micro-climate and in part
the macro-climate of adjoining areas. Wind turbulence is generated
which gives increased atmospheric exchange and reduced atmospheric dust burden. This is brought about by the dust inter-
cepting properties of the leaves of trees and shrubs.

The second function includes the active promotion of all factors making for landscape and green area improvement. Action in this field is taken with a view to improving the quality of existing tree populations. SVR funds made available annually are distributed in such a way as to form part only of the total finance required. The balance is borne by the landowner himself. The subsidy granted by the Ruhr Regional Planning Authority may vary between 30 and 80 per cent of the total cost according to the importance attached to the area in question in the regional plan. The following principle has been established with a view to the sharing of finance: in the densely populated central conurbations as much encouragement as possible is given to green areas; in the rural areas conservation of existing green areas by wise development is emphasised. In this way landscape planning makes a valuable contribution to environmental maintenance and protection.

In spite of this, woodlands in the area coming under SVR jurisdiction decreased between 1927 and 1970. But this decrease was far less in comparison to other conurbations of high demographic density over the same period of time. Note should be taken also of the fact that much effort was devoted, to improving forest areas and to making them available to the population as recreational resources.

The efficiency of agricultural areas is also being raised by the planning strategy of the SVR.

ENCOURAGING THE PROVISION OF GREEN AREAS

Individual projects involving financial subsidies and expert advice include:

1. establishment of vegetation on areas occupied by industrial plants

2. shelter planting in connection with sewage treatment plants, pumping stations, transformers, and water catchment areas

3. establishment of vegetation on areas occupied by housing estates, open air swimming pools, camping sites and weekend houses

4. planting of trees and shrubs alongside water courses, on slopes of various kinds, on embankments and along cuttings and roads if a recreational interest is involved

5. development of hiking paths, parking space, and leisure areas

6. provision of park benches and shelter huts

7. provision of nature trails

8. promotion of private and municipal woodlands by means of waste land reclamation and re-afforestation of denuded forest areas

9. conversion to resistant mixed woodland where forests have suffered from air pollution and mining activities

10. costly operations to maintain young tree populations with a high species diversity

11. setting up of forest playgrounds and other facilities for physical exercise

12. soil protection plantings and shelter plantings

13. reclamation of quarry areas.

Special efforts have been made under 'ACTION GREEN RUHR' and 'ACTION GREEN GERMAN RAILWAYS' in the Ruhr. 'ACTION GREEN RUHR', sponsored by the Ruhr Regional Planning Authority aimed at establishing vegetation on all industrial and coal mine spoil heaps. To that end landowners were provided with the necessary planting stock free of charge. The landowner, in return was obliged to establish trees and shrubs according to SVR specifications. Thus from 1952 to 1971 a total of 12 million trees were planted including 500 ha spoil heaps, 250 ha embankments along roads, railway lines and water courses, 200 ha tipping areas, 600 ha derelict land, and 50 ha woodland damaged by mining operations. In this way some 1,600 ha new green areas have come into being.

Since 1970, within the framework 'ACTION GREEN GERMAN RAIL-WAYS', in association with the Federal German Railway Company, the State of Northrhine-Westphalia and the Ruhr Regional Planning Authority, a total of one million trees and shrubs have been established in the Ruhr along the main railway lines totalling 90 rail kilometres of railway. This amounts to 100 ha having been planted.

RECLAMATION TO PROVIDE GREEN AREAS

The landscape architecture and forestry section is also concerned with the reclamation of spoiled land. Before approval is granted for the siting of mine spoil banks certain procedures must be followed:

1. investigation into the siting of the spoil dumps and their extension

2. evaluation of spoil material quantities and the coordination of the dumping of spoil material from several mines

3. the recording of the location of the spoil dump and its siting in relation to its environs on maps and in development plans

4. information is made available for plans of the future development, shape, completion of successive sectors and plans for the removal or disposal of surface water

5. the nature of the colliery spoil and the proportion of combustible constituents must be ascertained and the likely water contamination effects and the possibility of damage to vegetation

6. reliable data for the site on which colliery spoil is to be dumped, including subsoil studies, the fixing of appropriate angles of slope and safety distances to the nearest buildings

7. descriptions and plans of machinery and devices employed in compacting colliery spoil and the measures to be taken to protect people working on the site against accidents

8. indication of measures adopted to cope with groundwater and surface water generated on the site

9. information on the nature and extent of immissions that are likely to occur as well as on equipment provided to combat immissions

10. information on measures against the outbreak of fires or their early detection

11. plans covering the utilization of the spoil dump area using earth material on which vegetation will develop

12. coordination of all the procedures with the State Control Board of Mines and the embodiment of procedures in an operational plan (Betriebsplan).

CONCLUSION

Forward planning, both for the establishment and maintenance of green areas and their creation by reclamation of waste material following sensibly planned waste disposal procedures, is the

cornerstone of environmental management by the landscape architecture and forestry section of the Siedlungsverband Ruhrkohlenbezirk.

COAL WASTE MANAGEMENT

COAL WASTE MANAGEMENT - IDENTIFICATION OF PROBLEMS

The procedures adopted in mining coal lead to a large number of environmental problems:

Deep-mining gives rise to surface subsidence with subsequent flooding. Examples of nature reserves that have been created in such subsidence areas can be found in a number of industrial countries. Large expanses of water provide refuges for waterfowl and islands may be created by judicious tipping of colliery spoil.

One of the problems of tipping waste is tip stability. In the United Kingdom the stability of each tip must be certified by a civil engineer before vegetating procedures are begun. Moisture content monitoring is carried out as a precaution. In the Ruhr region of the Federal Republic of Germany colliery tips are required, by law, to be terraced even though this can give rise to reduced rates of tree growth.

Water pollution arises from the contamination of waterways by acid mine drainage water and drainage and erosion from tips and strip-mined areas. Even slopes with gradients of as little as 2.5 per cent may suffer from gullying, and droplet splashing can give rise to loss of particulate matter to water courses.

Air pollution has occurred due to combustion of colliery spoil materials. This feature also has significance for colliery spoil as a subsequent medium for plant growth. The degree of combustion will depend upon:

I   Coal rank (the degree of coalification determining reactivity with oxygen)

II Coal content of the tip

III Tip permeability to air

Tips will be very susceptible to burning if they are relatively rich in coal, loosely compacted and have steep slopes which intercept wind currents. Tips often possessed these features in the past. However, combustion is prevented in modern tips by reducing air permeability fo 10^{-9} cm.sec^{-1}) by compacting to 20 per cent voids or less. As some of the voids will be water filled this corresponds to less than 5 per cent air-filled and air penetration will only be to a few centimetres into the tip.

Colliery spoil ignition may be caused by:

I Spontaneous combustion

II Intentional ignition (to stabilize or produce 'red shale')

III Accidental causes (lightning or untended fires)

Burning may be controlled by digging or burning material or covering the tip with a moist impermeable "blinding layer" of finely-textured clay, compacted colliery spoil material or other suitable non-permeable layers.

The progress of combusion can be characterized as follows:

Temperature (°C)	Effect
20	Over 99 per cent of oxygen content absorbed; where gases escape the tip contains (v/v dry basis): < 0.005 per cent O_2 0-10 per cent CH_4 2-10 per cent CO_2 2-10 per cent N_2
100	As burning increases trace amounts of CO are released.
200	Depending on the oxidation state and flow rate of the gas stream, varying amounts of CH_4, CO, CO_2, C_2H_6, C_4H_8, H_2S and SO_2 are released.
300	Carbonization occurs and coals become fluid.
450	All interstitial water is eliminated.
500	Fumes of $(NH_4)_2SO_4$ and NH_4Cl are produced; NH_3, tar, CO_2, CO, SO_2, H_2S and N_2 released.
600	Micaceous material begins to break down
800	Clay content is eliminated and 'brick' produced.

Visual disamenity of both deep-mined and strip-mined areas is
generally dealt with by landscaping and vegetation establishment.
In the Federal Republic of Germany the practice is to reclaim the
lower slopes of terraced heaps before tipping to form the upper
slopes is completed. Planting on the steep slopes is accomplished
by men in parachute harness attached by a rope to a tractor which
proceeds slowly around the terrace above them.   In the United
Kingdom tipping is now generally subject to planning controls.
Top soil is removed and stored for subsequent replacement over the
tip and good establishment of herbage species usually is attained.
Where planting is carried directly into spoil material the features
listed below may cause problems for the establishment of vegetation.

    Combustion
    Instability
    Steep slopes
    Slope aspect
    Adverse textural composition
    Erosion
    Compaction
    Wind effects (blasting of plants)
    Chemical toxicity (Al, Mn, H, S, Cu, Zn, B)
    Chemical deficiency (N, P, K)
    Water stress
    Extreme surface temperatures
    Lack of appropriate plant species or genotypes
    Vandalism

THE DISPOSAL OF COAL MINE SPOIL IN THE UNITED KINGDOM

H.G. Glover

National Coal Board, Rotherham, United Kingdom

ABSTRACT

About 50 M tonnes of spoil are produced annually from underground coal mines in the United Kingdom and more than 2000 M tonnes
are known to be lying in disused tips throughout the coalfields of
England, Scotland and Wales. Opencast mine workings have not
produced any residual spoil tips, the land being fully restored aft
completion of the mining operation.

Freshly produced spoil consists mainly of argillaceous and
arenaceous rocks in the form of sandstones, siltstones and mudstones
together with lesser quantities of coal, inferior coal, ironstone,
pyrite, ankerite and minerals. The siltstones and mudstones degrade
under the action of handling and weathering, but the shear
strengths remain adequate for the safe construction of disposal
tips and for applications of the spoil in civil earthworks.
Chemical changes such as the oxidation of pyrite and coal, and the
leaching of salts, may be significant in loosely compacted spoil
but in most cases can be prevented by compaction.

Spoil which has burned may be stronger and more granular than
unburnt spoil but is liable to be variable in composition.

The construction of spoil tips in the United Kingdom is
regulated by several Statutes which control the location, shape
and height of a tip, the mechanical stability of the spoil and
pollution of the atmosphere and adjacent watercourses.

Considerable areas of land formerly covered by disused spoil
tips have been reclaimed for buildings, highways, recreation,

agriculture etc. Special methods of reclamation may be necessary
when the spoil is burning and/or acidic, but otherwise the spoil
may be treated as a normal engineering soil.

Applications for spoil are steadily increasing. The bulk of
the unburnt spoil taken from disused tips is used in civil earthworks such as highway embankments, and small quantities are used
for brick, cement and lightweight aggregate manufacture.

INTRODUCTION

Underground coal mining has provided a major source of basic
wealth to the United Kingdom for over two centuries. In the early
years of mining, the coal was relatively accessible and labour
was cheap, with the result that good quality coal could be selected
manually from the associated shales and other unwanted minerals in
the mine. Consequently, little or no mineral spoil came to the
surface. It is interesting to surmise that at the time, such
spoil as was produced was largely utilised in building up the land
to form foundations for the mine buildings, rail embankments etc.
It is understood that, at one nineteenth-century mine, the land
fill reached such a depth that the shafts were built up above the
original ground level to a height equal to the depth to which
they were sunk into the ground.

As the result of increasing mechanisation of the mining process
the proportion of unwanted minerals in the coal brought to the
surface (the run-of-mine) has increased to about 50% at the
present date. Much of this unwanted mineral is separated from
the coal before sale and has to be disposed of. Until recent years,
the normal method of disposal was to form a tip near to the mine
itself, and, until some twenty years ago, these tips were loosely
placed and had steep faces. Many of these older tips caught fire,
either by spontaneous heating or by accident, or even in some cases
by design, possibly in an attempt to increase the mechanical
stability of the tip or to generate burnt spoil. Such tips often
caused atmospheric and water pollution, rendered the land derelict,
and many were unattractive features of the landscape.

During the last twenty years, restrictive legislation and
the voluntary efforts of the coal mining industry has led to improved methods of tip construction. The new tips are mechanically
stable, do not become heated and do not cause significant water
pollution.   The modern tips also have a low profile and are
covered with vegetation so that they become an acceptable and often
useful part of the landscape.

The considerable heritage of disused spoil tips in the United
Kingdom is being systematically attacked by public authorities

in each coalfield, and the sites recovered for purposes such as
buildings, highways, agriculture, recreation, etc. Particularly
difficult sites may simply be beautified for the time being by
developing grass and tree covers on the spoil.

Spoil from the disused tips is also being used at an increasing
rate as a raw material for civil earthworks, brick manufacture and
other purposes.

It is now becoming more generally appreciated that coal mine
spoil is not a completely valueless waste material and that disused tips can be regraded to yield additional areas of land and
much of the spoil can be directly utilised. It may soon be
necessary to ask whether new spoil tips should be regarded as
temporary stocking grounds for a raw material rather than as permanent sites for the disposal of a solid industrial waste.

The following text is set out in the form of a review of the
properties of coal mine spoils, followed by methods of tip construction, the reclamation of land from disused tips, and, finally,
applications for the utilisation of spoil.

COAL AND SPOIL PRODUCTION

The coalfields of the United Kingdom are distributed throughout England, Wales and Scotland, as shown in Fig. 1. The coal
seams occur in the carboniferous strata, those of England and
Wales being largely in the Westphalian series (the Pennsylvanian
series of the United States of America) and those of Scotland
largely in the Westphalian and the Namurian (the Mississipian
series of the United States of America).

The coal seams have been mined in shallow workings since
Roman times in South Wales, the eastern and western slopes of the
Pennine hills of central and north-eastern England, and parts
of Scotland. These shallow deposits are now largely exhausted and
mining has extended to increasing depths, during the last 150
years, the deepest workings having reached over 1400 m. About
280 out of a total of some 500 underground coal mines in the United
Kingdom are operated by the National Coal Board, a nationally
owned, but independant Corporation, the other mines being privately
owned. About 99.5 percent of the total output of coal is produced
from the National Coal Board mines. Opencast (strip) mining of
coal was almost unknown until the year 1941, but today accounts
for about 9 percent of the total coal production of about 140 M
tonnes/year.

The solid wastes produced by coal mining have acquired a
variety of names within the different countries and regions of the

United Kingdom. These have included 'waste', 'refuse', 'dirt', 'slag' and 'spoil' and the deposits have been known as 'tips', 'bings', 'rucks', 'banks' and 'heaps'. For present purposes, the term 'spoil' will be used to describe all these materials, although it is evident that even this term is confusing since material temporarily displaced during opencast mining is also known as spoil. The term 'tip' will be used to describe deposits of all types of spoil, in conformity with recent legislation. The methods used for spoil disposal in the United Kingdom have recently been reviewed (Currie[1]).

Until some seventy years ago, the spoil discharged from coal mines in the United Kingdom consisted largely of waste rocks from shaft sinkings and headings, and small-sized coal screened from run-of-mine. The development and increased use of extractive machinery in the mine introduced increasing proportions of shales and sandstones from the roof and/or the floor of the coal seams into the run-of-mine. These unwanted components were rejected at first by manual selection and later by mechanical processes known as 'coal preparation' or 'washing'. The proportion of spoil compared to the saleable coal produced has risen from a few percent at the start of the twentieth century to some 30 to 50 percent at the present date.

The bulk of the spoil from inland underground mines has been deposited on the surface as near to each mine as was practicable, usually within a distance of a few kilometres. Spoil from mines near to the coast (amounting to some 9% of the total spoil production) has been deposited on the shore or in coastal waters within 5 kilometres of the shore. Negligible quantities of spoil have been re-stowed underground in the mine workings.

Opencast workings have not produced any residual spoil tips, the sites having been fully restored to their former use or to an alternative use selected by the landowner or by a public authority (Brent-Jones[2]).

The National Coal Board are currently in charge of some 2000 spoil tips containing about 2000 M tonnes of spoil. A further. possibly equal number of disused tips and comparative quantity of spoil are in either private or public ownership.

It was estimated in 1964 that disused coal mine spoil tips covered 8000 hectares of land in England and Wales, that active tips occupied a further 6000 hectares and that new land was being covered at the rate of 320 hectares/year ([3]). The rate of use of new land has more recently been estimated to be 250 hectares/year (Currie[1]).

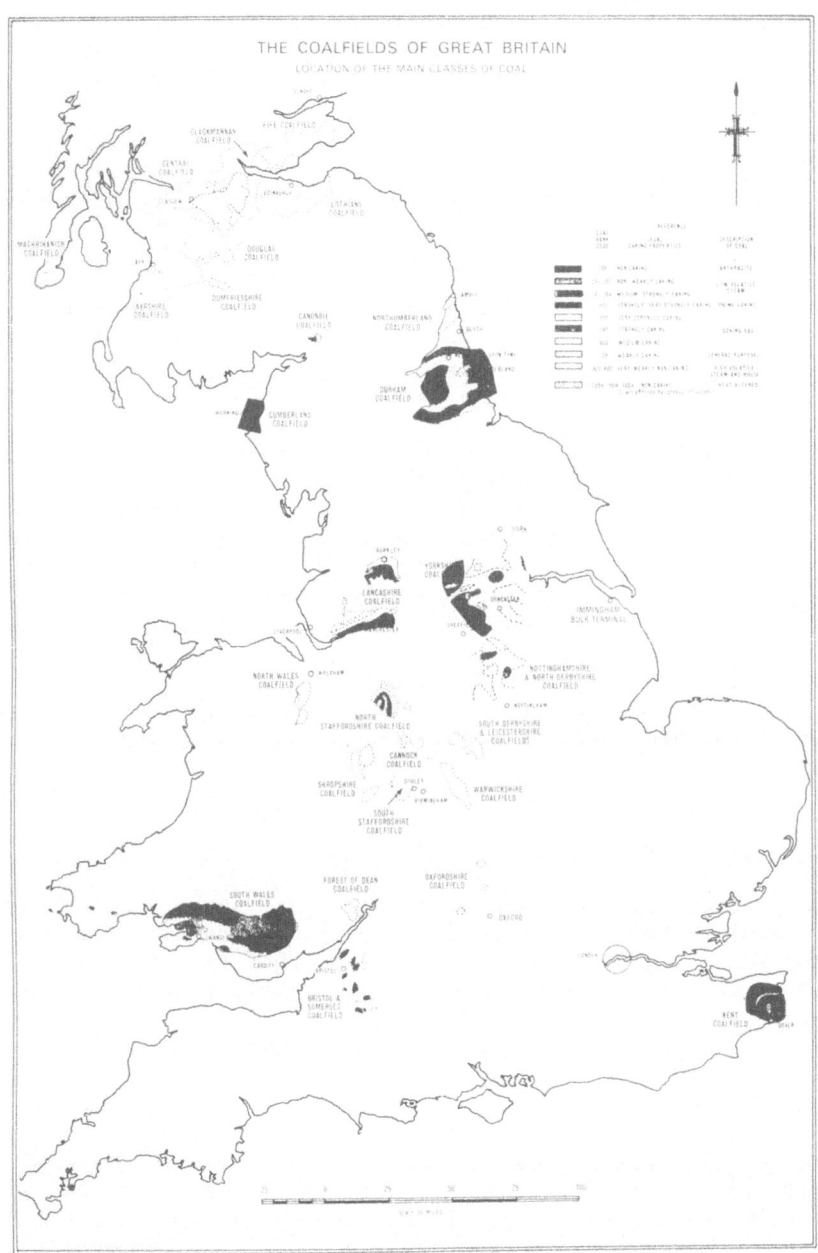

Figure 1. The Coalfields of Great Britain
(facsimile of the chart inserted at the end of the book)

THE PROPERTIES OF FRESHLY PRODUCED COAL MINE SPOIL

Spoil is discharged from coal mines or coal preparation plants in the following forms:-

Run-of-mine spoil   -   partially comminuted, usually dry, rocks
(also known as 'dry'   discharged directly from the mine workings.
 'white' or 'dirt')

Discard   -   solids larger than ½ mm discharged from
                              coal preparation processes. Discard
                              forms about 90% (dry weight) of the spoil
                              discharged from coal preparation plants.
                              Discards are usually saturated with water.

Slurry and Reject   -   suspensions of, or sludges containing,
                              solid particles, including coal, which
                              are smaller than ½ mm and which are dis-
                              charged from coal preparation processes.

Tailings   -   as slurries, but containing almost no
                              coal.

In testing coal mine spoils it may be advisable to avoid the drying-out of the samples before test. Test procedures based on British Standard 1377:1967 have been developed by the National Coal Board ([4]).

Particle Size Distribution

The particle size distribution of a large number of freshly produced spoils from coal mines in the United Kingdom are shown diagrammatically in Fig. 2. It is evident that the freshly produced discards and run-of-mine spoils are essentially granular in nature.

Specific Gravity

The specific gravity of the particles of freshly produced spoils lie in the ranges 1.8 to 2.7 for discards and 1.5 to 2.5 for slurries and tailings. There is a close correlation between the specific gravity and the ash yield (the complement of loss-on-ignition) of the spoils.

Loss-on-Ignition

The loss-on-ignition of oven-dried spoil containing no coal

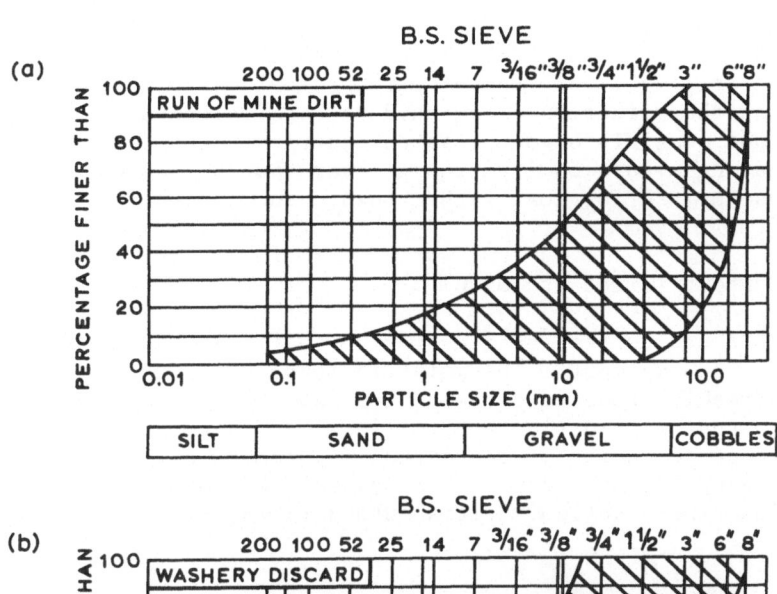

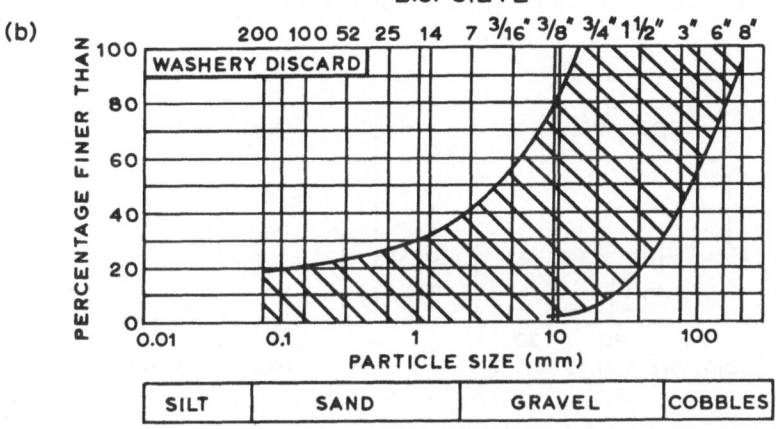

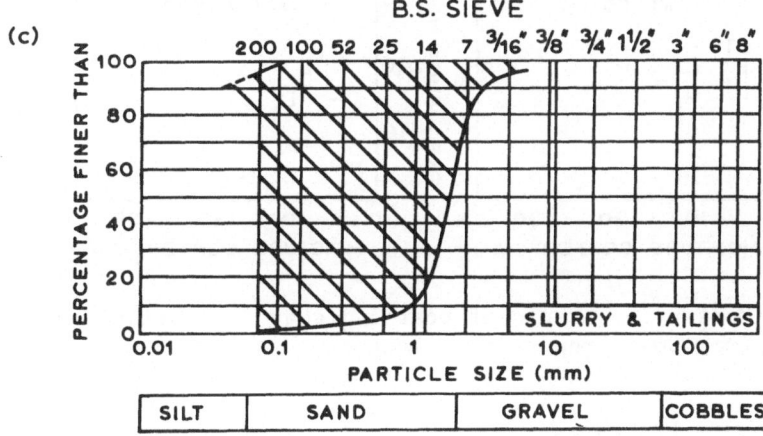

Figure 2. General ranges of particle size distribution for colliery discards as delivered from the Coal Preparation Plant

is about 10 to 20 percent, this loss being attributed to the dehydration of clays, the loss of carbon dioxide from carbonate minerals and the oxidation of sulphides. Spoils containing coal show a proportionately higher loss-on-ignition. In general, material having a loss-on-ignition greater than 60 percent is regarded as an inferior coal rather than a spoil.

Moisture Content (dried at 105°C)

The moisture content of run-of-mine spoils are usually in the range 1 to 3 percent. The moisture contents of a large number of freshly produced discards are shown in Fig. 3. The moisture content of slurries and tailings may range from 25 to 400 percent depending on the method of production. (Moisture contents are quoted on a dry-solids basis in conformity with civil engineering practice, British Standard 1377:1967).

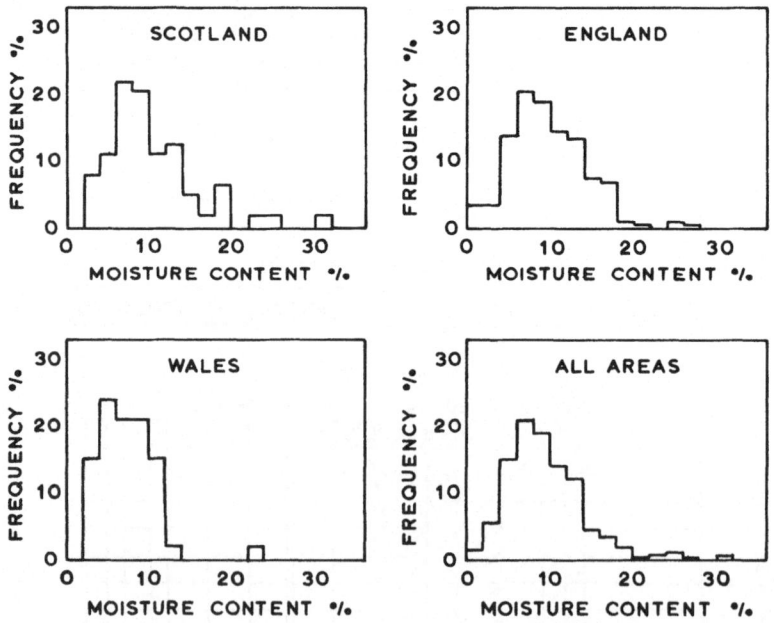

Figure 3.   Moisture contents of coarse discards as delivered from Coal Preparation Plant

Compaction Characteristics

The compaction characteristics of freshly produced discards
are shown in Fig. 4. It would appear from the graphs that
freshly produced spoil is slightly too dry to obtain maximum
compaction, but this is not so, the moisture content of fresh
spoil usually being a few percent too high. This apparent
anomaly is caused by the rejection of larger particles from
laboratory compaction tests.

Mineralogical Composition

The mineralogical composition of coal mine spoils in several
regions of the United Kingdom have been described by several
authors. The major components of the spoils are quartz and clays,
admixed in all proportions. The quartz grains vary from less
than 2 μm to 2 mm in size, and various cementing agents, including carbonate minerals, may be present. The clays are usually
kaolinite, with illite and lesser quantities of chlorite (Taylor[5];
Lawrence[6]; Dixon & Skipsey[7]). The other major components of
many spoils are coal, inferior coals and siderite.

The more arenaceous coarse grained rocks are termed sandstones; the finer grained rocks, siltstones; and the finest
grained rocks, mudstones. The mudstones and siltstones are
often collectively termed shales, although this term should
perhaps be reserved for the highly laminated argillaceous
rocks.

Minor but possibly important components of the spoils include iron pyrite, ankerite, calcite, apatite, garnet, feldspar,
rutile, sphene, tourmaline, and zircon (Lawrence[6]; Dixon & Skipsey[7]).

Chemical Composition

The chemical composition of the spoils reflects the
mineralogical composition ([5,6,7,8,9]). Free silica (quartz) may
be present in concentrations from 3 to almost 100%, and combined
silica in the form of clays may occur in the range from a few %
to about 60%. Aluminium concentrations may range from a few %
to 40% (as oxide) with calcium, magnesium, iron, sodium, potassium
and titanium in concentrations of a few %. Lower concentrations
of manganese and phosphorus may also be present, with trace
concentrations of the heavy metals, copper, nickel, and zinc.
Carbonates may be present in concentrations up to several percent.
The sulphur content of freshly produced spoil may be as high as
10% but is commonly less than 1%. The sulphur is distributed as
organic sulphur in the coal fraction of the spoil, and as iron

44

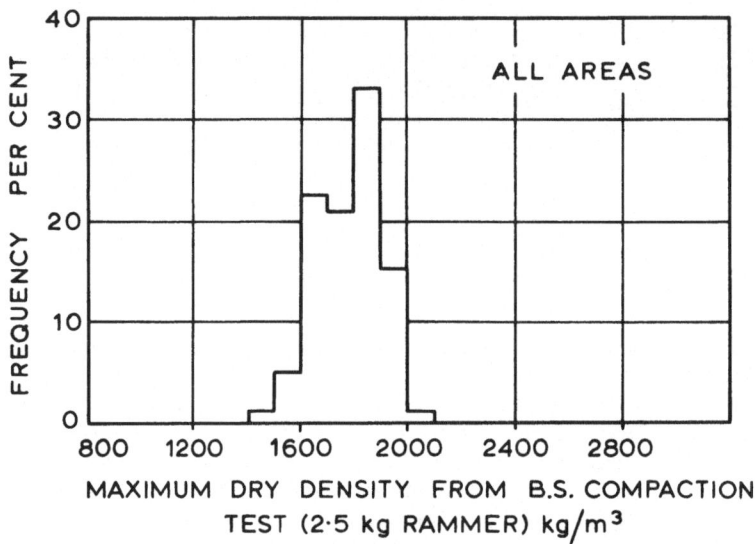

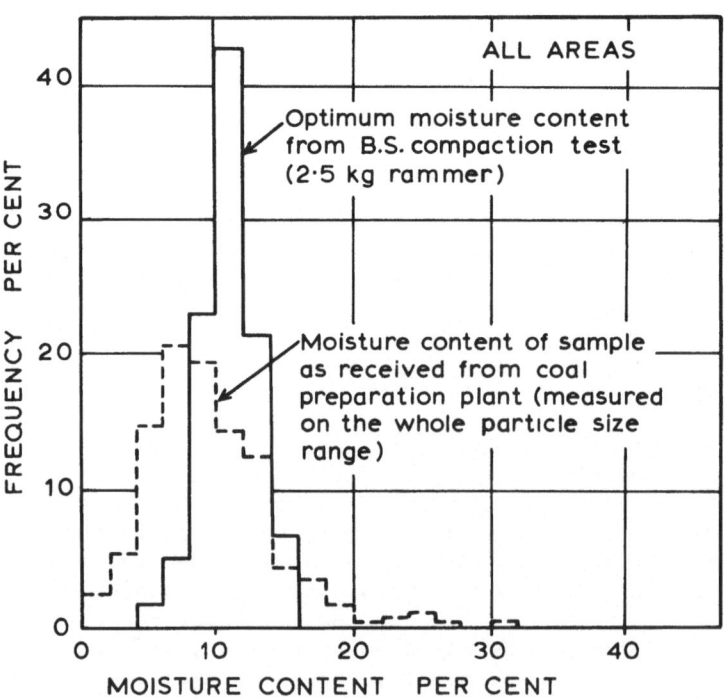

Figure 4. Compaction characteristics of coarse discards as delivered from coal preparation plant

pyrite. Freshly produced spoils contain no sulphur in the form of sulphates, other than barium sulphate, and a small amount dissolved in adhering water

Permeability

The permeability to water of the freshly produced discards is generally high (probably greater than 10^{-1} cm/sec) in the un-compacted state, and about 10^{-2} to 10^{-4} cm/sec) in the compacted state, provided that size degradation has not occurred during placing and compaction (see next section). Should size degradation have occurred, the permeability may be as low as 10^{-7} cm/sec.

The permeability to air of freshly produced spoils has not been studied in detail. It may be inferred from the few tests which have been reported that, provided the air-voids content of a well-compacted spoil is less than, say 5%, the permeability to air is effectively nil.

Shear Strength

The shear strengths of coal mine spoils have been the subject of considerable research during recent years, details of which are beyond the scope of this text. A summary of recent research on the properties of coal mine spoils has recently been published [10]. The angle of drained shearing resistance for a large number of spoils was reported to be in the range from 25.5 to $39°$ in England and Scotland, and from 25.5 to $41.5°$ in Wales, with the drained cohesive strength approximating to zero.

CHANGES IN THE PROPERTIES OF SPOILS

Coal mine spoils, having been recovered from a regime of low oxidation potential, high confining pressure and low moisture content, are both chemically and mechanically unstable with respect to conditions on the surface of the earth. The freshly produced spoils are, in effect, rocks which undergo a more or less rapid transformation into engineering soils, and given the correct conditions, into agricultural soils.

Oxidation and the loss of adsorbed gases start as soon as the minerals are exposed to air in the mine, and these changes, together with hydration and comminution, continue during subsequent transport and separation from the saleable coal. These changes, which may have important effects on the mechanical stability of the tip, the risks of fires within the tip, the

risks of water pollution by drainage from the tip, and on the
fertility to plants of the surface layers of the tip, will be
considered under appropriate sub-headings. The discussion relates almost entirely to discards which form the bulk of the
coarse spoil. As far as is known, the fine spoils (tailings and
slurries) do not change appreciably on exposure other than in
conditions of desiccation.

Particle Size Degradation Due to Wetting

The effects of water on the spoil component of the freshly
produced run-of-mine are important from the point of view of
the coal preparation process, and hence, the quality and quantity
of the spoil subsequently produced. Studies of the spoils produced in the different coalfields of the United Kingdom have
shown that the shales associated with low rank coals have a
greater tendency to break down in water than the shales associated
with the higher rank coals. The type of clay minerals present
did not show any correlation with the tendency of a shale to
breakdown in water, but the base exchange capacity, the grind-
ability, and the free moisture content did correlate with the ease
of break down ([11],[12]). Later, some doubt was cast on the validity
of the results of these tests, by the observation that the lower
the moisture content of a given shale at the moment of wetting
the greater the degradation (Cook[13]). A similar observation was
made on carboniferous shales from shallow deposits, these shales
tending to fracture if allowed to dry, but retaining their in-
tegrity if placed in water immediately after excavation ([14]). The
mechanism of shale degradation in water has been attributed to
the compression of air in the pores of the rock, to the swelling
of clay minerals, and to the osmotic effects of water soluble
salts in the rocks (Lawrence[6]; Cook[13]; Taylor & Spears[15]).

Tests made on a spoil which was known to degrade considerably
on handling show that no particle size degradation occurred during
storage in plastic bags of wet, freshly produced samples over
a period of three months (Whorton[16]).

Particle Size Degradation During Placing on the Tip

During the construction of trial embankments with freshly
produced discard, it was observed that considerable particle
size degradation occurred during the placing and compaction of the
spoil ([10]). At one tip, the percentage passing the 75 μm sieve
increased from about 7 to 25% during this operation. The bulk
of the degradation was attributed to the action of tracked bull-
dozers pushing the spoil across the surface of the tips for
distances of up to 300 m. The degradation observed on the tip

could be simulated in the laboratory, the effects of compaction
being appreciably less than the effects of abrasion which increased the percentage passing the 75 μm sieve from 9 to 50%
(Whorton[16]).

These observations are consistent with the general conclusion that freshly produced discards placed in thick layers
with little handling or compaction may behave as granular materials
of low to moderate permeability, whereas similar spoils subjected
to severe handling and heavy compaction may become cohesive and
of very low permeability.

Swelling

Individual particles of freshly mined spoil may swell on
wetting. This phenomenon may be associated with the size degradation previously discussed (Lawrence[6]; Taylor & Spears[15];
Spears[17]). The extent of swelling of water-sensitive specimens
of shale has been reported to be 0.1 to 1.2% ([18]), and swellings
of up to 20% in particular planes have also been reported (Annen
& Stalmann[19]). However, apart from the possible size degradation,
the swelling of freshly produced spoil should not have an appreciable effect on the construction of spoil tips or on most
types of civil earthwork, since sufficient voids are probably
present even in compacted spoil to accommodate any swelling. It
has been claimed that in highly compacted embankments, the effects
of swelling would be significant only in the top 2 m layer, and
even then would be uniform and generally acceptable ([19]).

Tests made by the National Coal Board have failed to observe
any swelling in highly compacted specimens reconstituted from
spoil taken from existing tips.

Effect of Freezing

Cycles of freezing and thawing have been found to cause size
degradation of the more argillaceous components of coal mine
spoils, the larger particles breaking down within a few cycles
into sand and fine-gravel sized particles with relatively little
increase in the silt and clay size fraction. The sandstones, coals,
inferior coals and iron-stones are relatively unaffected by
freezing.

The phenomenon known as 'frost heave' occurs when a compacted spoil lying on the ground is frozen via the exposed
surface. At the lower limit of penetration of the frost, an
ice lens tends to form which becomes progressively thicker due to
the suction of water from greater depths into a region of

effectively zero pore pressure. Ice lenses many centimetres
thick have been formed by this process under highways constructed from unsuitable materials. It is obviously necessary
to select materials having low frost-heave characteristics for
the parts of civil earthworks which will be exposed to frost
(Croney & Jacobs[20]).

The results of many tests made by or on behalf of the National
Coal Board have shown that many of the unburnt spoils have a
satisfactory frost-heave resistance, presumably because of the
low permeability of the compacted material.

Particle Size Degradation During Exposure on a Tip

It is necessary to distinguish between conditions on the
outer surface of a tip and conditions within the tip. Once spoils
have been buried at depths of more than about 1 metre within a
tip, little further changes occur due to the absence of oxygen and
the low rate of movement of water (Spears, Taylor & Till[21]).
These conclusions are applicable only to tips which have been compacted to a reasonably high density by machinery or have con-
solidated to a similar density. Some of the older tips containing
more granular spoils placed at low densities obviously suffer from
weathering effects such as oxidation at considerable depths.

The spoils exposed in the outer layers of tips undergo
considerable size degradation and oxidation ([6,15,17,19]). Tests
have shown that the mudstones and siltstones are the most sensitive
components whereas the sandstones, the ironstones, the coals and
the inferior coals are the least sensitive. The immediate effects
of weathering are the fissuring of the larger particles into sand
and fine-gravel sized particles. Some of the shales form flaky
particles which may partially protect the weathered layer. The
production of silt and clay-sized particles is generally negligible
at first, with the possible exception of some of the seat-earths.
Prolonged weathering of a tip eventually produces a layer of up
to 0.5 m thickness containing finer particles which may begin to
form crumbs suggestive of an agricultural soil. The weathered
layer may have a low permeability, and on steep slopes or on large
exposed areas may be subjected to erosion. Tips containing more
granular and resistant materials may permit the surface degradation
products to wash through the spoil into the base of the tip ([6]).

Salt Leaching

Neutral salts such as calcium and magnesium sulphates which
can inhibit plant growth, may be produced in the spoils by the
action of pyrite oxidation products on the carbonate and clay

minerals in the spoil. Calcium, magnesium, barium and strontium chlorides may also be present in the freshly produced spoil.

Little information has been published on the total soluble salt content of the United Kingdom spoils, or of the rate of leaching under various conditions of exposure. Toe drainages from spoil tips may contain as much as 5000 mg/l of chloride (as Cl) and 5000 mg/l of sulphates (as SO4), but these are often directly attributable to the presence of tailings and slurry lagoons constructed on the tips.

The Oxidation of Iron Pyrite

The oxidation of iron pyrite is of importance for the following reasons:

It may produce conditions toxic to plant life on the surface of a tip.

It may lead to the discharge of a polluting drainage.

It may affect the mechanical properties of the materials within a tip.

It may reduce the suitability of the spoil for subsequent utilisation.

It may increase the temperature of the spoil.

Oxidation begins at the moment of exposure of the pyrite to atmospheric oxygen within the mine, but the effective rate of oxidation is at first low. Oxidation products which may form in the mine, and during subsequent transport to the surface, are washed off the spoil during the coal preparation process, so that the spoil discharged from the coal preparation plant is usually free from acids and sulphates, although there may be a little sulphate present in the adhering water film. Providing the pyrite is placed, within a few hours or days, at depths greater than about 1 metre inside a tip, and the spoil is moderately compacted, it is unlikely that any further pyrite oxidation will occur.

Pyrite which is exposed for longer than a few weeks on the surface of a tip will oxidise at an appreciable and increasing rate to form the water-soluble products, ferrous and ferric sulphate, which may hydrolyse to sulphuric acid and ferric oxide. These primary oxidation products are almost immediately trapped by the adjacent clay and carbonate minerals to form calcium, magnesium and manganese sulphates, and provided that the oxidation continues, aluminium sulphate. As oxidation progresses, the zone of acidity around each pyrite particle continues to spread, depending on the relative abundance of the pyrite and the neutralising minerals, and the degree of leaching, and may eventually cover the entire surface and finally penetrate through to the toe of the heap. Alternatively, the pyrite may be exhausted and the oxidation products leach away to leave a potentially fertile soil. The shelf purifying capacity of spoil may be judged by the fact that some weathered spoils contain 5% of calcium sulphate, none of which could have been present in the fresh spoil.

The mechanism of pyrite oxidation has been studied in detail in recent years ([22],[23]). Two inorganic oxidation processes based on electron transfer at the pyrite crystal face have been identified. The significance of acidophilic ferrous oxidising bacteria in pyrite oxidation has not been established unequivocally. It has been claimed that in practice, bacterial cell densities are insufficient to account for the observed rates of pyrite oxidation ([24]). No practicable means of preventing pyrite oxidation, other than exclusion of atmospheric oxygen, has been discovered.

The Oxidation of Coal

The oxidation of coal is important since, if sufficient heat is evolved, the spoil may ignite. In the older spoil tips containing uncompacted granular spoils, fires often developed which were started by accident, natural causes, spontaneous heating, or deliberate initiation. In the more modern compacted tip and in civil earthworks, spontaneous heating is extremely unlikely and the rate of propagation of fires started by external sources of ignition should be negligible, provided that the spoil has not been dried-out after compaction, for example, by an external source of heat such as an industrial furnace.

Recent studies have shown that the rank of the coal is related to the reactivity to oxygen, the anthracites being least reactive ([25],[26]). Coal particles exposed to the atmosphere slowly lose their original affinity for oxygen. In compacted spoils it is evident that the factor which limits coal oxidation is the availability of air to the spoil inside the tip or embankment. The air supply is determined by convective and wind pressures and by the gaseous permeability of the spoil. The permeability is in turn a function of the particle size distribution, the moisture content and the dry density. The air permeability of highly compacted spoils is so low as to be unmeasurable in the laboratory, and calculations show that even under extreme conditions of wind pressure, the rate of oxidation within the spoil must be negligible.

THE PROPERTIES OF WEATHERED SPOIL

It will not be necessary to discuss the properties of weathered spoils in detail since the properties of freshly produced spoils and the effects of weathering have already been described. It may be concluded that weathered spoils differ from fresh spoils in the following respects:-

1. The particle sizes of spoil inside a tip are likely to be finer than that of the original spoil (Fig. 5, compare with Fig. 2), but the extent of degradation will not necessarily be related to the age of the tip.

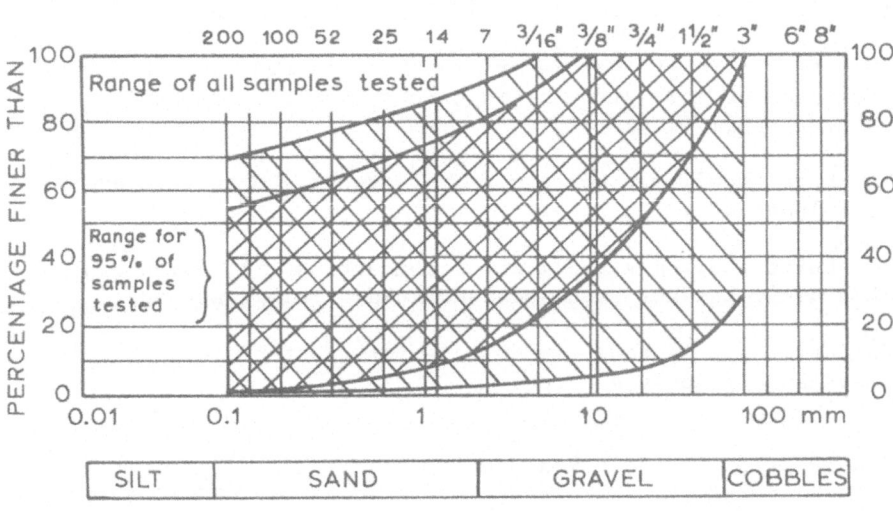

Figure 5.   General ranges of particle size distribution for coarse discards as found in existing tips

2. The particle size distribution of the spoil in the outer layers of a tip, including spoil washed from the surface into the toe, may be completely different to that of the original spoil.

3. Compacted weathered spoils are unlikely to swell even if subsequently soaked in water.

4. Weathered spoils may contain up to 0.5% of sulphates in the soil water (1:1 water extract) and up to 5% of acid soluble sulphates (as SO_3) (Sherwood & Riley[27]).

5. Weathered spoils, particularly spoil from the surface of
   a tip may be toxic to plant life because of the content
   of pyrite oxidation products. The pH value may be as
   low as 2 to 3. It is usual for the acidity to be confined to a zone around each pyrite particle.

6. The coal particles present in weathered spoils may have
   lost much of their original affinity for oxygen.

7. Should the oxidation of spoil in a tip have generated
   sufficient heat to cause ignition, the spoil will be in
   a more or less burnt condition. Burnt and partly burnt
   spoils have special characteristics as follows:-

   The minerals may be more or less completely modified,
   the clays being particularly susceptible to change (Taylor[5]).

   The shear strength parameters may be enhanced, and the
   spoil may be more granular in nature than the original
   unburnt spoil (Taylor[5]).

   The burnt spoil may contain high (percent) concentrations
   of sulphates.

   The burnt spoil may swell after compaction ([28]) (although
   this claim has not been substantiated by other authors.

   The burnt spoils have a wide range of frost heave
   resistance (Croney & Jacobs[20]).

THE CONSTRUCTION OF SPOIL HEAPS

   Until recent years, the only practicable means of disposal of
coal mine spoil in the United Kingdom was the construction of tips,
which were assumed to be the ultimate fate of the spoil. The
availability of large scale earth-moving equipment, and the demand
for large quantities of engineering soils for major civil earthworks is now leading to the concept that the spoil disposal tips
may be temporary features and that ultimately all of the spoil will
be put to beneficial use. However, the present rate of production
of spoil, some 50 M tonnes/year, is still considerably in excess
of the rate of utilisation (some 10 M tonnes/year) and it is assumed
in designing and constructing tips that they may remain undisturbed
indefinitely. The methods used by the coal industry in complying
with laws governing the construction of spoil tips have recently
been reviewed (Taylor[29]). The relevant Acts are:-

   Mines & Quarries (Tips) Act
   Town & Country Planning Act

Clean Air Act
Rivers (Prevention of Pollution) Acts
Deposit of Poisonous Wastes Act

The Mines & Quarries (Tips) Act

This Act was passed following the collapse of a coal mine spoil tip at Aberfan in 1966. Previously there was almost no legal control over the security of coal mine spoil tips. The Act and its corresponding Regulations specify the responsibilities and liabilities of the owner and of the manager of a spoil tip. The Act differentiates between active, closed and disused tips. Part 1 of the Act relates to the acitve and closed tips which are associated with a working mine, and part 2 of the Act relates to tips not associated with a working mine.

For the purposes of the Act, a tip is defined as, "An accumulation or deposit of refuse from a mine or quarry (whether in a solid state or in solution or suspension) other than an accumulation or deposit situated underground, and where any wall or other structure retains or confines a tip, then, whether or not that wall or structure is itself composed of refuse, it shall be deemed to form part of the tip for the purposes of this Act."

The chain of responsibility for the design of a new tip, supervision of the construction and maintenance of an active tip and the inspection and maintenance of closed and disused tips have been defined in detail by the National Coal Board ([30]).

After a considerable examination of the subject by the Board's Civil Engineers and by various consultants, a comprehensive technical handbook on tip construction has been published ([31]). This document gives advice firstly, on the continued construction of tips, including factors affecting the mechanical stability, site investigations, landscaping and the risk of fires. A second section deals with the design of new tips, including heaps of discard and lagoons, and a third section describes methods and safety aspects of tip construction.

Historically, tips were constructed in the United Kingdom on land adjacent to the mine by tipping from trucks or wagons (giving tips of low profile), or by tipping from an aerial ropeway, flight or tippler, giving tips of conical or ridge shape. In general, the object of tipping was to place as much spoil as possible on the available land, although there was a common law liability for material which overflowed on to other persons' or public lands.

The Town & Country Planning Acts

The location, the dimensions and the general appearance of
new spoil tips are controlled by public authorities under the
Town & Country Planning Act. The Act requires the mine owner to
apply to the planning authority for permission to construct a tip.
The authority may place restrictions on the actual area of land
to be covered, the dimensions of the tip, the method of construction
in so far as it may affect local amenities, and the appearance
of the tip during and after construction. Normally, the maximum
height will be chosen to blend into the local landscape, and the
tip may be designed so as to conceal an unsightly industrial
structure. The face gradients are usually limited to a maximum
slope of 1 in 3, and may be required to be as shallow as 1 in 7.
It is normally necessary to lift the surface soil before construction, and to place the soil on the finished surface of the
tip. Before final acceptance, the planning authority must publish
the proposals, and members of the public at large have the opportunity to object.

An important exception to the powers of the planning authority
under the Acts, is the general Development Order which permits the
continued construction of tips already in use in the year 1948.
Many of these sites are now filled and spoils are largely being
placed on controlled areas.

The Clean Air Acts

Under these acts, the owner of a mine is required to employ
all practicable means of preventing the combustion of refuse deposited from the mine, and of preventing or minimising the emission
of smoke and fumes from the refuse. These controls do not apply to
tips which were neither in use nor under the control of the present
owner in the year 1956.

It has been found that adequate control of fires can be obtained by compacting the spoil at the time of placing on the tip.
The generally lower face gradients which have been used in recent
years, the finer particle size distribution, and the higher
moisture content of the spoils are seen as additional safety factors.

The Rivers (Prevention of Pollution) Acts

These Acts, which are administered by the River Authorities,
control the deposition of solid waste in a water course, and
the pollution of surface water courses by any form of discharge,
and require that consent should be obtained for discharges of
trade effluents. In general the drainages from coal mine spoil

tips are not considered to be trade effluents within the meaning of the Acts, but where pollution by such drainages has been persistent and severe, the River Authorities have asked that an application for consent should be made.

The greatest quantities of polluting substances discharged from spoil tips are probably the fine solids which erode from the surface of the spoil in storm conditions. However, under such conditions, the adjacent surface watercourses are normally in spate, and are carrying solids washed off highways and agricultural land, so that the extra contribution from spoil tips is probably not significant.

Since the quantity of toe drainages from the tips is usually a small fraction of the total volume of rain falling on the tip, it is assumed that the majority of the rainfall either runs off the surface, soaks into the surface and is subsequently re-evaporated or soaks into the ground. Some of the seepages from the spoil tips are polluting in nature, but the quantities of pollutants involved are usually low and have not given rise to widespread complaint. Discharges from a very small number of spoil tips containing high concentrations of pyrite which have been constructed with steep slopes and little if any compaction, are more polluting, and in some cases it has been necessary to instal chemical treatment plants to purify the toe drainages. Treatment is considered to be only a temporary solution and it is anticipated that a control over the rate of pyrite oxidation and the rate of leaching of such oxidation products will eventually be obtained by regrading of the tips and the sealing of the spoil with materials of low permeability to air and water.

The Deposit of Poisonous Wastes Act

This Act makes it an offence to deposit on land, any poisonous, noxious or polluting waste in circumstances in which it can give rise to an "environmental hazard". For the purposes of the Act, it is not necessary to give notice of the intention to make deposits of coal mine spoils, but any pollution caused by the deposits may be liable under the Act, with the exception of discharges for which consent has been obtained under the Rivers (Prevention of Pollution) Acts.

The Deposit of Poisonous Wastes Act has also introduced the further problem that some coal mine spoil tips have been used as depositing grounds for noxious or polluting substances such as certain wastes from coke manufacture, and have also been used, sometimes without the permission of the owner for the deposition of other toxic wastes. The control of these activities is now a further problem for the mine owner.

THE RECLAMATION OF LAND FROM DISUSED SPOIL TIPS

Once a coal mine tip has become disused, the owner, or a
public authority may wish to reclaim the site for the purpose of
building, recreation, agriculture, or simply to improve the
appearance. It will not normally be necessary to reclaim tips
which have been constructed subject to Planning Act controls, since
these should conform to the planned use of the land.

The older tips which were constructed before the passing of
the Town & Country Planning Acts, and those which have been constructed since under the General Development Orders, may be
owned by the National Coal Board, or privately, or by a public
authority. The National Coal Board are not required to reclaim
any of these tips which are within their control, but may offer
to do so in connection with an application for the extension of
a tip or the construction of a new tip, or to reclaim land for
industrial use.

The costs of reclamation may be found by the owner of a tip,
particularly where the value of the reclaimed land would repay
the costs of reclamation. More commonly, the land is passed to
a public authority and then may become eligible for a grant from
central government funds of up to 85% of the costs, under Statutes
such as the Industrial Development Act, the Local Authority Act,
and the Local Employment Act.

The various methods which are available for reclamation
depend on the quality of the spoil, the projected use of the site,
the topography and other factors ([32,33,34,35]). These will be
considered briefly.

Complete Removal of the Spoil

This is a most drastic form of land reclamation which is seldom
feasible, and may even be undesirable, for example at a site where
subsidence caused by undergound mining has depressed the original
land surface below the drainage level. Situations in which total
removal of the spoil are practicable occur, for example, when
the spoil can be transferred directly into an adjacent opencast
mine working, or a disused quarry, or where the spoil can be
transferred for utilisation in a civil earthwork. Several examples
of the complete clearance of sites, each involving hundreds of
thousands of tonnes of spoil, have been observed in recent years.
It is a measure of the progress of technology that a mound of
spoil which may have taken hundreds of men several decades to build
up can now be moved away by a few men in a few weeks.

Recovery of Coal

A few of the older spoil tips contain coal which was un-
saleable, or could not be separated from the spoil, at the time
of deposition. Spoils of high coal content are more common in
the anthracite and steam coal areas of South Wales than elsewhere
since other coals are more reactive and are more likely to have
burned away after deposition. It may be practicable and profitable
to recover coal from some of these tips, particularly as part of
a general scheme of reclamation. In general, it is necessary that
the spoil should contain at least 25% of recoverable coal, and
that about 0.5 to 1 M tonnes of spoil should be available for
recovery (Hambleton[36]). Reclamation of as little as 12% of
recoverable coal may be profitable in particular situations. The
separation of the coal may create a secondary problem of spoil
disposal since much of the spoil will be in a very wet and more
finely divided state after recovery of the coal, but usually this
problem is preferable to the original problem of placing safely,
a spoil of high coal content which may be granular and may possibly have dried out by internal heating.

The Reclamation of Land for Buildings, Highways etc.

Since the conditions inside an existing spoil tip can seldom
be known accurately without making a comprehensive investigation,
the costs of which may not be justified, the only way to reclaim
land from a spoil tip with complete certainty as to its mechanical
strength and chemical stability is to lift and replace all of the
spoil. Fortunately, this is seldom necessary, even for applications
such as the support of motorways or industrial buildings. However,
the uncertainties as to the quality and degree of compaction of
any spoil which is left in place during reclamation pose difficult
design problems in the direct utilisation of coal mine spoil tips
in civil earthworks. The normal practice is to design the reclamation so that an appreciable quantity of spoil is taken from
the highest parts of the tip, knowing that, if necessary, the
transferred spoil can be inspected and properly placed. The
remaining spoil will hve undergone consolidation during the life
of the tip, and the density of the exposed surface can be measured,
assuming that, if anything, the density of the spoil in the deeper
layers will be even greater. If necessary, further site investigations can be made after regrading.

The dry density of the spoil in existing tips has been studied
extensively and has been found to vary considerably about a mean
value equivalent to 90% of the density attained in the standard
compaction test specified in B.S. 1377:1967 (Fig. 6). Spoils
having such densities probably have adequate strength for many
applications and several old tips have been used in this way as

58

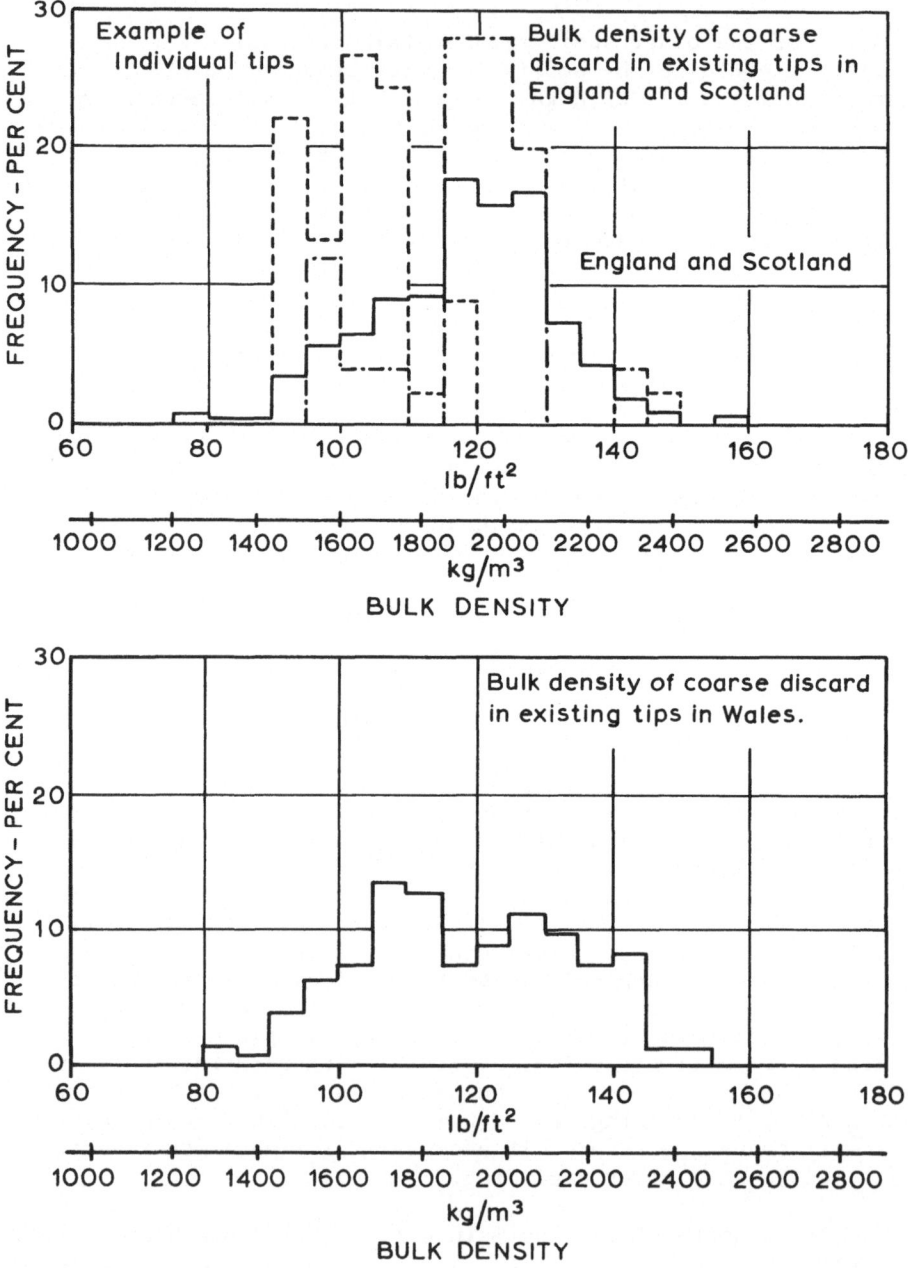

Figure 6.   Bulk densities of coarse discards as found in existing
            spoil heaps

foundations for new motorways, industrial estates, etc.

Should greater strengths and/or chemical stability than is
found in an existing tip be required it may be sufficient to cut
the spoil to a metre or two below formation level, and to replace
the spoil, after any necessary correction of the moisture content,
with appropriate compaction in thin layers. This process would
develop a strong layer which would be resistant to the penetration
of water and air and so would minimise the risks of heating, the
production of acidic conditions, and the discharge of polluting
seepages.

Alternatively, it may be practicable to reduce the permeability
of a spoil by the injection of a grout or sealant, but the costs
and durability of these methods do not seem to be attractive at
present [37,38].

The Reclamation of Land for Amenity Purposes, Agriculture etc.

The reclamation of tips for these purposes is generally less
critical than reclamation for buildings and civil earthworks, since
the bearing capacities of the surfaces are less critical and
greater surface gradients may be acceptable or even desirable.

The majority of tips present no particular materials handling problems other than the expense of regrading and, possibly,
the provision of adjacent land to reduce the gradients on the
original site. It is obviously advantageous if the spoil is sufficiently fertile to support plant life with no more than a
minimum of pH correction and fertiliser application, and many spoils,
both burnt and unburnt, have been found to have this property.
However, a significant number of spoils have been relatively infertile and some have been very toxic to plant life. Many studies
of spoil toxicity have been reported [35,40,41,42,43,44,45,46,47,
48].

In general, the freshly produced spoils are free from all
forms of life, are slightly alkaline in reaction, and would support
plant life, given the addition of certain essential nutrients. As
the spoils weather, the particles of iron pyrite oxidise, to
form small acidic zones which may spread to an extent determined
by factors such as the relative quantities of pyrite and alkaline
minerals, the particle size distribution, the surface gradient,
and the rate of erosion. The surface of a well-weathered spoil
tip may thus contain only small toxic patches, or large toxic
sections or may be completely toxic to plant life. In the long
term, all of the exposed pyrite will be oxidised and the oxidation
products leached away sufficiently to give a potentially fertile
soil.

A further problem of plant toxicity may be caused by neutral salts present in freshly produced spoil. These salts include sodium, calcium, magnesium, barium and strontium chlorides and are present in the strata from which the spoils are extracted. The concentrations of chlorides in the strata water may reach over 2000,000 mg/l (as Cl). Such spoils are only expected to be discharged from the deeper workings of the central English coalfields, and occasionally from the deeper workings in other coalfields.

During reclamation works involving the regrading of an existing tip, it might be necessary to make frequent inspections of the spoil exposed by the excavations in order to select the most suitable, potentially fertile spoils to be used to form the new surface layers.

Should suitable, potentially fertile spoils not be available during a reclamation scheme, it may be necessary to import a fertile soil. This is normally an expensive operation unless the scheme can be combined with, for example, an adjacent civil earthwork which yields suitable material.

An existing spoil tip which is burning or which is so acidic that polluting drainages are being discharged will normally require special consideration. These problems are caused by active oxidation of the spoil, and may be controlled by the formation, in the manner previously described, of a sealing layer over the affected parts of the tip. It may be necessary for this purpose to select well graded (normally unburnt) spoil which will compact to a low air-voids content. Such a sealing layer would have a low water permeability, and would be liable to erosion if left unprotected over any extensive area.

A suitable form of protection would be a layer of fertile soil which would provide a plant growth medium and would reduce the risk of drying out or disturbance by root action or frost of the seedling layer.

Sealing layers have been used successfully for the control of fires and polluting drainages at several sites in the United Kingdom in recent years, and the use of compaction as a means of excluding air forms an integral part of modern spoil tip design.

The lack of success of previously reported attempts to control fires by means of sealing layers (McNay[39]) may have been due to the use of unsuitable materials, to the layers being of inadequate thickness, or to the design having failed to prevent the drying out of the sealing layers with a consequent increase of permeability to air.

THE UTILISATION OF COAL MINE SPOIL

The applications to be considered under this heading are distinct from those of the previous section in that the spoils are transported from the site to be used elsewhere, rather than re-formed for some beneficial purpose on the original site.

Apart from the successful use of spoil for brick manufacture, the early history of spoil utilisation was not without problems, for example, the attack of concrete by sulphates present in burnt spoils which had been laid without regard to good practice, i.e., the provision of a suitable membrane between the spoil and the concrete. These unfortunate experiences did not encourage the wider use of coal mine spoils (O'Brien[28]; Eldridge[49]) and it was assumed for many years that the unburnt spoils would not only cause sulphate problems, but would also be liable to spontaneous heating, and would possibly soften or swell in water. During the last fifteen years, it has been found, largely as the result of experience with the modern spoil tips, and with large civil earthworks such as dams constructed from carboniferous shales ([14]), that these fears are unfounded, and the outlook for the unburnt spoils has changed considerably.

The incentive to find new applications for coal mine spoils received new impetus following the collapse of a tip at Aberfan in South Wales in 1966 which increased public interest in the disposal of coal mine spoils and other forms of solid wastes. The National Coal Board have considered spoil utilisation to be so important that a complete marketing organisation known as the 'Minestone Executive' has been established.

Utilisation applications for spoil, either raw or processed, which are now in use or under review have been described in recent publications (Hambleton[36]; Tanfield[50]; Nicolson[51]) and will be considered briefly under appropriate headings.

Common Fill for Civil Earthworks

This title refers to the filling of land and the construction of embankments generally up to the deepest anticipated frost level. Applications include highway construction, and the levelling of sites for buildings ([19]). Some ten to fifteen years ago the rate of utilisation for these purposes was probably 5 M tonnes/year, most of the spoil being burnt. The rate of consumption is now about 10 M tonnes/year in the United Kingdom, nearly 6 M tonnes of which is supplied from National Coal Board tips. Most of the present consumption is unburnt spoil. The total length of highway embankments in the United Kingdom which have been constructed on unburnt spoil fill is not known accurately, but is almost certainly greater than 20 Km. The very few problems which have arisen from these applications have been confined to the incorrect use of burnt spoils.

The unburnt spoils have been found to give adequate strengths when compacted in the normal manner, for engineering soils used as common fill, and, despite early forebodings, to have been easier to handle in wet weather than many of the competitive conventional engineering soils ([14]).

It is a fortunate coincidence that the moisture content of the spoils in the unburnt tips has almost always corresponded to the optimum for compaction so that no additional water has been required at the time of placing other than in very dry weather. The unburnt spoils have compacted so well that it is obvious that the water and air permeability is very low, and highway embankments consisting of spoil alone have been constructed to depths of more than 15 metres.

Studies at the Road Research Laboratory indicated that the sulphate content of coal mine spoil was not a limitation to its use for bulk fill, subject to precautions being taken to control contact between the fill and cement products (Sherwood & Ryley[27]).

No evidence of swelling of spoils after compaction has been reported, even though, at a few sites, freshly produced spoils have been used, and at one site, a light factory building was constructed directly on a mixture of old and freshly produced discard within a few weeks of placing.

No problems have been encountered with spontaneous heatings or fires in any compacted unburnt spoil used for a civil earthwork. The National Coal Board have constructed a demonstration embankment to motorway specifications from some 8000 tonnes of spoil having a high coal content. Despite the fact that the embankment was constructed in a well exposed position, with a steep slope facing into the prevailing wind, no temperature rise occurred over a period of three years, as indicated by 180 thermocouples buried in the embankment.

At several sites, spoil has been taken directly from a burning tip and placed in motorway embankments, the only restriction being that the spoil was placed in thin layers which were allowed to cool before being covered.

Low Duty Roads, Stocking Areas, Highway Sub-Base etc.

Coal mine spoils are seldom suitable for applications

requiring strong aggregates, although some of the burnt spoils may
find particular applications where high strength is not required.
From the point of view of frost-heave resistance, the burnt spoils
have been found to be very variable (Croney & Jacobs[20]). The
unburnt spoils are less variable and many may prove to be suitable
for direct use in applications such as 'superior common fill' above
the nominal frost level.

It is possible to up-grade spoils by stabilisation either
with lime or cement. The costs of adding as much as 10% of
cement to the spoil to give materials having high compressive
strengths may be justified in some circumstances. A large area
of cement stabilised unburnt spoil has been laid recently in
Scotland to provide a working area for the assembly of oil drilling
platforms.

In placing unburnt spoils near to the surface, it may be
necessary to take precautions to avoid excessive loss of moisture
which could lead to an increase of air permeability and a subsequent risk of fire. Obvious situations to be avoided are the
proximity to the spoil of industrial furnaces and heated pipe lines.
In critical situations, it may also be desirable to avoid planting
large trees near to paved areas or to buildings.

Burnt Clay Brick Manufacture

The use of coal mine spoils for brick manufacture is probably
nearly as old as the coal mining industry, and even today, some
of the carboniferous fireclays are mined specifically for the
purpose of making refractory bricks. At present, some 500,000
tonnes/year of coal mine spoil is used directly for the manufacture
of building bricks.

Cement and Sulphuric Acid Manufacture

About 130,000 tonnes/year of unburnt spoil is used in the
manufacture of cement by the conventional clay/limestone process
(Tanfield[50]), and in the combined cement/sulphuric acid process
(Smith & Gutt[51]).

Lightweight Aggregate Manufacture

It has been known for many years that coal mine spoils can
be converted into lightweight aggregates, either by heating to
induce bloating or by sintering of agglomerates of the spoil. The
coal content of the spoil may be used as the source of process
heat but if the coals concerned are of low rank, special steps

have to be taken to maintain the emission of smoke (tar fume) at
an acceptable level. Full scale plants have been constructed in
the U.S.A., in the U.K. and in Europe (Tanfield[50]; Nicolson[52];
Hanquez, Boutry & Chauvin[53]; Toubeau[54]). Some of these plants
consume spoil in hundred thousand tonne quantities annually.

Investigations made recently by the National Coal Board have
shown that tailings and slurries, which are produced at the rate
of some 5 M tonnes/year in the United Kingdom, can be dried and
burned in a fluidised bed combustor possibly without the need for
extra fuel. Pilot tests were made in a 1 m^2 combustor, and a
larger reactor is being constructed in Derbyshire. The ash may
find applications in the manufacture of lightweight building blocks,
and building boards, or may be used as a carrier for fertilisers
(Gibson[55]).

Road Surfacing Aggregates

A few of the spoils, both burnt and unburnt may contain
aggregates sufficiently strong to be used directly as road surface
dressings. These spoils are exceptional and little bulk market
can be expected to develop in this field. Trials have shown that
calcined spoils containing inclusions of granular calcined bauxite
offer promise as road surfacing aggregates having adequate strength
and high skid resisting properties (Gibson[55]).

Concrete Aggregates

A few of the spoils, in particular, the burnt spoils, may
contain aggregates of sufficient strength to justify their use
in the manufacture of concrete for low-duty applications. One
of the main outlets for this type of material could be the construction of packs in underground coal mines (Gibson[55]).

Tailings Digest Manufacture

Investigations made by the National Coal Board have shown
that tailings can be converted into a low grade plastic by digestion with heavy coal tar oil. The product is a bituminous
material which has low grade properties compared with synthetic
resins but is not as brittle as pitch (Gibson[55]).

ACKNOWLEDGEMENTS

The author wishes to thank the National Coal Board for permission to publish this paper and his several colleagues for assistance with the manuscript. Any opinions expressed are those of the author and not necessarily those of the National Coal Board.

Figs. 1 to 6 are reproduced by kind permission of the National Coal Board.

REFERENCES

1. Currie, W.J., Deep coal mining and the environment, Colliery Guardian, A. Rev. 1972, 56 & 63.

2. Brent-Jones, E., Methods and costs of land restoration, J. Inst. of Quarrying, 55. 341, 1971.

3. Anon., Derelict Land. A Study of Industrial Dereliction and How it May be Redeemed, Civic Trust, London, 1964.

4. Anon., Application of British Standard 1377:1967 to the testing of Colliery spoil, Natn. Coal Bd, London, 1971.

5. Taylor, R.K., Compositional and geotechnical characteristics of a 100 year old colliery spoil heap, Trans. Inst. Min. & Metall., 82, 1, 1973.

6. Lawrence, J.A., Some properties of South Wales colliery discards. Colliery Guardian, 1972, 220, 270 & 329.

7. Dixon, K. & Skipsey, E., The distribution and composition of inorganic matter in British coals I, Initial study of seams from the East Midlands Division of the Natn. Coal Bd, J. Inst. Fuel, 37, 485, 1964.

8. Dixon, K. & Skipsey, E., The distribution and composition of inorganic matter in British coals II, J. Inst. Fuel, 43, 124, 1970.

9. Dixon, K. & Skipsey, E., The distribution and composition of inorganic matter in British coals III, J. Inst. Fuel, 43, 229, 1970.

10. McKechnie Thomson, G. & Rodin, S., Colliery spoil tips - after Aberfan, 1972, Civ. Engrs, 1972.

11. Berkovitch, I., Manackerman, M. & Potter, N.M., The shale breakdown problem in coal washing I. Assessing the breakdown of shales in water, J. Inst. Fuel 32, 579, 1959.

12. Horton, A.E., Manackerman, M. & Raybould, W.E., The shale breakdown problem in coal washing II. Some causes for shale breakdown and means for its control, J. Inst. Fuel 37, 52, 1964.

13. Crook, M.D., The effect of changes in the moisture content of shale on its breakdown characteristics, Natn. Coal Bd, West Midlands Division, D6 B44/559, 1964.

14. Kennard, M.F., Knill, J.L. & Vaughan, P.R., The geotechnical properties and behaviour of Carboniferous shale at the Balderhead Dam, Q.J. Engng. Geol., 1, 3, 1967.

15. Taylor, R.K. & Spears, D.A., The breakdown of British coal measure rocks, Int. J. Rock. Mech. Sci, 7, 481, 1970.

16. Whorton, B., Laboratory investigation of the particle size distribution of coarse washery discard from Askern Colliery, Natn. Coal Bd, Scient. Dep., Yorkshire Regional Laboratory Report No. YRL 2037, 1972.

17. Spears, D.A., A laminated marine shale of carboniferous age from Yorkshire, England, J. Sedimentary Petrology, 39, 106, 1969.

18. Belin, J., Jeger, C. & Dubart J-Ch., A study of the effect of humidification on the mechanical properties of some rocks, Revue De L'industrie Minerale, 53, 21, 1971.

19. Annen, G. & Stalmann, V., Washery discard in dike and embankment construction, Natn. Coal Bd. Transl. A. 2794/AL, 1336, 1969.

20. Croney, D. & Jacobs, J.C., The frost susceptibility of soils and road materials, Minist. Transport, Road Res. Lab. Report LR 90, 1967.

21. Spears, D.A., Taylor, R.K. & Till, R., A mineralogical investigation of a spoil heap at Yorkshire Main Colliery, J. Engng. Geol., 3, 239, 1970.

22. Anon., Acid Mine Drainage Formation and Abatement, U.S. Environ. Prot. Agency, Water Quality Office, Wat. Pollut. Contr. Res. Ser., DAST -42, 14010 FPR 04/71, 1971.

23. Morth, A.H., Smith, E.E. & Shumate, K.S., Pyritic systems, a
     mathematical model, U.S. Environ. Prot. Agency. Office
     of Res. and Monitoring, Environ. Prot. Technol. Ser.,
     EPA-R2-72-002, 1972.

24. Lau, C.M., Shumate, K.S. & Smith, E.E., The role of bacteria
     in pyrite oxidation kinetics, Proc. Third Symp. on Coal
     Mine Drainage Res. Mellon Inst. Pittsburgh, Pennsylvania,
     1970.

25. Chamberlain, E.A.C., Hall, D.A. & Thirlaway, J., The ambient
     temperature oxidation of coal in relation to the early
     detection of spontaneous heatings I, Min. Engr., 130,
     1970.

26. Chamberlain, E.A.C., Hall, D.A. & Thirlaway, J., The ambient
     temperature oxidation of coal in relation to the early
     detection of spontaneous heatings II, Min. Engr., 132,
     387, 1973.

27. Sherwood, P.T. & Ryley, M.D., The effect of sulphates in
     colliery shale on its use for roadmaking, Minist. Transp.
     Road Res. Lab. report LR 324, 34, 1970.

28. O'Brien, T., Solid floor on a shale fill, why did it fail?
     Building, 222, 102, 1972.

29. Taylor, A.R., The management and disposal of coal mining
     refuse practice at British Collieries. World Hlth. Org.,
     Training Course on the Management and Disposal of Indust-
     rial Solid Wastes Mining Residuals and Sludges, Katowice,
     Poland, 20, 1972.

30. Anon. Tips - Codes & Rules, Natn. Coal Bd., 82, 1971.

31. Anon., Spoil heaps and lagoons - Technical handbook, Natn.
     Coal Bd., 1970.

32. Ranwell, D.S., Landscape improvement advice and research,
     J. Ecol., 55, 1, 1967.

33. Goodman, G.T., The nature of the technical advice required
     when creating land affected by industry, J. Ecol., 55,
     27-34, 1967.

34. Nicholson, M., Landscape surgery at Stoke-on-Trent. New
     Scientist, 51, 752, 1971.

35. Anon., Landscape Reclamation, vols I & II, I.P.C., Guildford,
     1971 & 1972.

36. Hambleton, G.C., Reclamation from coal mine waste dumps. Can. Min. and metall. Bull., 65, 83, 1972.

37. Anon., Use of latex as a soil sealant to control acid mine drainage. U.S. Environ. Prot. Agency, Wat. Pollut. Contr. Res. Ser. 14010, EFK 06/72, 1972.

38. Anon., Silicate treatment for acid mine drainage prevention, U.S. Environ. Prot. Agency, Wat. Poll. Contr. Res. Ser. 14010 DLI 02/71, 1971.

39. McNay, L.M., Coal refuse fires, an environmental hazard, U.S. Bur. Mines. Inf. Circ. 8515, 1971.

40. Ellsworth, K.J., Plant growth on pit heaps - a literature survey, Natn. Coal Bd, Res. & Devel. Dep., Intelligence Group, 1967, supplement, 1972.

41. Wood, R.F. & Thirgood, J.V., Tree planting on colliery spoil heaps, For. Comm. Res. Branch, Pap. 17, 1955.

42. Frank, R.M., A guide for screen and cover planting of trees on anthracite mine-spoil areas. U.S. Forest Serv. Res. Pap. NE-22, 1964.

43. Cornwell, S.M. & Stone, E.L., Availability of nitrogen to plants in acid coal mine spoils, Nature, Lond. 217, 768, 1968.

44. Berg, W.A. & Vogel, W.G., Manganese toxicity of legumes seeded in Kentucky strip-mine spoils, U.S. Forest Serv. Res. Pap. NE-119, 1968.

45. Chadwick, M.J., Cornwell, S.M. & Palmer, M.E., Exchangeable acidity in unburnt colliery spoil. Nature, Lond., 222, 161, 1969.

46. Beyer, L.E. & Hutnik, R.J., Acid and aluminium toxicity as related to strip-mine spoil banks in Western Pennsylvania, Penn. State Univ., Sch. Forest Resources, Coll. of Agric., Special Res. Rep. SR-72, 1969.

47. Davis, G., Ward, W.W. & McDermot, R.E., Coal Mine Spoil Reclamation, Scientific Planning for Regional Beauty and Prosperity, Proc. Sch. Forest Resources Symp. Penn. State Univ., Coll. Agric., 1965.

48. Barthauer, G.L., Kosowski, Z.V. & Ramsey, J.P., Control
    of mine drainage from coal mine mineral wastes, Phase 1,
    Hydrology and related experiments. U.S. Environ. Prot.
    Agency, Wat. Pollut. Contr. Res. Ser. 14010 DDH 08/71,
    1971.

49. Eldridge, H.J. , Concrete floors on shale hardcore. Mun.
    J., 72, 2893 & 2896, 1964.

50. Tanfield, D.A., Construction uses for colliery spoil. Con-
    tract J. 14 & 21, Natn. Coal Bd. Minestone Exec., 1971.

51. Smith, M.A., & Gutt, W., The use of colliery shale as a raw
    material and reductant in the cement/sulphuric acid
    process. Cement. Technol., 4, 3, 1973.

52. Nicolson, S.R., Utilisation of colliery waste in the building
    and construction industries and in agriculture. Min.
    miner. Engng. 7, 27, 1971.

53. Hanquez, E., Boutry, C. & Chauvin, R., Colliery shale as a
    raw material for building purposes, Sixth Int. Conf. on
    Coal Prep., Paris, Pap. 7E, 1973.

54. Toubeau, G., Manufacture of lightweight aggregates from
    washery screen discard. Sixth Int. Conf. on Coal Prep.,
    Paris, Pap. 8E, 1973.

55. Gibson, J., An appraisal of the utilisation of potential
    colliery waste. Coal Prep. Soc. Min. miner. Engng., 1970.

RECLAMATION OF COAL MINED LAND IN THE UNITED STATES AS COMPARED
WITH THE RUHR

R.J. Hutnik and Grant Davis

Pennsylvania State University and United States
Department of Agriculture, Forest Service, U.S.A.

ABSTRACT

There are differences in reclamation practices between the
United States and the Ruhr district of Germany mostly because of
the dissimilarities in geography and the stage of development of
mining laws. Mining in the United States is conducted over wide
ranges of climate, physiography, and geology as constrasted to
the rather homogeneous geographic area of the Ruhr. Surface
mining, underground mining, and open-pit operations are found
throughout the coal fields in the United States, but underground
mining is dominant in the Ruhr. Reclamation practice is generally
more sophisticated in the Ruhr, especially in spoil placement,
water-quality control, and land-use planning. More intensive
reclamation in the Ruhr is probably a result of a longer history
of mining to allow development of adequate reclamation laws
and because of greater demands on the land in this densely
populated region.

INTRODUCTION

The comparison of coal mining and reclamation in the United
States with that in the Ruhr district of Germany must first take
into account the dimensions of distance and time. Much of the
difference between the two areas can be explained in these terms.
Even if we consider only the 48 contiguous states, the area of
the United States is 1700 times that of the Ruhr. This almost
ensures that there will be important differences in the climate,
physiography, and geology among the coal regions of the United
States. In addition, the United States contains many different

types of coal, ranging from lignite to anthracite. In contrast,
the Ruhr district is characterized by small differences in
climate, physiography, geology and coal types. Anthracite mining
in northeastern Pennsylvania, for example, has no counterpart in
the Ruhr, nor has coal mining in the arid or high-altitude regions
of western United States.

Another related spatial consideration is that of population
density. The Ruhr has one of the highest concentrations in the
world, exceeding 1200 inhabitants per square kilometre. Nowhere
in the coal mining regions of the United States does the population
density approach this value for a similar-sized area. Indeed,
much of the coal mining in the United States is in remote areas
far from high concentrations of people.

The significance of this difference in population density
is that mining or reclamation operations of similar size and
nature would directly affect many more people in the Ruhr than
anywhere in the United States. It follows that the mining
industry in the Ruhr should be subject to greater public pressures
than it is in the United States. In the United States, however,
substantial public pressures are generated by those living outside the coal regions; often organizations are established for
just that purpose.

Problems are more complex in the Ruhr district because it is
a centre of industrial activity other than coal mining. In
fact, the environmental impact of these other industries may well
be substantially greater than that of coal mining. For example,
species of plants used in reclamation must not only be tolerant
of the adverse conditions created by mining, but must also be
tolerant of the conditions caused by other industries, such as
the air pollution they generate.

The other important difference between the two areas is in
their mining history. Coal mining in the United States did not
commence on any appreciable scale until the middle of the last
century. And only in the past few decades has there been any
regulation of the industry as far as environmental effects are
concerned. In contrast, early reclamation work in the Ruhr dates
back to the end of the last century and since 1932 the Ruhr
Planning Authority has been concerned with the location and
reclamation of colliery spoil heaps to ensure that they are taken
fully into account in the total regional plan.

In the broader perspective, the differences in reclamation
are traceable to differences in the philosophical approach to
land use. Over the centuries in Germany, land has been recognized
as a valuable asset to be used wisely so that future use will not
be adversely affected. Until recently, the philosophy in the

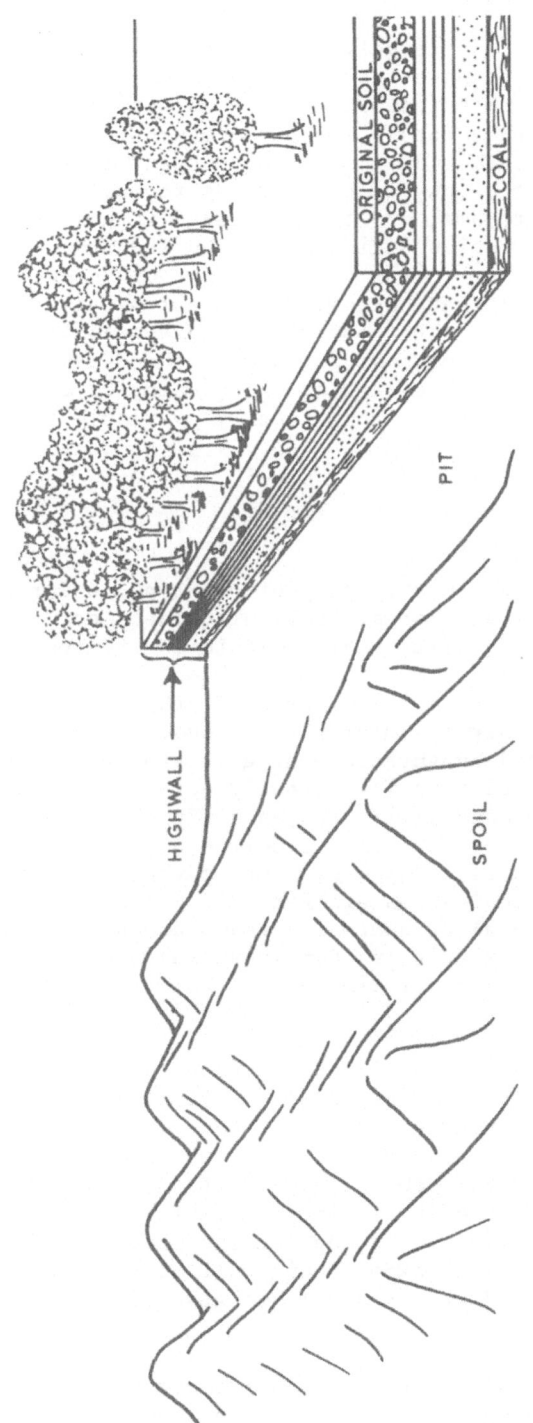

Figure 1. Typical cross-section and isometric view of area strip mine showing spoil pits and horizontal rock strata exposed by the highwall

United States has been one largely of land exploitation. To
extract from the land the greatest return on investment was
considered not only permissible but also good business practice.
It was inconceivable that future economic returns could be
sufficiently high to permit any money to be spent on reclamation.
The legacy of this philosophy is readily apparent in the mining
scars, the polluted streams, and the depressed economy so
prevalent in the Appalachian region of the United States.

Forecasts indicate that coal mining is not a dying industry
as was once believed. In both countries, there will be a continually increasing need for coal so long as the demand for
electrical energy keeps increasing at its present rate and until
other sources of energy become more readily available than at
present.

MINING METHODS

Coal mining in the Ruhr is only of one type, underground
mining. All the mines are large. Only bituminous coal is
produced, although its quality varies within the region.

In the United States on the other hand, there is about as
much coal produced by surface mining as by underground mining.
The proportion produced by surface mining is gradually increasing
with time. Mining operations vary greatly both in size and in
method in both underground mining and in surface mining. Most
of the surface mining for coal in the United States can be
classified as either area strip-mining or contour strip-mining.
In regions where the coal seams are very thick (as in the lignite
deposits in South Dakota) or where the seams are highly pitched
(as in the anthracite deposits in Pennsylvania) the mining
practice is closer to open-pit mining than to either area or
contour strip-mining.

Area strip-mining is used if the terrain is relatively flat
and the coal seam is nearly horizontal. In this method a trench
is made through the overburden, exposing the coal seam. After
the coal is removed from the initial cut, the adjacent strip is
excavated, with the overburden being deposited in the previous
cut. Each adjacent succeeding strip is made in a similar manner.
In the final cut (usually at the edge of the property) an
open trench is left with an exposed highwall as the boundary.
Unless graded, the mined area resembles a gigantic washboard
(Fig. 1).

It is often possible to remove a large amount of coal in a
single area strip-mining operation. For this huge equipment is
most effective and it may be economically feasible to remove 60 m

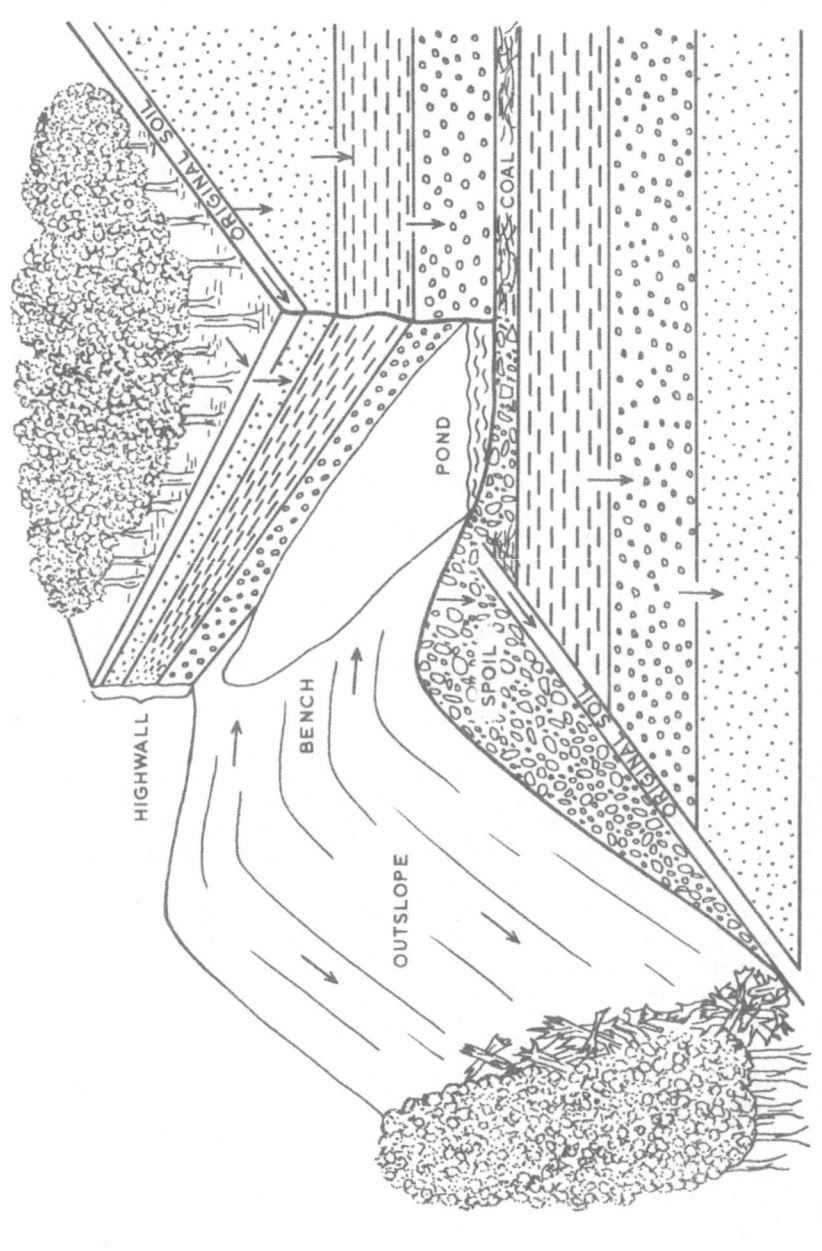

Figure 2.   Typical cross-section and isometric view of contour strip mine showing graded bench, pond, outslope, and windrow of vegetation at toe of spoil.   Arrows represent the direction of water movement

or more of overburden.

In hilly or mountainous terrain, contour strip-mining is employed. Usually the coal seam is almost horizontal. The overburden is removed and piled on the down-hill side, exposing the coal. The width of the strip depends on the slope of the land and the amount of overburden which can be removed economically. Eventually a point is reached when it is no longer economical to remove the overburden for the amount of coal obtained. The finished operation leaves a nearly vertical highwall, a bench from which the coal has been removed, and spoil heaps on the outer slope (Fig. 2).

Additional coal may be removed by using augers which are bored into the coal seam at the highwall for distances up to 50 m. This is very efficient in terms of cost per ton of coal removed but very inefficient in terms of proportion of coal removed. Drift mines can also be established leading from the headwall.

The development of more efficient equipment permits the economical removal of larger amounts of overburden and generally results in a greater degree of land disturbance. It is now economical to return to many of the previously strip-mined areas and remove additional coal. Such operations often destroy the results of reclamation carried out following the initial strip mining. On the other hand, previously non-reclaimed or poorly reclaimed sites mined for a second time must now be reclaimed under newer and stricter regulations.

Land disturbance is not restricted to surface mining. Underground mines generate considerable amounts of waste material, which are often piled near the mine openings or the coal-processing sites. The amount of disturbed land is much smaller than that resulting from strip-mining. However, the environmental problems created are relatively more serious. Such wastes are more often located close to residential areas, and the piles are visible over a wide area. Usually they are more physically unstable and contain greater amounts of toxic materials. Such problems are dealt with in greater detail in the next section.

ENVIRONMENTAL PROBLEMS CREATED BY COAL MINING

Many of the adverse environmental conditions resulting from coal mining cannot be improved by revegetation schemes. These include such problems as: land subsidence; mine fires; noxious fumes from burning mine dumps; dust and erosion from haul roads; flying debris, noise and vibration from blasting operations and polluted water drainage from both active and abandoned underground mines. Although these are serious problems that need

to be solved by the mining industry and governmental agencies,
they lie outside the scope of this paper. Instead, emphasis
will be placed on the problems which are alleviated by revegetation of the disturbed land.

Perhaps the most obvious problem, the blighted landscape
resulting from mining, cannot be evaluated economically. Certainly,
most people are offended by the scarred landscape following
mining. The closer the scars are to towns, cities, and major
highways, the more visible they are to greater numbers of people.
Thus the aesthetic problems are potentially more serious in the
Ruhr region than in the United States. On the other hand,
reclamation and planning have progressed much further in the
Ruhr region, due to public pressures and it does not contain the
backlog of extremely ugly areas such as those found in Appalachia.
Towns containing these scars of past mining operations are
usually economically depressed, and new industries are reluctant
to move into such unattractive surroundings.

Physical instability of the banks can be a tragic consequence
of bad spoil-heap management as witnessed by the disaster at
Aberfan, Wales, in 1966. The outslopes of contour strip-mines
in steep topography are also subject to large-scale landslides.
Although less noticable, the smaller-scale instability is much
more common. Streams become choked with sediments from nearby
eroding banks. On windy days, clouds of dust originate from
piles or from filled sediment basins containing the fine waste
material separated from the coal in the cleaning operations.

An even more serious form of water pollution in some areas
is chemical pollution. Sulphuric acid results from the oxidation
of pyrites and other sulphur-bearing minerals. The acid water
enters streams, often killing almost all of the stream biota for
several kilometres downstream. Associated with the low pH is
the precipitating out of dissolved substances in the water. For
the most part these substances are iron compounds, especially
hydrated ferric oxide and iron sulphate. Aluminium, manganese,
calcium and magnesium sulphates may also be present. The water
and the rocks in and along the stream become discoloured, and
the precipitates, especially hydrated ferric oxide, can build
up to a considerable level. Although these chemical pollutants
emanate mostly from underground mines, they may also be present
in the surface runoff and seepage waters from both underground-
mine and strip-mine banks.

ENVIRONMENTAL FACTORS PREVENTING PLANT ESTABLISHMENT

Although many of the environmental problems created by coal
mining could be alleviated by establishing a vegetation cover on

the spoil banks, natural colonization is slow and many banks are essentially bare even after more than 50 years. Several factors contribute to the failure of plants to become established on these banks. The spoil material represents a substrate that is far different in many important physical and chemical characteristics from those of the native soils to which the vegetation is adapted. Furthermore, the spoil material varies from place to place because of the many different seams of coal and overlying rock strata.

Physical instability limits plant establishment on steep banks. Recurring landslides, spoil slippage, and sheet and gully erosion prevent seedlings from becoming established. Even on level areas, plants can be killed or seriously injured by the blowing of unstable fine surface waste materials.

The surface layers of the banks dry rapidly because of the coarse texture and high porosity of the spoil material and because of its exposure to sun and wind. This effectively prevents the establishment of almost all the newly germinated seedlings. However, once seedlings are established, soil moisture is seldom limiting since there is usually adequate moisture at the deeper layers.

The dryness of the surface layers, combined with their customary dark coloration and exposure to solar radiation, are conducive to rapid changes in surface temperature. Surface temperatures high enough to kill young seedlings are common during the Summer. Older seedlings are better able to survive such high surface temperatures because they have thicker bark, they furnish more shade to the surface at their base, and a relatively smaller part of the plant is exposed to the lethal temperature zone at and near the surface.

Spoils containing significant amounts of fine material have higher moisture, and hence better surface temperature, conditions. However, young seedlings on such banks may encounter two additional problems, compaction and frost heaving. Frequently, grading operations compact the soil enough to drastically reduce infiltration and root penetration. Consequently sheet erosion is accelerated and seedlings cannot become established. Frost heaving, a less common problem, can occur on banks with high surface moisture contents during the Winter and early Spring.

In addition to the potentially adverse physical characteristics described above, vegetation of coal-mine spoil banks face a host of problems of a chemical nature. Foremost among these is the high acidity resulting from the oxidation of iron sulphides such as pyrite. On spoils having pH of 3.5 or less, seedling growth is severely retarded and mortality is high. Although a high concentration of sulphuric acid is in itself highly toxic, the low pH also affects the solubility of other compounds. Nutrients, such as phosphorus, are commonly fixed into insoluble compounds, and toxic ions, such as aluminium and manganese, become soluble.

Even in the absence of high acidity, the spoil material is characteristically deficient in nutrients. Nitrogen and phosphorus are almost always deficient. Contents of the other nutrients are highly variable from bank to bank.

Total salts, primarily sulphates, are comparatively high. The contents may be high enough on some banks to prevent seedlings from becoming established.

Although any one of these chemical factors by itself may be sufficiently severe to kill plants on the coal-mine spoil, more likely they reduce the growth and vigour sufficiently to make the plants susceptible to other environmental stresses (such as drought, high temperatures, and diseases) for a much longer period of time.

RECLAMATION LAWS

In spite of the environmental problems associated with mining the American public did not seem to be greatly concerned about regulating surface-mining operations until after World War II. It was not until the late 1940's when surface mining for coal went through a period of rapid expansion that several eastern coal states began to enact laws requiring some reclamation.

In the United States, the primary responsibility for regulation of mining and reclamation has been assumed by the individual states. As the citizens within a state became concerned about environmental damages from mining, they pressed for regulations. Various environmental groups were formed, and attempts to regulate the mining industry became a highly emotional issue which resulted in many heated debates. Because there was little interstate organization among the environmental groups proposing legislation and because mining conditions differed from state to state, every state has different reclamation laws. One state may place great emphasis on some phases of mining or reclamation that would be virtually ignored in other states. Administration of laws is further complicated because enforcement of regulations in some states is delegated to several agencies.

Whereas Germany seems to have highly developed mining and reclamation laws, the situation in the United States is very

dynamic. The pattern of legislation in the United States has
been to start with minimal requirements and strengthen the laws
as the public demands. Some states have already strengthened
their basic mining laws several times, while other states are
just now contemplating initial legislation. The latter is the
case in several western states where coal reserves are being
tapped at an increasing rate because of the energy crisis.

The environmental movement has had some influence on the
development of state laws. There has been more communication
among conservationists, legislators and reclamation agencies
with their counterparts from other states. Reclamation practices
that work well in other states are often included in the legislation when a state has revised or updated its laws. However,
the greatest impact the environmental movement has had is in
changing public attitudes and policies. The environmentalists
have insisted on a national policy that considers both internal
and external costs in the total cost of producing material and
energy.

Federal bills are still being developed but it is likely
that administration of these laws will remain with the states
and they will have to meet the federal standards within a set
time frame. There is also an attempt to include some land-use
planning in the federal law. This aspect of reclamation is well
developed in Germany but it is stressed in the laws of a few
states only.

RECLAMATION PRACTICES

The emphasis in the United States is to achieve what is
known as basic reclamation. Simply stated, basic reclamation
requires that the land be stabilized and revegetated so that no
further damage to the environment occurs. Returning the land to
a high rate of productivity is not stressed in the United States
as much as it is in Germany. However, there has been some
reclamation in the United States to achieve this particular
purpose.

Basic Reclamation

The regulations of most states require some spoil shaping,
water quality control, and revegetation as basic reclamation
practices.

Spoil Placement and Grading. Segregation and burying of
toxic material is required in many states. In some cases, topsoil
is stockpiled and replaced. However, analysis of overburden strata

to determine proper placement of spoil is not yet common practice
in the United States as it is in Germany. In the past few years,
many of the United States coal operators have converted their
mining equipment from shovels and draglines to front-end loaders
and trucks which gives their operations greater flexibility in
segregation and placement of spoil.

Grading spoil to a specific shape is required in most states.
This varies from minimal grading to placing the land back to its
original contour. In some instances, the highwall must be
reduced or partially covered.

Several practices are required to reduce spoil slides or
slumps in mountainous areas. These include spreading spoil down
slopes to reduce depth of spoil, placing more spoil on solid
benches, and storing excess spoil in the heads of valleys rather
than dumping it over steep outslopes.

Water Quality Control. Most state laws outline procedures
that must be used to reduce sedimentation and chemical pollution
of streams. Erosion and sediment are problems throughout the
coal fields, and several practices are required. Ditches may
be constructed above the highwalls to keep water from entering
the operations. Drainage from terraces and benches is required,
but rarely are the drainways designed or constructed as well as
is common in Germany. Impoundments to trap sediment in the upper
reaches of streams draining mined areas have been very effective
in some states. Road location, construction and abandonment
are usually regulated to reduce the amount of erosion and sediment.

Acid drainage and chemical pollutants are problems associated
with some coal seams. Portable equipment to treat water pumped
from pits or draining into streams is used to meet water quality
standards when required. Treated water must pass through sediment
basins to allow the chemical precipitates to settle out.

Revegetation. All states now require revegetation of strip-
mined areas. Trees or herbaceous vegetation or both may be
required. The United States has had a comparatively long history
of revegetation research on mined areas and as a result, a range
of species selected for growth on spoils is available. Methods
for testing spoil to determine lime and fertilizer requirements
have been developed. Most states either regulate the procedures
for spoil testing and applications of lime, fertilizer, or mulch
or require that a minimum cover be established within a specified
time. In order to obtain rapid vegetative cover on mine spoils,
planting of trees at close spacings is practiced in the Ruhr in
contrast to the combination of herbaceous cover and wider spacing
of trees in the United States.

Underground-Mine Refuse Piles. Reclamation of waste rock
from underground-mine operations is usually regulated separately
from surface-mine operations. This waste must be compacted to
prevent burning and also graded for aesthetic purposes. Some
states require that it be revegetated or covered with soil. The
detailed planning, careful location, construction and revegetation
procedures used in Germany are in sharp contrast to the rather
"broad-brush" treatment of extensive areas used in the United
States. In the Ruhr, mine waste is placed in huge piles with
terraced outslopes, and no attempt is made to blend these man-
made mountains into the landscape. Some of the underground waste
in the Ruhr is disposed of in gravel pits or used as land fill.

Special Purpose Reclamation

Surface mined land in the United States has been reclaimed
for a wide variety of special uses beyond the basic reclamation
requirements.

In the mountainous areas of Appalachia the coal fields were
found under forest cover and therefore, much of the spoil has
been planted to trees for potential timber production. Wildlife
food and cover species are also planted on mountain strippings.
If the mined areas were farmland or adjacent to farms, they may
be put under cultivation after reclamation. The most common
reclamation for agricultural use is for grazing of cattle.

Reclaiming mined land for residential or industrial use
depends very much on location and ownership. Areas adjacent to
highways have been developed for houses, mobile homes, factories,
and airports. The larger mines resulting from area stripping are
generally more suited for intensive development. Lakes have
been constructed on some of the larger areas and recreation
complexes developed around the water. Some of the areas near
cities have been used for solid waste disposal because they offer
convenient sites for domestic refuse.

Land-Use Planning. The lack of land-use planning in the
United States has meant that reclamation for special uses has
been done on an individual case basis. There is no planning commission as there is in Germany to assist in the development of
mined areas for special uses. In the United States, often the
coal is owned by one party, the surface by another, and the mining
is conducted by a third party. Although some mined areas have a
good potential for development, most cannot be highly developed
under present economic conditions.

The German practice of removing and storing topsoil before
any land is disturbed for industrial development, highway

construction or mining is a conservation measure worthwhile for
the Americans to consider in land-use planning. Only when
mining operations are coordinated with overall land-use planning
will the full potential use of mined land be realized.

ACIDITY AND NUTRIENT AVAILABILITY IN COLLIERY SPOIL

M.E. Palmer*

Department of Biology, University of York, York, U.K.

ABSTRACT

Variability of spoil materials is a significant problem in re-clamation and revegetation of colliery spoil heaps. In samples of spoil from the weathered surface of colliery tips in the East Pennine coalfield of Britain the pH values ranged from pH 8.0 down to pH 2.4. Extreme acidity in these materials is due to oxidation of iron pyrites, (FeS_2), in the debris. The resulting acid sulphate medium is buffered by reaction with the clay mineral component of the debris. Weathering of the host minerals for phosphate, magne-sium and manganese is extremely rapid under these conditions. At low pH values potassium is precipitated as Jarosite $(KFe_3(OH)_6$ $(SO_4)_2)$, and the level of plant available potassium is greatly reduced.

Spontaneous combustion of coal debris in spoil heaps raises the temperature within burning spoil to 800°C or more. The composition and subsequent weathering of the spoil material is significantly affected by heating.

INTRODUCTION

To a large extent the success of a reclamation scheme will depend ultimately on the establishment and growth of an imposed vegetation that is appropriate to the projected land use. In

*Present address: Environment Department, South Yorkshire County Council, Barnsley, U.K.

order to enhance the success of this, investment in a soil cover may
be made but this will depend upon availability, expense and the
land use envisaged. Many schemes will thus have to rely on establishing a vegetational cover of plants rooted in spoil material.
Factors affecting spoil nutrient availability and spoil toxicity
are therefore of considerable importance to reclamation practise.

Hall[1] studied the range of plant types and species in the
colonising flora of unreclaimed derelict colliery spoil heaps in
most of the British coalfields, and concluded that colliery spoil
materials were broadly comparable with many of the poorer soils
of north-west Europe. If this conclusion is correct it seems
reasonable to suppose that where appropriate tests and analyses
are carried out, and ameliorants applied as indicated, the range
of useful species can be extended to include the commercially
available amenity species and productive agricultural grass
mixtures. The selection of plant material for reclamation sites
would therefore be determined by the end use of the site.

The selection of species which the tip material may not support
without intensive amelioration, initially and perhaps continuously
over a long period of time, has been questioned by ecologists.
Without the use of a soil cover the implied alternative requires
that the choice of plant material, and perhaps therefore the end
use of the site, should be determined by the nature of the spoil
materials on site.

In practice, consideration of the properties of spoil materials
is seldom the basis for the choice of function or the selection of
the species used on reclamation sites, except perhaps in the case
of tree species. The consequences of the widely adopted strategy
of returning reclaimed colliery spoil heaps to intensive functions
such as productive agricultural grassland, or even public open
space, cannot be assessed without some understanding of those
factors which will determine the nutrient status of spoil materials.

The evaluation of possible alternative reclamation strategies
requires that the probable requirements for initial and subsequent
amelioration treatments be estimated. This paper deals only with
some chemical characteristics of colliery spoil materials. The
effects of spoil physical factors are not considered here.

Glover[2] has reviewed the qualitative aspects of the mineralogy
and composition of the spoil source materials (the roof and floor
strata and the dirt bands and non-carbonaceous mineral matter in
the coal itself). Quantitative relationships which may exist between the composition of these strata and that of the material on
a spoil heap are heavily obscured by the selective nature of the
coal mining and coal preparation techniques used at different
locations and at different times during this century. Many of the

collieries in this coalfield have worked more than one seam and
there is very little reliable indication of the relative volumes
of dirt output associated with separate workings. It is nevertheless convenient to recognise four types of spoil material which
represent the range of possible composition for freshly mined
spoils, neglecting the coal and carbonaceous matter content. These
four spoil types are as follows:

1. Debris containing detrital minerals only

2. Debris containing detrital minerals and diagenetic carbonates
    but not iron pyrites

3. Debris containing detrital minerals and iron pyrites, but
    not carbonates

4. Debris containing detrital minerals, carbonates and iron
    pyrites.

Spontaneous combustion was a common feature of spoil heaps in
the East Pennine coalfield, particularly in the period 1930-1950.
The effects of combustion on spoil composition are reviewed in
the Introduction to this Section and some specific characteristics
of burnt materials are considered together with those of unburnt
spoil.

pH OF COLLIERY SPOIL MATERIALS

The abrasion pH of the principal detrital minerals in the
rock debris (quartz, kaolin, illite and chlorite) is neutral or
slightly alkaline, and the equilibrium pH for carbonates of
calcium and magnesium is also alkaline with a maximum value of pH
8.3 for calcite at normal atmospheric pressure and CO_2 content.
Iron pyrites, FeS_2, is insoluble and therefore has no influence on
the pH of freshly mined spoils, which are always neutral or
alkaline in reaction.

The distribution of pH values in samples of weathered surface
spoils from unreclaimed sites is given in Fig. 1. The distribution
of pH values in samples of surface spoil from recontoured sites
before application of ameliorants is given in Fig. 2. Mean pH
values for samples of spoil from surfaces of differing ages are
presented in Table 1. The depth of spoil sampled at each site
was 0-10 cm. For Gilcar both sets of data relate to samples taken
from the same area of the site, and the initial sampling was
carried out within two months of completion of the re-contouring
work. The second sampling was carried out eighteen months later.

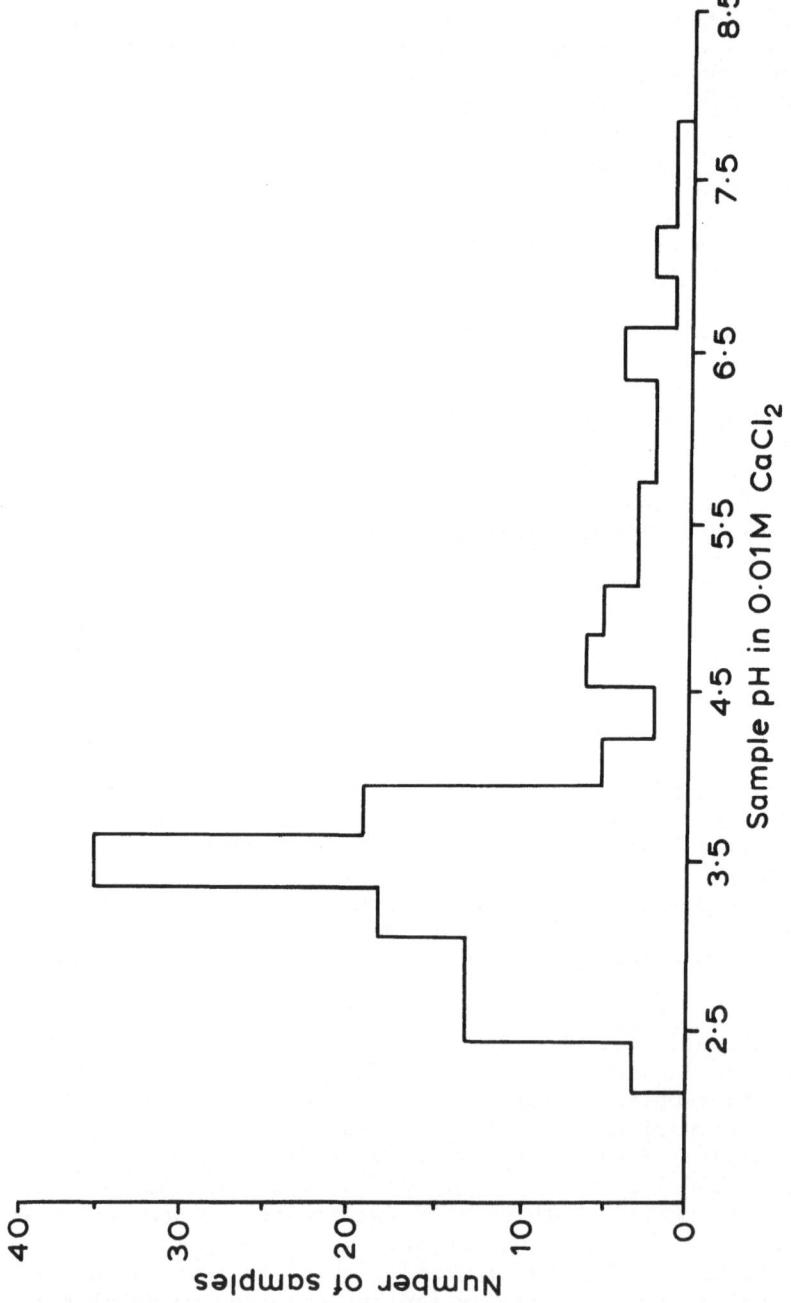

Figure 1. pH distribution in weathered colliery spoil.

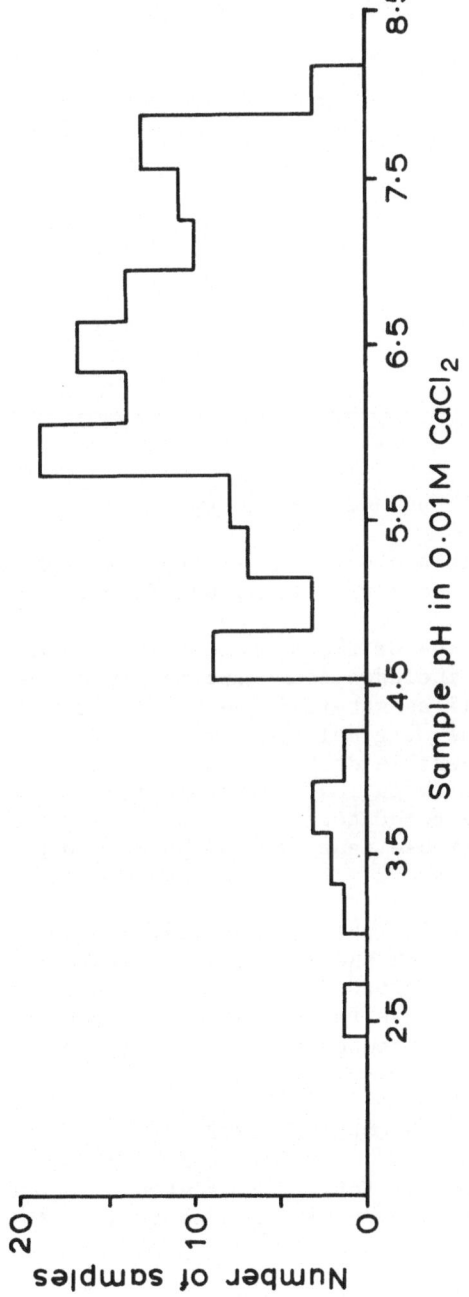

Figure 2. pH distribution for reclaimed colliery spoil sites.

Site	Approximate age in years	Mean pH in 0.01M $CaCl_2$	Number of samples
Woolley (R)	0	7.16±0.43	18
Gilcar (R)	0	6.21±0.67	9
Gilcar (R)	1.5	3.08±0.38	9
Water Haigh (R)	1.0	5.00±0.88	16
Bullcroft	10	2.92±0.30	12
Maltby Maine	15	3.52±0.56	24
Hound Hill	60	3.33±0.23	12

R = Re-contoured site

Table 1. Spoil pH and surface age

Data from this study suggests that the pH of surface spoil material may decrease with time, but from the values given in Table 1 it is clear that spoil pH is not in any way a simple function of the age of the surface. The surface material on re-contoured sites includes some material from the original weathered surfaces, as well as material previously buried at depths of up to 30 m below the original tip surface. Spoil which has been buried at considerable depths below the tip surface appears to retain the characteristics of fresh spoil. The pH of surface material on some re-contoured sites, for example Water Haigh, is extremely variable ranging from pH 2.8 to pH 7.2 within a few metres.

All the data in Table 1 and Figs. 1 and 2 was collected from an area in which the mean annual rainfall varies from 625 mm to 650 mm. The observed differences in pH of materials of different ages cannot be due to variation in leaching rates, and must therefore be a function of the composition of the spoil material.

PYRITES OXIDATION IN COLLIERY SPOIL

The oxidation of FeS_2, represented by equations (1) and (2) below, is clearly a potential source of acidity in colliery spoil materials:

$$FeS_2 + \frac{7}{2}O_2 + H_2O \rightarrow Fe^{2+} + 2SO_4^{2-} + 2H^+ \quad (1)$$

$$FeS_2 + \frac{15}{4}O_2 + \frac{7}{2}H_2O \rightarrow Fe(OH)_3 + 2SO_4^{2-} + 4H^+ \quad (2)$$

The potential acidity of pyritic soils, termed acid-sulphate soils or 'cat-clays', was discussed by Van Beers[3], and Knabe[4] examined the analagous properties of pyritic wastes, including mining spoils. The thermodynamic instability of FeS_2 in aerated surface environments is inferred from measured oxidation-reduction potentials such as those plotted by Baas Becking, Kaplan & Moore[5], which are generally above the calculated limits of the pyrite stability field, (Garrels & Christ[6]).

Fragments of massive framboidal pyrites, weighing up to 500 g and more than 90 per cent by weight FeS_2, associated with fusain coal and other carbonaceous material are of common occurrence on many spoil heaps in the East Pennine coalfield. Pyritic material exposed on the tip surface is frequently coated in white crystalline efflorescence of hydrated ferrous sulphate, and around fragments buried below the surface of the tip there is often a deposit of yellow, basic ferric sulphate.

Thin sections from sites Maltby Maine and Hound Hill were prepared by impregnating air-dry samples of undisturbed spoil material in a clear resin. Of the various forms of pyrites in coal described by Whelan[7] the following were recognised in sections of Maltby Main material:

1. Concretions of variable size, but mostly 20 to 200 microns diameter

2. Ramifying veins 20 to 200 microns in width

3. Infilling fractured coal structures

4. Filling fusain cell cavities.

All of these pyritic bodies appeared to be composed of framboidal pyrites, and in addition to the pyrites in coal fragments, framboidal pyritic aggregates ranging from 0.1 mm to over 3 mm in diameter were also observed in the matrix of rock debris.

Not all the pyritic material in these sections supported visible evidence of oxidation, but around most pyritic aggregates in coal and in the rock debris there was a substantial deposit of brown ferric hydroxide. The samples of material from Hound Hill appeared to be almost devoid of pyrites other than occasional small aggregates contained in fragments of coal debris. There were no recognisable ferric iron pseudomorphs of pyritic aggregates in the sections of Hound Hill material.

The total pyrites content of these spoils was not determined, but it has been suggested that values ranging from less than 1 per cent to 10 per cent FeS_2 could be expected. Estimates of the

amount of acid produced in weathered pyritic spoils were made
from determinations of the total buffering capacity to field pH
values. These calculations gave values ranging up to 4.2 g FeS_2
oxidised 100 g^{-1} of spoil assuming complete oxidation as in
equation 2, but it is recognised that values derived in this way
take no account of the losses in leaching and drainage water and
must always be lower than the true values. The rate of acid
production in spoil from Gilcar was estimated in this way for the
period during which the mean surface pH fell from pH 6.2 to
pH 3.1, and the average value obtained was equivalent to 0.47 g
FeS_2 oxidised 100 g^{-1} of spoil. Details of the methods used
to estimate buffering capacity of the spoil material are given
elsewhere in this paper. Reduced to an annual rate, the calculated
oxidation rate for the spoil on Gilcar is 0.32g FeS_2 oxidised
100 g^{-1} spoil $annum^{-1}$.

Total oxidation of 1g FeS_2 in 100g of spoil would release
33 milliequivalents of acidity 100 g^{-1} of spoil, equivalent to
36.5 tonnes of $CaCO_3$ ha^{-1} . If field oxidation rates of 0.5 g
FeS_2 100 g^{-1} of spoil $annum^{-1}$are maintained for more than a
relatively short time, say two years, the frequency of lime
applications required in a poorly buffered spoil would present
almost insuperable difficulties for reclamation work. Hart[8]
found that field oxidation rates in acid sulphate soils declined
rapidly after an initial burst of activity, and a notable
feature of all the pyritic spoils on undisturbed sites examined
in this study is the abundance of unoxidised pyrites in the tip
surface material. In thin sections of Maltby Main spoil material,
most pyritic aggregates which showed signs of oxidation were still
apparently more than 50 per cent intact. With the exception of
some areas of limited extent on a few sites, the problem of
high recurrent lime requirements has not yet arisen on these
reclamation sites, and there is a strong presumption that very
high rates of acid production may be of short duration in these
colliery spoils.

Many of the laboratory studies that have been carried out
with a view to determining the reaction mechanism of pyrites of
oxidation are difficult to relate to field situations, and
there are conflicting hypotheses regarding the specific mechanism
and the rate limiting step. Smith, Svanks & Halko[9] have
suggested that there are two entirely independent oxidation
mechanisms, one involving gaseous oxygen and the other ferric
ions in solution as the specific oxidant. The oxidants are said
to be adsorbed on reactive sites on the surface of the pyrites,
these sites are specific to one or other of the two oxidants
which can therefore be adsorbed simultaneously and without
competition. The rate limiting step for the reaction involving
oxygen is the electron transfer which takes place between ferrous

iron and oxygen. The oxidation by ferric iron is dependent upon the rate of Fe^{3+} adsorption which is in turn dependent upon total Fe^{3+} concentration and solution Eh. According to Singer & Stumm[10] the specific rate of oxidation by ferric iron is entirely independent of the surface properties of the pyrite and the rate limiting step is the oxidation of Fe^{2+} to Fe^{3+} in solution. In their model of the pyrites oxidation reaction the role of bacteria, such as the obligate chemoautotroph Ferrobacillus ferrooxidans described by Silverman & Lundgren[11], is limited to that of increasing the rate of ferrous iron oxidation in solution. Lorenz & Trapley[12], Silverman[13] and Carrucio[14] have all reported that the addition of iron oxidising bacteria to the reaction system results in a marked increase in the rate of acid production. The significance of these bacteria to field oxidation rates is questionable in view of the observation by Silverman[13] that concentrations of chloride greater than 5×10^{-3} M are completely inhibitory. In 1:2 spoil in water extracts of material from Gilcar the measured chloride concentrations range up to 8×10^{-3} M, and higher values have been recorded from samples of freshly exposed spoil on other sites.

There is no apparent reason to believe other than that the reaction mechanisms for pyrites oxidation under field conditions will be common to all pyritic spoils, and since all the proposed mechanisms are surface phenomena the overall reaction rate is dependent upon the particle size and surface area of the pyrites. This has been demonstrated in a number of experimental systems using crushed pyrites, for example Braley[15] and Scott[16]. The dependence of acid production rate on the particle size distribution of sulphide grains in pyritic spoils has been demonstrated by Smith[17] and Caruccio[14] who found that acid production rates were greatest in materials where the dominant form of FeS_2 was as grains of 10 microns diameter or less. Whelan[7] states that in British coals containing less than 10 per cent total FeS_2, nearly all the sulphide occurs as grains 0.5-40 microns in diameter, and as concretions 20 to 100 microns in diameter. Wandless[18] has examined the distribution of pyritic sulphur in the productive seams of this coalfield and gives values ranging from 1 per cent to 13 per cent FeS_2, but the variation within seams is almost as great as that between seams. It has been estimated that prior to 1940 the amount of free coal discarded on spoil heaps in this coalfield was commonly 10-15 per cent of the total spoil material. It is practically impossible to distinguish quantitatively the percentage of coal as opposed to other carbonaceous matter in the debris, but ashing samples of finely ground spoil at 380^0C gave values of up to 40 per cent total carbonaceous matter. Samples from size of the seven materials examined gave values between 8.4 per cent and 19.8 per cent total carbonaceous material. There is good reason to infer, therefore,

that these spoils contain a significant quantity of finely
disseminated pyrites which is responsible for the initial high
rate of acid production.

The oxidation of pyrites is ultimately dependent upon the
supply of gaseous oxygen whether the oxygen is a specific oxidant
or involved only in the oxidation of ferrous to ferric iron, with
or without the assistance of bacteria. The model for pyritic
systems constructed by Shumate, Smith & Brant[19] predicts that
the rate of acid production decreases as the oxygen path length
increases, particularly in moist porous materials where the pore
water acts as a barrier to oxygen diffusion. The pH profiles
from Maltby Main, Hound Hill and Gilcar are reproduced in
Fig. 3, and the profiles from Maltby Main and Gilcar, which
are typical of those from other acid pyritic spoils, suggest that
acid production decreases with increasing depth below the tip
surface. The effect of oxygen adsorption by coal on the
distribution of oxygen in the spoil atmosphere is not known but
it has been reported that the oxygen concentration in a sealed
borehole was reduced to almost zero shortly after the bore was
sealed off.

The pH profile from Hound Hill is very similar to the pH
profile of some weathered acid-sulphate soils described by Van
Breeman[20] but there was no evidence of the redistribution of ferric
iron which characterised these soils. In acid sulphate soils
the insoluble ferric sulphates precipitated in the acid surface
horizons during the period of active pyrites oxidation were said to
have undergone hydrolysis yielding ferric hydroxide and soluble
sulphate, and this change is accompanied by development of a
characteristic pattern of iron staining not recorded from the
Hound Hill profile. Thin sections of spoil from depths up to
120 cm below the surface appeared to be of uniformly low pyrites
content and there were no recognisable psuedomorphs of the forms
of pyrites noted in Hound Hill spoil material. The site is within
4 km of the mining town of Barnsley, and is only about 15 km
downwind of the major industrial area of Sheffield. It is
concluded that the initial pyrites content of this spoil was very
low, and that the pH profile on this site is the result of
leaching by acidic rainwater on a spoil of low initial buffering
capacity.

BUFFERING CAPACITY OF COLLIERY SPOILS

The pH of colliery spoils is not determined solely by the
amount of acid generated from the oxidation of pyrites. The rate
at which spoil pH changes in response to the addition of acid,
whether from pyrites or from external sources such as polluted
rainfall, is a function of the buffering capacity of the spoil
material. Van Breeman[20] has identified three principal components

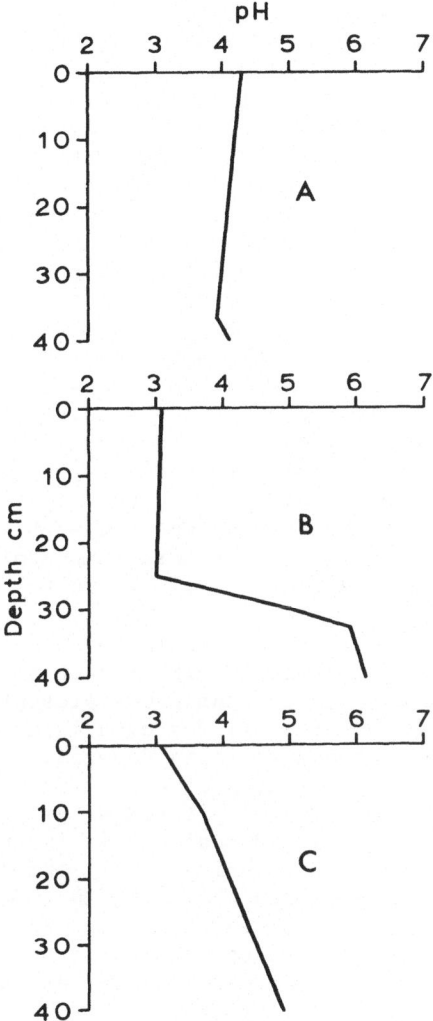

Figure 3.   pH profiles for three colliery spoil heaps in
            Yorkshire: A Maltby Main; B Hound Hill; C Gilcar.

of the buffering capacity in acid sulphate soils, and the
buffering capacity of the spoil materials examined here is
considered to be closely analogous to that of the acid sulphate
soils. For spoil materials containing only debris of the
detrital rock material the spoil pH is buffered by adsorption of
H_3O^+ onto the cation exchange surfaces at the expense of
exchangeable bases which are displaced into solution. The
capacity of this component of total buffering is determined as

the difference between total cation exchange capacity at the
initial pH of the material, and the residual exchangeable bases
at field pH. The cation exchange capacity at pH 7.0 was
estimated in ammonium acetate solution and values ranging from
8.0-22.5 meq 100 g^{-1} of the <2 mm fraction of spoil were obtained
for material from eight sites. Reliable values for the residual
exchangeable bases at field pH were difficult to obtain from
most of the spoil materials because of the presence of water
soluble salts. Estimates for some of these spoils were made as
follows. Samples were extracted in a normal solution of KCl, the
extract was titrated against NaOH solution to phenolphthalein
end-point after boiling to give a value for extractable acidity.
The cation exchange capacity was determined at field pH in an
embuffered solution of NaCl, and the difference between cation
exchange capacity at field pH and extractable acidity was
considered to represent the exchangeable bases at field pH.
Estimates derived in this way indicate that the buffering capacity
of the exchangeable bases in colliery spoil is between 5 and 12
meq 100 g^{-1} for samples of pH 3.0-pH 4.0, but it is not known
how this capacity is distributed over the range pH 7.0 to pH 3.0.
The value quoted by van Breeman[20] for this component of the
buffering capacity in acid sulphate soils is 10 to 30 meq 100 g^{-1}.

The presence of carbonate minerals in the spoil increases
the buffering capacity at neutral to alkaline pH values. The
buffering capacity for each 1 per cent $CaCO_3$ present is 20 meq
100 g^{-1}, and 24 meq 100 g^{-1} of spoil for each 1 per cent $MgCO_3$.
For $FeCO_3$, however, the net neutralizing capacity is nil if it
is assumed that all the ferrous iron liberated from the carbonate
is oxidised to ferric iron and precipitated as $Fe(OH)_3$. The
carbonate minerals in the rock and coal debris in the East
Pennine coalfield include both siderite $FeCO_3$, and mixed
carbonates of variable composition $Fe.Ca.Mg.Mn .CO_3$. Total
carbonate content is therefore an inadequate parameter of
potential buffering capacity. An approximation to the buffering
capacity of carbonates in colliery spoil material was derived
from determinations of the total calcium and magnesium extracted
in 30 per cent HCl. The difference between acid extractable
cations and the soluble plus exchangeable cations in ammonium
acetate extracts at pH 7.), expressed as meq 100 g^{-1}, is given
for samples of material from the surfaces of recontoured sites
in Table 2. These determinations were made only the <2 mm
fraction of the spoils.

The pH range over which the buffering capacity contributed
by carbonates can be effective is limited by the partial pressure
of CO_2 in the spoil atmosphere, assuming that the buffering
reactions are in equilibrium with solid carbonate in the spoil
(Garrels & Christ[6]). At values of pCO_2 close to the normal
earth atmosphere, $pCO_2=10^{-3.5}$, the pH will not fall below pH 7.

Site	Net Acid Soluble Bases:meq 100 g^{-1}		
	Ca	Mg	Total
Firbeck Ridge	22.24±2.85	32.01±1.98	54.25
Darfield Wombwell	1.55±1.46	17.88±4.46	19.43
Water Haigh	24.14±15.25	6.99±4.63	31.13
Gilcar	8.93±5.18	8.92±2.99	17.85
Granville	8.81±3.31	9.44±0.73	18.25

Table 2. Net acid soluble bases in colliery spoil

If pCO_2 values in the spoil are higher than the normal earth atmosphere value, the lower limit of the carbonate buffering range will fall below pH 7. Guney[21] has reviewed much of the work done on the low temperature oxidation reactions of coal. In view of the established relationship between these reactions and spontaneous combustion in coal and colliery spoil materials containing coal, it must be assumed that in freshly exposed spoils at least, the composition of the spoil atmosphere is modified by the absorption of O_2 and release of CO_2 during such reactions. Furthermore, the assumption that the reaction between carbonate minerals and acidic ions in the spoil takes place at equilibrium conditions must be questioned since in thin sections of Maltby Main spoil (pH 3.2) there were numerous fragments of crystalline carbonate from the clear partings of the coal. The analyses of Gilcar spoil material also indicate the presence of free carbonate in an acid spoil. The persistance of the cleat carbonate in acid conditions is thought to be due to the rather large particle size of these fragments, 1-3 mm in width and up to 12 mm in length, and to the coating of hydrated ferric oxide which surrounds most of the intact fragments. Ultimately, however, the carbonate must either be totally dissolved if the total acidity is greater than the buffering capacity of the carbonate, or the presence of carbonate in excess of total acidity will eventually drive the pH back up until the carbonate CO_2 is restored.

The third major component of the buffering capacity of these spoils is comprised of the reactions involving H_3O^+ ions and the aluminosilicate minerals. These reactions have been widely studied and the literature produced has reached huge proportions, however, discussion can be simplified by examining these reactions from two different aspects. Various concepts of soil acidity and buffering in acid soils were reviewed by Coleman & Thomas[22] and these concepts may be regarded as being concerned mainly with reaction mechanisms and with the intermediate compounds formed. The alternative approach is concerned with gross changes in the composition of aluminosilicates resulting from weathering reactions in an acid medium. Keller[23] and Garrels & Christ[6] have examined the equilibrium conditions for some of these reactions, and where the equilibrium conditions for a reaction are known the total buffering capacity of the system can be calculated.

When exchangeable bases on the cation exchange surfaces of clay minerals are exchanged for H_3O^+ ions from solution, the adsorbed H_3O^+ ion migrates into the lattice, displacing Al^{3+} ions from their position in the lattice. The behaviour of the displaced aluminium is one of the most controversial subjects in soil science, but it is generally accepted that many characteristics of acid soils are determined by the distribution of aluminium ions between the cation exchange surfaces and solution, and by the degree of hydrolysis of the aluminium ions in solution.

According to Schofield & Taylor[24] the pH of a solution containing dissolved aluminium is determined by the hydrolysis reactionsof aluminium ions, and the first stage hydrolysis reaction is represented as:

$$(Al \cdot 6H_2O)^{3+} + H_2O = (AlOH \cdot 5H_2O)^{2+} + H_3O \qquad (3)$$

Schofield & Taylor[24] estimated the value of the reaction constant to be 1.05×10^{-5} and most other values reported in the literature are close to 1×10^{-5}. The buffering capacity of the hydrolysis reaction depends upon the total amount of aluminium in solution. At a fixed concentration of total dissolved aluminium the buffering capacity, when additional H_3O^+ is added to the system, depends upon the amount of divalent aluminium available for reaction. The equilibrium of this reaction is determined by the ion activities rather than the concentrations of the ions in solution. Hem[25] gives values for the ion activity coefficients of aluminium ions at various levels of total ionic strength. The values given for trivalent Al^{3+} and divalent $AlOH^{2+}$ are 0.03 and 0.23 respectively in a solution of total ionic strength $I = 0.71$. At a fixed concentration of total dissolved aluminium, the effect of increasing the total ionic strength is to reduce the buffering capacity of the system to added H_3O^+, in proportion to the ratio of the ion activity coefficients for these aluminium ions.

The behaviour of dissolved aluminium in pyritic colliery spoil materials and acid sulphate soils is further modified by the formation of complex ions, $AlSO_4)_2^-$. Hem[25] gives the reactions for the formation of these ions, and the equilibrium constants for these reactions, as follows:

$$Al^{3+} + SO_4^{2-} = AlSO_4^+ \quad K = 3.2 \tag{4}$$

$$\text{and} \quad Al^{3+} + 2SO_4^{2-} \quad K = 5.1 \tag{5}$$

Typical values from saturation paste extracts of spoil from Bullcroft (pH 2.9, I = 0.10) are 2.5×10^{-2} M in SO_4^{2-}, and 1×10^{-3} M in total dissolved Al. The data given by Hem[25] predicts that the activity of Al^{3+} in solution would be approximately 1×10^{-4} for a solution of this composition, i.e. about 1 per cent of the total aluminium concentration in solution.

It has already been noted that the total dissolved aluminium concentration determines the overall buffering capacity of the hydrolysis reaction in solution, equation (3). Various reactions, by which the level of soluble aluminium in acid soils might be determined at any given value of soil pH, have been examined and reported in the literature. In many earlier studies it has been widely assumed that the level of soluble aluminium is controlled by the solubility product of gibbsite, $Al(OH)_3$, for which Hem & Robertson[26] gave the value of $K_{SP} = 2.24 \times 10^{-33}$:

$$Al(OH)_3 = Al^{3+} + 3(OH)^- \tag{6}$$

If this assumption is correct, the buffering capacity of the hydrolysis reaction in equation (3) increases by two orders of magnitude for each unit decrease in the pH of the system below pH 5. Lindsay, Peech & Clarke[27] have shown that in acid soils where the pH is close to pH 5, the level of soluble aluminium is close to the value predicted in equation (6), but at lower pH values the soil solutions were undersaturated with respect to gibbsite.

Congruent dissolution of the kaolin clay mineral, as in equation (7) below, may effectively limit the activity of dissolved Al^{3+} in solution:

$$Al_2Si_2O_5(OH)_4 + 6H^+ = 2Al^{3+} + 2H_4SiO_4 + H_2O \tag{7}$$

where $\quad \log K = \log Al^{3+} + \log H_4SiO_4 + 3pH$

In the presence of solid amorphous silica the activity of dissolved silica is limited to $10^{-2.6}$, according to Garrels & Christ[6]. The soluble aluminium increases as the pH falls, and the buffering capacity predicted from equation (7) increases by about two orders of magnitude for each unit decrease in pH.

The distribution of pH values for weathered, untreated spoils, given in Fig. 2 indicates that these colliery spoil materials are

poorly buffered in the range pH 5.0 to pH 3.5. This is confirmed
by estimates of the total exchangeable acidity in spoil materials
as determined in $BaCl_2$/triethanolamine buffer, after Peech,
Cowan & Barker[28]. The values obtained are plotted against
sample pH in Fig. 4. Values for total soluble aluminium in
saturation paste extracts of acid colliery spoils are plotted agains
against sample pH in Fig. 5, and it is apparent from the figure
that the relationship between total aluminium in solution and
sample pH that might be anticipated from equations (6) or (7), is
not realised in these materials.

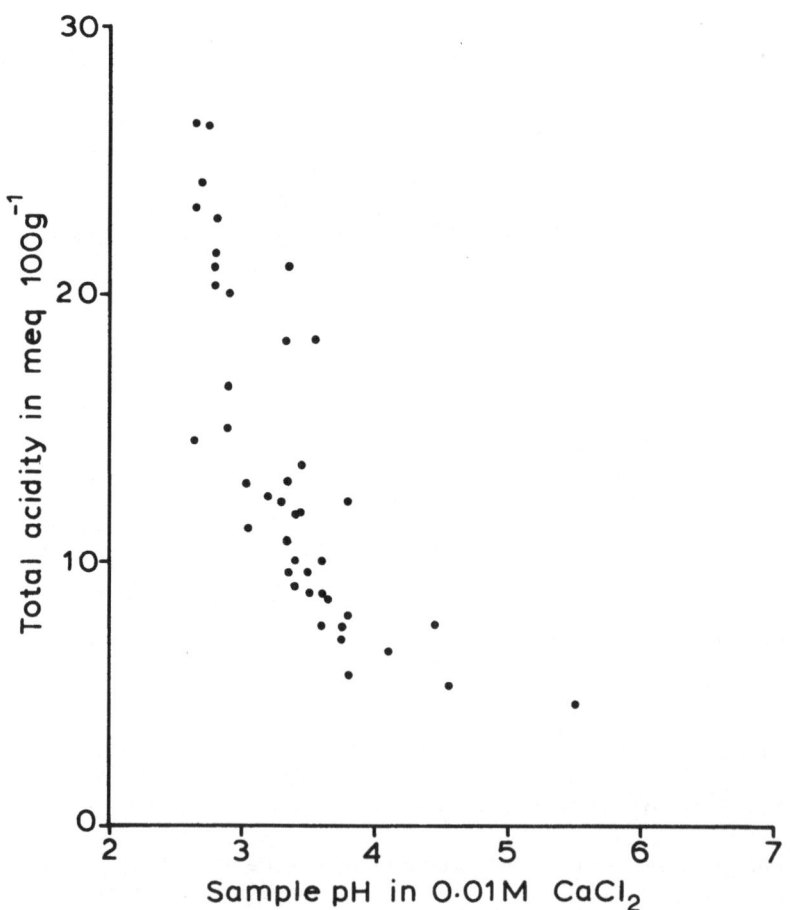

Figure 4. Total exchangeable acidity of weathered colliery
spoil.

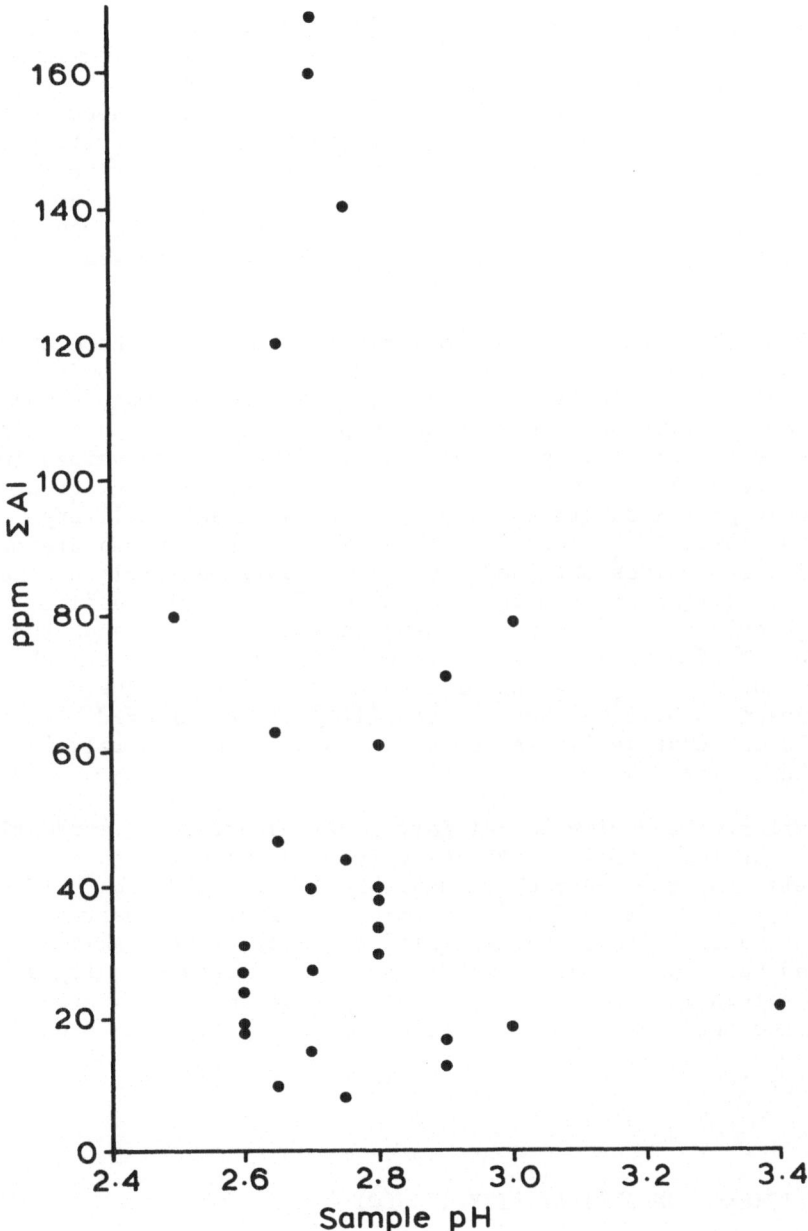

Figure 5. Total dissolved aluminium in saturation paste
extracts of weathered spoil from Bullcroft.

Van Breeman[20] has suggested that poor buffering in acid
sulphate soils may be due to the failure to maintain equilibrium
for the congruent dissolution of clay minerals weathering in a
strong acid medium. The solubility of the basic sulphate
$AlOHSO_4$ probably determines the level of soluble aluminium under
these conditions, according to van Breeman[20]. The activity of
the Al^{3+} ion would be a function of pH, total ionic strength and
sulphate ion activity, but the formation of complex ions and
ion pairs, such as $CaSO_4^0$ in solution, necessitates the use of
a computer to perform the iterative calculations required to
determine Al^{3+} activities.

Although it has not been possible to demonstrate here that
the data obtained from samples of acid colliery spoil conforms
to the solubility plot for $AlOHSO_4$ given by van Breeman[20], the
following observations are seen to support the extension of his
hypothesis to colliery spoil materials. Firstly, the values for
total soluble aluminium for a wide range of total ionic strength
and sample pH for saturation paste extracts of acid colliery
spoils are broadly comparable with data presented by van Breeman[20].
Secondly, the values obtained from burnt spoil materials are not
evidently different from those obtained for unburnt spoils,
although evidence is given elsewhere in this paper to show that
the amounts of aluminium extractable in KCl solution are much
greater for burnt spoils than for unburnt spoils and there is
good reason to believe that the stability of the altered clay
minerals in burnt spoils is less than that of the minerals in
the unburnt material.

Buffering capacity in colliery spoil materials is evidently
greatest at the opposite extremes of the pH range for these
materials. At the upper pH values, pH 7 to pH 8.3, the buffering
capacity is very largely a function of the spoil composition
i.e. carbonate content and composition. Spoils are weakly
buffered below pH 7, and the increase in buffering capacity at
low pH values is due to the reaction between the acid weathering
medium and the clay minerals content of the spoil. Van Breeman[20]
has also noted that other sulphate compounds of low stability
also contribute to the buffering capacity of acid sulphate
soils.

MACRONUTRIENT ION AVAILABILITY IN SPOIL

Calcium, Magnesium and Sodium.

Table 3 gives values for the soluble plus exchangeable cations
extracted in neutral ammonium acetate solution for samples of
surface from re-contoured and unreclaimed sites.

Site	Cations extracted at pH 7.0:meq 100 g^{-1}				pH in 0.01MCcCl$_2$	Number of Samples
	Ca	Mg	Na	K		
Firbeck Main	10.43	3.26	0.08	0.38	6.01	10
	±2.77	±0.37	±0.01	±0.04	±0.68	
Maltby Main	6.74	4.05	0.27	0.36	4.75	10
	±2.40	±0.79	±0.28	±0.06	±1.25	
Gilcar (R)	15.24	6.45	4.55	0.28	3.08	20
	±3.98	±2.25	±3.92	±0.16	±0.38	
Darfield Wombwell (R)	4.53	3.55	6.69	0.80	5.23	8
	±0.94	±1.57	±2.32	±0.14	±0.50	
Water Haigh (R)	4.97	2.64	1.10	0.42	5.36	16
	±4.13	±1.75	±1.53	±0.11	±1.77	
Mitchells Main	2.82	0.40	0.12	0.64	4.67	10
	±2.58	±0.24	±0.04	±0.37	±0.55	
Bilsthorpe	10.09	2.21	0.26	0.32	5.26	10
	±4.35	±0.43	±0.31	±0.06	±1.47	
Mansfield	3.25	1.66	0.04	0.22	3.59	10
	±4.35	±0.77	±0.006	±0.03	±0.30	
Tibshelf (R)	103.47	5.28	n.d.	0.19	6.3)	16
	±27.27	±2.61		±0.03	±0.40	
Firbeck Ridge (R)	6.80	2.50	6.03	0.63	7.50	8
	±0.42	±0.16	±0.88	±0.10	±0.20	

R = Recontoured site

Table 3. Soluble plus exchangeable cations in colliery spoils

From the data presented in Table 3 it is apparent that extreme variability is a significant characteristic of colliery spoil materials, particularly in respect of those elements which are contributed from the non-detrital component of the debris as in equation (8) below:

$$(Ca.Mg)CO_3 + 2H^+ + SO_4^{--} \rightarrow Ca^{++} + Mg^{++} + SO_4^- + H_2O \quad (8)$$

The water soluble salt component of the total extractable bases in colliery spoil is composed of chloride salts, which are occasionally present in significant quantities in the freshly mined debris, plus soluble sulphate salts which are the secondary weathering products of pyrites oxidation. Euhedral and microcrystalline forms of gypsum were recorded in most thin sections of Maltby Main material, and lathlike crystals of gypsum up to 15 mm in length have been observed adhering to rock debris in the majority of spoils examined. Under field conditions the distribution of free calcium ions between the cation exchange surfaces and the spoil solution, and the plant availability of calcium (as Ca^{2+} ions) is controlled by the ion product K_{SP} for gypsum.

$$\text{Where } K_{SP} = \gamma_{Ca}.mCa^{2+} \quad \gamma_{SO_4}.mSO_4^{--} \quad (9)$$

The activity coefficient γ_{Ca}, and the concentration of free cations in solution, mCa^{2+}, are determined by the composition and the total ionic strength of the spoil solution. For solutions in which SO_4^{2-} is the dominant anion the formation of ion pairs, e.g. $CaSO_4^0$ and $MgSO_4^0$ becomes increasingly significant as the ionic strength of the solution increases, and the concentrations of free ions are correspondingly lower than the measured concentrations. The measured calcium concentration in saturation paste of spoil samples extract rises to 17 $mMol.1^{-1}$ and the value of I, the total ionic strength of the extract solution, ranges up to 1.5. The corresponding values for a simple saturated solution of gypsum are $mCa = 11$ $mMol.1^{-1}$ and I = 0.04. The minimum amount of gypsum required to bring the spoil saturation paste extracts to the equilibrium calcium concentration is about 0.06 per cent by weight $CaSO_4.2H_2O$ (equivalent to approximately 0.75 meq Ca 100 g^{-1}).

In surface spoil materials containing more than the minimum amount of gypsum, the rate at which calcium is lost in drainage and run-off water is a function of mCa, the measured calcium concentration in the leachate. Over most of the coalfield the mean evapo-transpiration for the period October – March is 56 mm, and the mean rainfall for the same period is 305 mm. The potential rate of loss of calcium in drainage and run-off from surface spoil materials is estimated to be 5.5–8.5 meq Ca 100 g^{-1} $annum^{-1}$. The calcium status of pyritic spoils is therefore dependant on the rate of sulphate formation (i.e. pyrites oxidation rate), the value of mCa in the leachate, and the total acid soluble calcium content of the spoil. Calcium can only be accumulated as gypsum in surface spoil materials at high pyrites oxidation rates, although the rate of gypsum accumulation is not necessarily related to spoil pH. Subsequently, gypsum formation is limited by the decline in pyrites oxidation rate, and also by depletion of the total acid soluble calcium reserves in the spoil material.

The solubility products of the magnesium and sodium salts are considerably in excess of that of gypsum, and therefore the leaching rate for both these cations is at least equal to that of calcium in spoils of high water soluble salt content. No chloride was detected in water extracts of weathered spoil material from any of the ten sites examined in this study where the surface age was greater than ten years. Consequently it is concluded that the problem of high water soluble salt content is confined to some, but not all, recently exposed materials and that a high level of water soluble salts cannot be maintained against leaching for more than a few years, probably much less than ten years for most spoils.

In spoils of low pyrites content and in weathered surface spoils, the presence of a small quantity of gypsum buffers the spoil against loss of available calcium which normally occurs in non-calcareous soils as a result of displacement of exchangeable bases by H_3O^+ and Al^{3+} or other aluminium species. In weathered colliery spoils the available calcium status is therefore seldom a function of spoil pH. The correlation between available calcium status and soil pH which is common to most non-calcareous mineral soils, and which has some ecological significance according to Clarkson[29], is only encountered in colliery spoil materials of negligible pyrites content or negligible pyrites oxidation rate. The variable composition of the carbonate minerals in the spoil source materials suggest that even in spoils where the pH is buffered at pH 7 or greater by the carbonates, the ratio of available calcium to magnesium varies over such a wide range that the available calcium level in some of these spoils may be lower than in other spoils with lower pH values.

Potassium

The available potassium status of colliery spoils, estimated in a single extraction of the <2 mm spoil fraction in neutral ammonium acetate solutions, ranges from 20-700 ppm. Values of less than 150 ppm available K would be regarded as deficient in the context of productive agricultural grassland. The frequency distribution of values obtained in ammonium acetate extractions is plotted in Fig. 6.

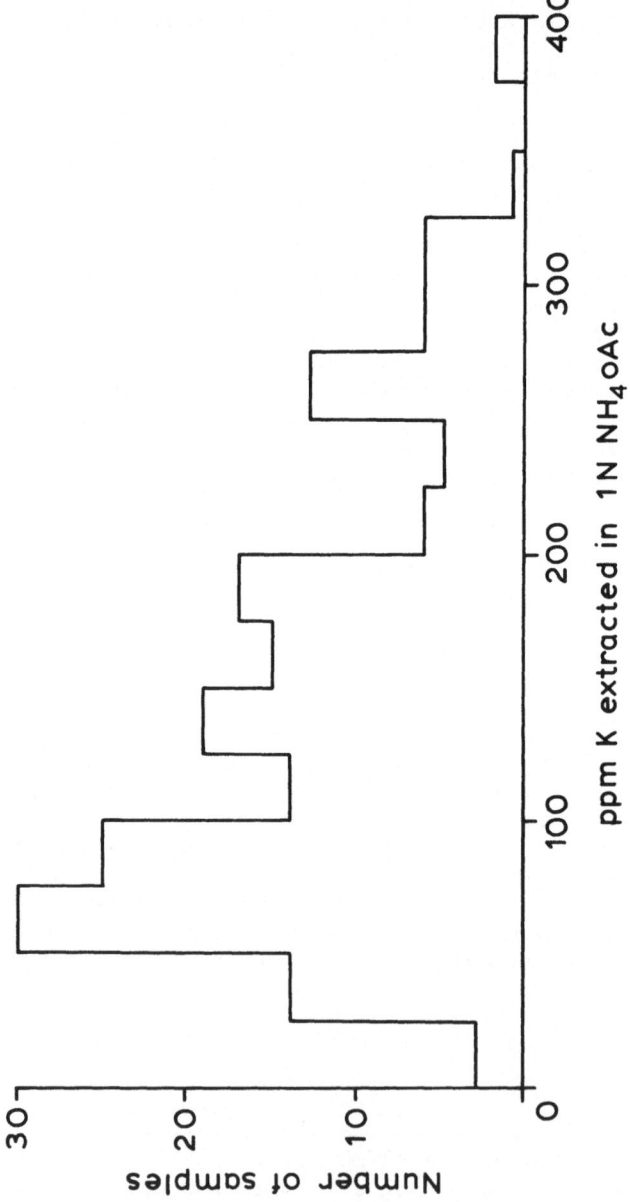

Figure 6. Distribution of extractable K from colliery spoil sites in the East Pennine coalfield (1N NH₄oAc).

The host minerals for K^+ in freshly exposed colliery spoil are illite and hydrous mica from the detrital clay mineral fraction of the debris. Flakes of a micaceous mineral measuring up to 150 microns across, and corresponding to the minerals described by Dixon, Skipsey & Watts[30] and by Carr, Grimshaw & Roberts[31] as hydrous mica, were identified in all the spoil materials examined in this study. Newman[32] concluded that the rate of release of K^+ from its position as in interlayer cation in various micas is controlled ultimately by the lattice characteristics of the host mineral. Potassium is released from the micaceous host minerals into solutions containing other cations, principally Ca^{2+}, Mg^{2+}, and Na^+, which are capable of expanding the host mineral structure. The amount of K^+ released into solution is limited by the equilibrium K^+ concentration for the interlayer cation exchange reaction. For a given mica, the equilibrium K^+ concentration increases with increasing concentration of the replacement ion in solution. Newman[32] found that for those micas which readily absorb protons from solution into the lattice structure, the equilibrium K^+ concentration was much higher in acid solutions. For those micas in which proton uptake was relatively slow, the equilibrium K^+ concentration was barely affected by solution pH.

The micaceous minerals in spoils have not been as well characterized as the experimental material used by Newman[32]. If it is assumed that the host minerals in spoil material from one location have relatively uniform properties it can be shown that a number of quite separate factors contribute to the variability in available potassium status of colliery spoils.

The K^+ concentrations in spoil water extracts of material from the recently exposed surface of Gilcar, and of material from the weathered surface of Maltby Main, are plotted against sample pH in Fig. 7. The replacement cations in these water extracts are Ca^{2+}, Mg^{2+}, and Na^+ in varying proportions. If the values for K^+ concentration in the spoil water extracts may be regarded quasi equilibrium concentrations, then it must be concluded that for these spoil materials there is no marked tendency for K^+ release to increase with spoil acidity. Indeed for material where the sample pH is less than about pH 3.5, it is apparent that the K^+ concentration in solution is determined either by circumstances which prevent the interlayer cation exchange reaction from reaching the normal equilibrium condition, or by a second reaction involving K^+ which has a much smaller equilibrium constant than the ion exchange reaction.

In an attempt to distinguish between the two possible alternatives, samples of acid spoil material from weathered surfaces were allowed to equilibrate with additional potassium in a 1:1 paste of spoil in 0.01 N K_2SO_4 solution. Replicate

samples of each material were extracted in neutral ammonium
acetate solution at intervals over a 28 day period. It was
found that within 14-21 days the total extractable K^+ decreased
to a constant minimum value for each material. The loss of
additional potassium, as determined from the constant minimum
extractable K^+ value by subtracting the extractable K^+ of
control sample with no added K^+, is given in Table 4. The amount
of K^+ immobilised was increased when ferric chloride or ferrous
sulphate was added to the suspension. In the case of the
ferrous salt it was found that there was an increase of up to
12 days in the length of time taken to immobilise the added K^+.
Addition of citric acid blocked the reaction between added Fe^{2+}
or Fe^{3+} and K^+.

Sample pH in 0.01M $CaCl_2$	meq K^+ added to 5g spoil	Composition of initial treatment	% of additional K^+ recovered after 28 days
2.35	0.250	0.050N K_2SO_4	19.0%
2.35	0.050	0.010N K_2SO_4	Nil
2.35	0.050	0.01N K_2SO_4 + 5% citrate	86.0%
2.65	0.050	0.010N K_2SO_4	0.5%
2.80	0.050	0.010N K_2SO_4	3.0%
2.90	0.050	0.010N K_2SO_4	25.0%
3.05	0.050	0.010N K_2SO_4	60.0%
3.05	0.050	0.010N K_2SO_4 + 0.05M Fe^{3+}	30.0%
3.10	0.050	0.010N K_2SO_4	96.0%
3.10	0.050	0.010N K_2SO_4 + 0.05M Fe^{2+}	3.0%

Table 4. Immobilisation of added K^+ in acid colliery spoils

For the same spoil materials comparison of the specific
rate of K^+ release was estimated from determinations of the
equilibrium K^+ concentration in successive extracts of the 100
mesh fraction of the spoil materials in 0.20N $CaCl_2$ solution.

The method used to determine the equilibrium K^+ concentration
was similar to that described by Newman[32] except that Ca^{2+} rather
than Na^+ was employed as the replacement cation. The spoil
materials were washed in water to remove water-soluble salts
before commencing the extractions in $CaCl_2$ solution. For each
extraction, 2.5g of spoil material was equilibrated for 72 hours
in 60 mls of $CaCl_2$ solution. The equilibrium K^+ concentration
for the initial extract in $CaCl_2$ solution was generally greater
than that of subsequent extracts, but from the fifth extract
onwards the values obtained remained relatively constant for
each of the spoils. Values for the equilibrium K^+ concentration
in the fifth extract are reported in Table 4.

From the experimental data in Table 4 it is apparent that
the very low K^+ concentration in water extracts of some acid
spoils is not a function of the rate of K^+ release from the
micaceous host minerals in spoil, but is determined by a
subsequent reaction involving only K^+ which has been released
from the interlayer position in a micaceous host mineral.

The basic ferric sulphate mineral jarosite, $KFe_3(OH)_6(SO_4)_2$,
is of common occurrence in acid sulphate soils (van Breemen[20]),
and it has been identified in weathered shales (Hartley[33]) and
also in weathered colliery spoils from other British coalfields
(Doubleday[34]). The gm:ion ratio of K:Fe in jarosite is 1:4.3.
From Table 4 the ratio of K immobilised to added Fe in these
experiments on acid spoils is between 1:7.1 and 1:10.2.

In the majority of pyritic spoils examined in this study,
yellow deposits of basic ferric sulphate were found in the
matrix of spoil debris and around pyritic fragments buried below
the tip surface. Formation of jarosite is almost certainly
the correct interpretation of the data in Fig. 7 and Table 4,
but jarosite cannot be identified positively except by physical
techniques, such as X-ray diffraction, which were not available.

Van Breemen[20] states that the upper pH limit of the jarosite
stability field is pH 3.7 in the presence of hydrated ferric
oxide. The concentration of K^+ in the spoil solution and in
water extracts of acid spoil at pH 3.7 or less is regulated by
the reaction:

$$KFe_3(OH)_6(SO_4)_2 + 6H^+ = K^+ + 3Fe^{3+} + 2SO_4^{2-} + 6H_2O \qquad (10)$$

where $\log K = -12.5 = \log K^+ + 3\log Fe^{3+} + 2\log SO_4^{2-} + 6pH$
$$\qquad (11)$$

The length of time taken to immobilise added K^+ in the
experiments for which results are given in Table 4, suggests that
the equilibrium for this reaction is approached relatively slowly

110

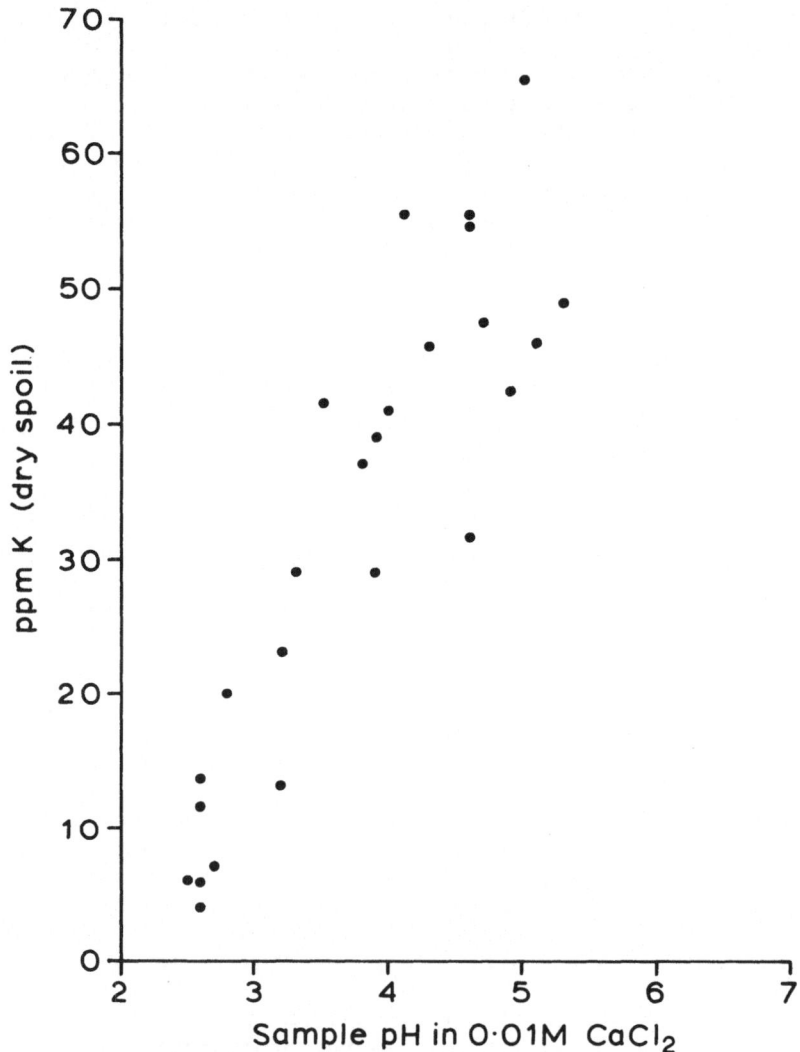

Figure 7. K⁺ in water extracts of acid colliery spoil.
Data from Gilcar (I = approx. 0.10).

in spoil materials, and that the rate of K^+ immobilisation is
also dependent upon the kinetics of the oxidation reaction
$Fe^{2+} \rightarrow Fe^{3+} + e$.

For a spoil containing 6 per cent FeS_2, the potential
sulphate content is almost 10 per cent by weight of the spoil.
The total potassium content of the majority of colliery spoils
ranges from 2.5 per cent to 4 per cent by weight. The gm:ion

ratio of K:SO$_4$ in jarosite is 1:4.9, and it is therefore
improbable even in the most acid spoils that the entire potassium
content of the spoil could be transferred from the micaceous host
to jarosite. Nevertheless, where jarosite is being formed in
acid spoils the rate of loss of potassium from the micaceous host
mineral is accelerated because the K$^+$ concentration in solution
is maintained at a lower level than the equilibrium K$^+$
concentration for the interlayer cation exchange reaction.

Variability in the available potassium status of colliery
spoil materials is a complex function of differences in the
initial spoil composition and the effects of subsequent weathering
reactions. In non-pyritic spoils, the rate of release of K$^+$ is
determined by the characteristics of the micaceous host minerals,
and by the water soluble salt content of the spoil. In pyritic
spoils the initial effect of pyrites oxidation may be to increase
the rate of K$^+$ release by increasing the water soluble salt
content of the spoil. If, as a result of pyrites oxidation, the
spoil pH falls below pH 3.7 the availability of K$^+$ is reduced
by jarosite formation. Liming to raise the pH of acid spoils
affects only the potassium availability of pyritic spoils where
jarosite has been formed.

In the initial stages of K$^+$ release from some of the micas
studied by Newman[32] the rate of K$^+$ release declined sharply
between nil and 10 per cent total K$^+$ exchanged, and a similar
trend was observed for most of the spoil materials extracted in
CaCl$_2$ solution. The availability of K$^+$ in colliery spoils is
therefore dependent in varying degrees on the previous history
of the material that is upon how much of the 10 per cent of the
total spoil K has already been immobilised as jarosite or lost
through leaching.

From the experimental work described by Newman[32] it was
shown that in acidified salt solutions the lattice of the
micaceous minerals was not significantly altered as K$^+$ was removed,
and that there was negligible increase in the rate of potassium
release due to collapse of the host mineral. Data from the
experiments carried out in this study on K$^+$ release in spoil
materials and summarised in Table 5, shows that for materials
containing burnt spoil the increased rate of potassium release
is accompanied by a correspondingly higher rate of Al loss from
the sample. Samples containing burnt spoil did not show the
same tendency for the rate of potassium release in later
extracts to become constant as for samples of unburnt spoil.
These observations are interpreted as evidence that in burnt
spoils the rate of potassium release is greatly enhanced by
collapse of the host mineral because of irreversible alterations
to the lattice which occur at elevated temperatures. As a result
of these differences between burnt and unburnt materials the

frequency distribution of estimates of the available potassium
in colliery spoils tends to be bi-modal as in Fig. 6.

Sample	ppm K^+ extracted in N l^{-1} NH_4oAc	ppm K^+ in 0.20N $CaCl_2$ 1st extract	ppm K^+ in 0.20N $CaCl_2$ 5th extract	g Al in 0.20N $CaCl_2$ 5th extract
Woolley (b)	320	384	211	1166
Gregory Springs (b)	90	122	68	914
Hound Hill	45	38	9	264
Bullcroft	8	19	14	70
Mitchells Main	80	157	51	141

(b) = burnt spoil

Table 5. K^+ release from acid colliery spoils

Phosphorus

Nicholls & Loring[34] and Spears[35] concluded that the phosphate
in rocks of the Coal Measurers sequence is present as hydroxyapatite
mineral, and according to Wandless[36] and Nelson[37] nearly all the
phosphate in coal can be attributed to inclusion of similar
minerals in the coal. Fractionation of the phosphorus in samples
of unweathered roof and floor rock and of weathered spoil
material passing a 100 mesh seive, together with subsoil material
from a soil profile developed on Coal Measure outcrop, using the
method described by Williams, Syers & Walker[38] gave results which
are summarised in Table 6.

Data provided by fractionation methods can only be interpreted
as establishing the major trends of the behaviour of phosphorus
in soils (Syers & Walker[39]). The results given in Table 6 are
as would be predicted from data on solubility of phosphate
minerals given by Lindsay & Moreno[40]. As the spoil becomes acid
the Ca-phosphate compounds dissolve but the phosphate released
is adsorbed by Fe and Al oxyhydroxides or precipitated as Al and
Fe-phosphates. Fractionation methods do not distinguish
successfully between Al and Fe-P compounds because some or all
of the phosphate released by the first extracting reagent (NH_4F
solution) is promptly adsorbed by ferric oxyhydroxide and is

then released by the second reagent (NaOH solution), and Bromfield[41] has shown that none of the methods proposed to compensate for this can be regarded as reliable.

Sample	Fe + Al-P Fraction	'Bound' Phosphate	NaOH-Soluble Phosphate	Ca-P Fraction
Seatearth rock	0.8	40	14	8
Shale rock	16.0	40	29	134
Bullcroft	256	50	26	11
Woolley (b)	205	67	72	30
Mitchells Main	176	30	29	27
Hound Hill	80	64	14	22
Maltby Main	80	66	26	58
Outcrop soil	40	80	28	40

(b) = burnt spoil; values as ppm P of 100 mesh Fraction of material

Table 6. Fractionation of phosphate in spoil materials

The results obtained from fractionation studies of the phosphorus in spoil materials shows that despite the comparatively severe weathering environment in acid spoils the proportion of phosphorus in the least accessible fraction, the reductant soluble P, has increased only slightly relative to the unweathered materials. The variability in size of the other phosphate fractions is considered to be due to variations in the initial phosphorus content of the fresh debris. These fractions include all the potentially available phosphorus in the spoils. Attempts were made to estimate the relative availability of phosphorus in colliery spoils by using mild extracting reagents such as 0.05N $NaHCO_3$ solution (Olsen's method) and 0.1N HCl plus 0.05N NH_4F solution (Bray's solution), and by anion exchange resins (Hislop & Cooke[42]). All of these methods gave results which were too close to zero to be of value. Subsequent experiments with the extracting reagents have shown that these methods are subject to severe limitations because readsorption of phosphate released by the reagent takes place to varying extent in different spoil materials.

From field trials and pot experiments carried out on a number of spoils from this coalfield it is apparent that lack of plant available phosphate is one of the principal limiting factors to the growth of sown species on these spoils, (University of York, Colliery Spoil Reclamation Research Unit, Annual Reports for 1970 - 71 and 1971 - 72). However, foliar analysis of Betula pendula carried out on samples collected in August 1971 from six sites in this coalfield, gave mean values of total P content in the leaves ranging from 1356 ppm to 1938 ppm dry weight. For the same species growing on soil over magnesian limestone, the mean value obtained was 1516 ppm dry weight.

Doubleday[34] and Fitter[42] have shown that if the capacity for phosphate adsorption is properly taken into consideration, the availability of phosphate in spoils and the response to added phosphate for sown species can be predicted with an acceptable degree of accuracy.

MICRONUTRIENT ION AVAILABILITY IN SPOIL

Manganese

The principal sources of manganese in the spoil materials from the coalfield are assumed to be the diagenetic carbonates of mixed composition occurring in the coal (Pringle & Bradburn[43]) and in the overlying rock strata (Spears[35]). Fragments of carbonate mineral from the cleat partings in coal were positively identified in thin sections of Maltby Main spoil, and fragments of similar material are of common occurrence in many other spoils in the coalfield.

Estimates of the total manganese content of spoil were made by extracting samples of the <2 mm material in cold 30 per cent HCl. The range of values obtained by this method was 60-1050 ppm. Table 7 gives the mean values obtained for manganese content, and the mean value of the ratio of manganese to magnesium in the acid extracts. Although the composition of the carbonate host mineral is evidently variable, the correlation between manganese and magnesium content extractable in strong acid confirms the supposition that these two elements share a common host. According to Spears[35] the manganese content of the detrital material is insignificant. The correlation between manganese and calcium in the acid extracts of spoil is generally not significant. Dixon & Leighton[44] stated that in the adventitious mineral matter of coal, of which the cleat carbonate is a major component, the calcium and iron content were much more variable than for the other elements.

Site	Mn ppm	Mn:Mg Ratio	Mn v. Mg
Water Haigh	239±13.2	0.115	n.s.
Firbeck Main	674±16.5	0.160	*
Tibshelf	315±14.1	0.119	n.s.
Gilcar	176±12.3	0.108	***
Granville	369±44.8	0.292	**
Darfield	509±870	0.696	**

Table 7. Acid extractable manganese in spoils from the East
          Pennine Coalfield.

If manganese and magnesium are presumed to have a common
carbonate host mineral, it would be anticipated that the two
elements would be weathered out at the same rates in acid spoils.
Data from samples collected at 0.20 cm depth from Gilcar shows
this to be the case. For magnesium the soluble plus exchangeable
ion determined by extraction in neutral ammonium acetate solution
was taken to be the weathered out fraction of the total acid
extractable magnesium content. Manganese extracted in 2.5 per
cent acetic acid was taken to be the weathered out fraction of
the total acid extractable manganese content.

Over the range pH 2.5-5.1, the correlation between the
relative amounts of manganese and magnesium estimated to be
weathered out of the carbonate host mineral was significant at
$P = 0.001$, $r = 0.91$. The linear regression equation calculated
from data summarised in Table 8 is $Y = 0.516X + 0.514 \pm 0.07$.

The data from these samples shows a trend to higher values
of Y at the upper end of the pH range. Consideration of the
thermochemical data given by Hem[45] shows that over the entire pH
range considered here, the carbonate mineral is not stable with
respect to the Mn^{2+} ion in solution. Therefore the value of Y
should not be greater than 1.0 in any of these samples and it
must be concluded that the reaction

$$(Mg.Mn)CO_3 + 2H^+ + SO_4^{2-} \rightarrow Mn^{2+} + Mg^{2+} + H_2O + CO_2 + SO_4^{2-}$$

has not yet reached equilibrium in this spoil.

Magnesium ppm dry spoil		Manganese ppm dry spoil	
a: HCl	b: NH_4 acetate	c: HCl	d: Acetic Acid
1486±224	584±284	167±51	92±48

a/b = X = 3.02±0.91   c/d = Y = 2.07±0.78

Table 8.   Weathering of Mn^{2+} and Mg^{2+} in recently exposed soil for Gilcar.

The manganese extracted in 2.5 per cent acetic acid is considered to be present as Mn^{2+} in solution and adsorbed on the cation exchange surfaces of the debris.   Extraction of the same samples in water shows that in water extracts of approximately equal total ionic strength the ratio of water soluble:soluble plus exchangeable ions is 0.78 for Mg^{2+} and 0.72 for Mn^{2+}.   The fraction retained on the ion exchange surfaces increases inversely as the total ionic strength of the water extract.   In older, weathered surface spoils the relationships between Mn^{2+} in water extracts and Mn^{2+} v. Mg^{2+} is similar to that observed in the recently exposed spoil.   The data in Table 9 is taken from analyses of saturation paste extracts of surface spoils ranging in age from 10 years to 60 years, and pH 2.6 to pH 4.5.   For all samples, the correlation between ppm Mn and total ionic strength is significant at P = 0.01.

Site	Saturation paste extracts ppm Mn^{2+}	Mn:Mg	Correlation Mn v. Mg	Mean pH in 0.01M $CaCl_2$
Bullcroft	12.23±	0.20	0.545 **	2.92±0.30
Maltby Main	2.68±	0.11	0.447 **	3.52±0.56
Hound Hill	0.74±	0.15	0.304 n.s.	3.33±0.23
Dennington No. 2	1.15±	0.16	0.421 **	3.62±0.41

Table 9. Magnesium amd Manganese in weathered spoils.

It is apparent that the relationship between the two elements, which originate from the same host mineral in spoils, is maintained over a long period of time and it may be supposed that the behaviour of the two elements is similar. Certainly there is no reason to believe that Mn^{2+} is made less available to plants by oxidation and precipitation as MnO_2, the most common form of manganese in soils. It seems more probable that the two cations behave similarly with respect to their distribution between the cation exchange surfaces and the spoil solution and are therefore lost in run-off and drainage in direct proportion to the amounts contributed by the host mineral.

The very high levels of soluble plus exchangeable Mn^{2+} in some of these spoils would seem to indicate that manganese should be considered as a potentially limiting factor to plant growth. However, because of the close relationship with the reactive diagenetic minerals in spoil, high manganese levels are associated with high total soluble salt content and therefore high ionic strength of spoil water extracts. In such extracts the activity coefficient of Mn^{2+} may be as low as 0.35, and the effective concentration of Mn^{2+} in solution under field conditions is probably even further reduced by the formation of ion pairs with SO_4^{2-}. According to Hem[45] the stability of the $MnSO_4^0$ ion pair is relatively high:

$$\frac{MnSO_4^0}{Mn^{2+} \, SO_4^{2-}} = 1.9 \times 10^2$$

Furthermore the association between manganese and magnesium, and the high levels of available Ca^{2+} in acid spoils as compared with acid soils may modify the toxicity of manganese. Berg & Vogel[46] suggested that spoil pH was the best guide for predicting manganese toxicity of legumes in Kentucky strip mine spoils. From the data obtained from samples of spoil from the East Pennine coalfield the relationship between pH and plant available manganese is unlikely to provide a reliable indicator of potential toxicity. The amount of available manganese in these spoils is determined by the size of the 'reservoir', (i.e. the amount and composition of the carbonate host mineral, and the extent to which the carbonate is dissolved by the acid sulphate product of pyrites oxidation, and by the rate of loss in run-off and drainage).

Copper and Zinc

Relatively little appears to be known with measurable certainty about the manner in which trace elements are held in the rocks of the productive measures of this coalfield. In general, recent geochemical studies do not follow earlier concepts which stressed the soil-plant relationship of the seat earth and coal seam, (Moore[47]). The trace element distribution of a section through a prominent marine horizon has been analysed by Curtis[48] who reported that the gallium, chromium, vanadium, boron, and strontium levels were related only to the clay mineral content of the strata. The elements barium, cobalt, nickel, copper and lead were concentrated in the horizons of highest organic matter content. Although Curtis[49] observed that the trace element distribution in these rocks was consistent with the hypothesis that the transition elements were originally incorporated into the sediments as stable organic complexes, correlations between total organic matter and trace element content reported by Curtis[48] and Nicholls & Loring[34] are not entirely satisfactory as indicators of the form in which the trace elements are held in the rock. The organic matter content shows covariance with other parameters of the conditions in which deposition took place, and early diagenetic processes appear to be important in fixing the ultimate distribution of trace elements between the organic phase, the diagenetic mineral phase and the clay mineral phase of the rock. Mobilisation of trace elements in poorly drained soils was discussed by Mitchell[50], and it may be significant that reducing conditions in the lower horizons of gley soils are apparently similar to the environment which is believed to have existed in the Coal Measure sediments during early diagenis (Spears[35]).

Samples of the <2 mm fraction of colliery spoil materials from weathered and recently exposed surfaces were extracted in 0.05M EDTA solution and 2.5 per cent acetic acid. Both solutions have been used extensively in studies of trace element availability and mobility in soils (Mitchell[50]). The spoil extracts were analysed by atomic absorption spectrophotometry for copper, zinc and lead. The extractable lead content of most spoils was below the limit of detection for the method used, (less than 2 ppm lead in dry spoil material). Values consistently above the limit of detection were obtained, from EDTA extracts only, in samples of surface spoil from a site where tipping is believed to have ceased about $4\frac{1}{4}$ years ago and which now supports a diverse colonising flora, including birch (Betula pendula). The mean extractable lead content was 15 ppm, and from soil samples taken from adjacent woodland on magnesium limestone the mean extractable lead content was 68 ppm. There is no evident reason to suppose that atmospheric pollution is significant at either site, and it is concluded that the extractable lead is located in recently accumulated plant debris rather than the colliery spoil material itself. Mobilisation and redistribution of the trace elements in colliery spoil as a result of plant uptake may become significant in a relatively short time under conditions where leaf fall is able to accumulate.

Values for the EDTA extractable copper and zinc are plotted
in Fig. 8 for samples from weathered surfaces of undisturbed
sites, and from recently formed surfaces on reclamation sites.
The values given in Fig. 8 are comparable with these obtained
from a wide range of soils (Mitchell[50]) but in relation to
colliery spoil materials the conventional use of EDTA as an
extracant for elements bound to organic matter may be misleading.

If it is correct to assume that both copper and zinc are
located in the octohedral layer of an aluminosilicate, the
results of these extractions do not indicate whether illite,
kaolin, or chlorite, is the host. An observation on the arbitrary
nature of the concepts employed by soil scientists makes the
following distinction, "all ions are exchangeable, but some are
more (easily) exchangeable than others". Mitchell[50] noted that
both copper and zinc have been reported to substitute for
aluminium in micaceous minerals including illite or hydrous
micas, and states that they may be expected to occur as "rather
difficulty exchangeable cations". As exchangeable cations on
the cation exchange surfaces of clay minerals or resins copper
is more difficult to displace than zinc. When the copper and
zinc concentrations in water extracts of colliery spoil materials
are expressed as a fraction of the acetic acid extractable element
it is apparent that the water soluble zinc fraction is greater
than that of copper. Values obtained from water extracts of
spoil materials, of approximately equal total ionic strength
for the spoil extracts, are plotted against sample pH in Fig. 9.
These results confirm that the behaviour of copper and zinc as
cations in colliery spoil materials corresponds with that in
soils and other ion exchange materials.

No attempt was made to determine the effects of high
temperatures on the extractable copper and zinc values for spoil
materials, or to estimate the total metal content of these
materials. Comparison of the limited available data from burnt
and unburnt materials showed that whereas aluminium was released
more freely in acid burnt spoils, the extractable copper and
zinc values were somewhat lower at all pH's than for unburnt
spoils. The analyses given by Curtis[48] indicate that the
probable total copper content of these spoils is 50–100 ppm.
The water soluble content of acid spoils, pH 3 or less, is commonly
about 5 ppm ranging up to 12 ppm. In relation to total copper,
the amounts lost in run-off and drainage water may therefore be
significant in acid spoils. There is no significant correlation
between acetic acid or EDTA extractable copper or zinc and
extractable manganese in these spoil materials, and it is
therefore considered improbable that the copper and zinc are
associated with manganese in the diagenetic mineral fraction of
the debris. Samples of pyritic material, all showing signs of
active oxidation, were analysed for copper and zinc content by

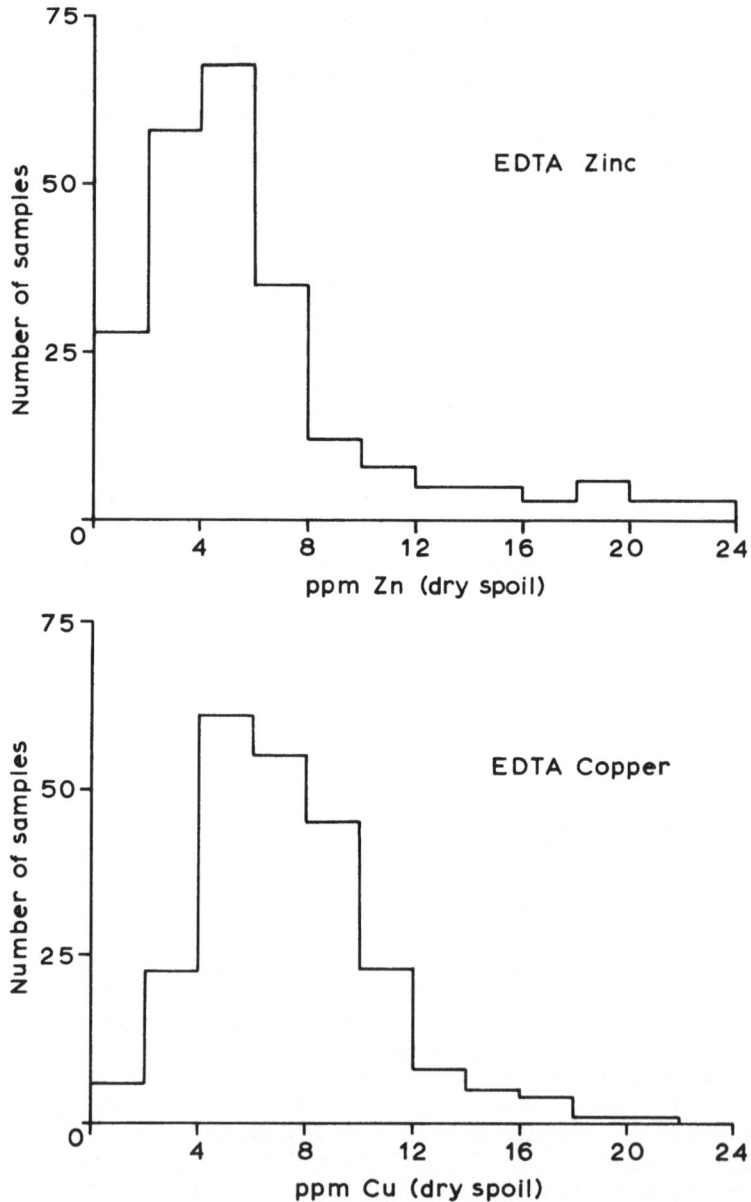

Figure 8. EDTA extractable Cu and Zn in colliery spoils.
Data from all sites.

digesting replicate subsamples in strong HCl and HNO_3. The FeS_2
content of each sample was estimated from the difference in iron
content of the two acid digests. The copper content was found to

range from 24–50 µg g^{-1} FeS$_2$ and the zinc content was 28–56
µg g^{-1} FeS$_2$. The HCl soluble copper and zinc were also determined
and the values related to non-pyritic iron on the samples. Copper
ranged up to 20 µg g^{-1} iron, and zinc up to 50 µg g^{-1} iron. These
values do not indicate any significant capture of copper and zinc
by co-precipitation with ferric iron. The amounts of copper and
zinc in the samples of pyritic material analysed are not
significant in relation to the extractable copper and zinc in the
<2 mm fraction of these spoils. However, the trace element
content of the finely disseminated pyrites in the coaly material
in spoil was not determined, and according to Whelan[7] this form
of sulphide is a product of early diagenesis and might therefore
be considered as a possible source of the extractable copper
and zinc, particularly in acid spoils.

Values of acetic acid extractable copper and zinc from
samples of Gilcar spoil material are plotted against sample pH
in Fig. 9. It is clear that as pH of the sample falls the
extractable copper and zinc increases, markedly, so for samples
of pH 3 or less for copper. Although the data presented here
could be used to support the view that copper and zinc are
contributed from different sources within these spoils it is at
least equally probable that the observed differences are due
to the characteristics of the extractants and of the two elements,
and that both elements are contributed from the same source.

The ionic radii for copper and zinc are 0.083 nm and 0.074
nm respectively, and the radius ratios in octahedral co-ordination
with oxygen (radius 0.145 nm) are 0.572 and 0.510 respectively.
The optimum radius ratio for octahedral co-ordination with
oxygen is 0.414. If both elements were incorporated in the
octahedral layer of an aluminosilicate mineral it would be
expected that the stabilities of the two ions would be
approximately equal. The removal of trace elements by acetic
acid is assumed to be by displacement of the least stable ions
by the smaller H$_3$O$^+$ ion, and the results presented in Table 10
show that copper and zinc are displaced in almost equal amounts
from most spoil materials. In samples of very low pH, pH 3 or
less, the penetration of the lattice by H$_3$O$^+$ ions in the
weathering medium is more extensive, Al^{3+} is displaced from
the octahedral layer (Barshad[51]) and the stability of the lattice
thereby reduced. In such circumstances it would be expected that
the stability of oversize ions incorporated within the lattice
would be further reduced. The increase in extractable copper and
zinc at about the same pH is the strongest indication that both
elements are contributed from the same host.

122

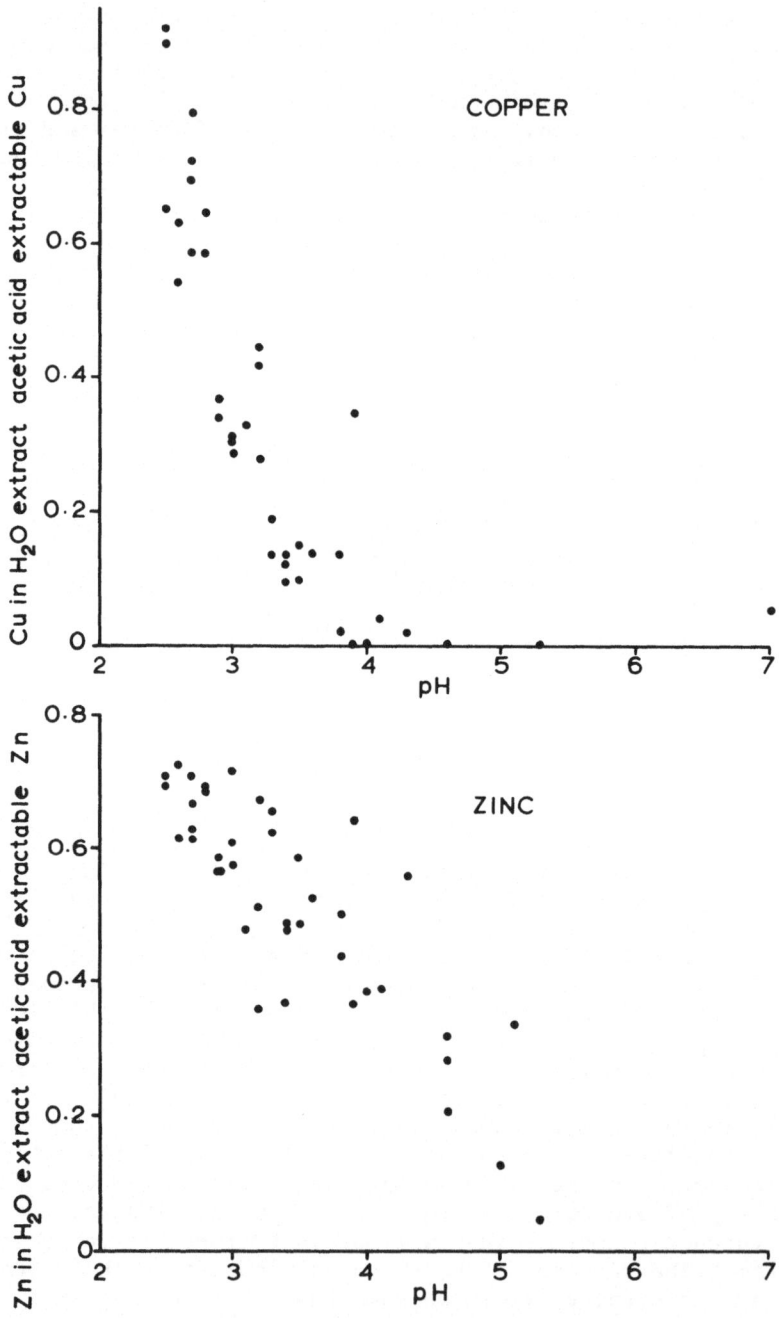

Figure 9. Relationship between water soluble trace metals and sample pH for colliery spoil.

REFERENCES

1. Hall, I.G., The ecology of disused pit heaps in England,
    J. Ecol., 45, 689, 1957.

2. Glover, H.G., The disposal of coal mine spoil in the United
    Kingdom, In Environmental Management of Mineral Wastes,
    (ed. G.T. Goodman & M.J. Chadwick), Noordhoff, Leyden, 1977.

3. van Beers, W.F.J., Acid Sulphate Soils, Bull. 3, Int. Inst.
    Land Reclam. & Improv., Wageningen, 1962.

4. Knabe, W., Observation on world wide efforts to reclaim
    industrial waste land, In Ecology and the Industrial
    Society, (ed. G.T. Goodman, R.W. Edwards & J.M. Lambert),
    Blackwell, Oxford, 1965.

5. Baas Becking, L.G.M., Kaplan, I.R. & Moore, D., Limits of the
    natural environment in terms of pH and oxidation-reduction
    potentials, J. Geol., 68, 243, 1960.

6. Garrels, R.M. & Christ, C.L., Solutions, Minerals and
    Equilibria, Harper International, New York, 1965.

7. Whelan, P.K., Finely disseminated sulphur compounds in
    British coals, J. Inst. Fuel, 27, 455, 1954.

8. Hart, M.G.R., Observations on the source of acid in empoldered
    mangrove soils II, Oxidation of soil polysulphide, Pl. Soil,
    19, 106, 1963.

9. Smith, E.E., Svanks, K. & Halko, E., Aerobic-anaerobic
    oxidation of pyrite, Symp. Poll. Contr. Fuel Combust. Proc.
    Min., Am. chem. Soc. Div. Fuel Chem., 13, 1969

10. Singer, P.C. & Stumm, W., The rate determining step in the
    production of acidic mine wastes, Symp. Poll. Contr. Fuel
    Comb. Proc. Min., Am. chem. Soc. Div. Fuel Chem., 13, 80,
    1969.

11. Silverman, M.P. & Lundgren, D.G., Studies on the chemauto-
    trophic iron bacterium Ferrobacillus ferrooxidans I, An
    improved medium and harvesting procedure for securing high
    cell yields, J. Bact., 77, 642, 1959.

12. Lorenz, W.C. & Tarpley, E.C., Oxidation of coal mine pyrites,
    U.S.D.I. Bur. of Mines Rep., 6247, 1963.

13. Silverman, M.P., Mechanisms of bacterial pyrite oxidation,
    J. Bact., 94, 1046, 1967.

14. Caruccio, F.T., An evaluation of factors affecting acid mine drainage production and the ground water interactions in selected areas of western Pennsylvania, 2nd Symp. on Coal Mine Drainage Res., Mellon Inst., Pittsburgh, Pennsylvania, 1968.

15. Braley, S.A., The oxidation of pyritic conglomerates, Industr. Wastes, 3, 1960.

16. Scott, C.S., Oxidation of coal mine pyrites, J. sanit. Eng. Div., ASCE, 92, 127, 1966.

17. Smith, E.E., Acid mine drainage research at the Ohio State University, 22nd Ann. Conf. Industr. Waste, Purdue, 1967.

18. Wandless, A.M., The occurrence of sulphur in British coals, J. Inst. Fuel, 32, 258, 1959.

19. Shumate, K.S., Smith, E.E. & Brant, R.A., A model for pyritic systems, Symp. Poll. Contr. Fuel Combust. Proc. Min., Am. chem. Soc. Div. Fuel Chem., 13, 50, 1969.

20. van Breeman, N., Soil forming processes in acid sulphate soils, In Acid Sulphate Soils, 66, Wageningen, 1972.

21. Guney, M., Oxidation and spontaneous combustion of coal, Review of individual factors, Colliery Guardian, 216, 105, 1968.

22. Coleman, N.T. & Thomas, G.W., Basic chemistry of soil acidity, In Soil Acidity and Liming, (ed. R.W. Pearson & J. Adams), Amer. Soc. Agron., Madison, 1967.

23. Keller, W.D., The Principals of Chemical Weathering, Lucas, Columbia, Missouri, 1957.

24. Schofield, R.K. & Taylor, A.W., The hydrolysis of aluminium salt solutions, J. Chem. Soc., 4445, 1954.

25. Hem, J.D., Graphcial methods for studies of aqueous aluminium hydroxide, fluoride, and sulphate complexes, U.S.D.I. Geol. Sur. Water-Supply Pap., 1827-B, 1968

26. Hem, J.D. & Robertson, C.E., Form and stability of aluminium hydroxide complexes in dilute solution, U.S.D.I. Geol. Surv. Water-Supply Pap., 1827, 1967.

27. Lindsay, W.L., Peech, M. & Clark, J.S., Determination of aluminium ion activity in soil extracts, Soil Sci. Soc. Am. Proc., 23, 266, 1959.

28. Peech, M., Cowan, R.L. & Baker, J.H., A critical study of the barium chloride/triethanolamine and the ammonium acetate methods for determining exchangeable hydrogen content of soils, Soil Sci. Soc. Am. Proc., 26, 37, 1962.

29. Clarkson, D.T., Calcium uptake by calcicole and calcifuge species in the genus Agrostis, J. Ecol., 53, 427, 1965.

30. Dixon, K., Skipsey, E & Watts, J.T., The distribution and composition of inorganic matter in British coals, I, Initial study of seams from the East Midlands Division of the National Coal Board, J. Inst. Fuel, 37, 485, 1964.

31. Carr, K., Grimshaw, R.W. & Roberts, A.L., A hydrous mica from Yorkshire fireclay, Min. Mag., 30, 139, 1953.

32. Newman, A.C.D., Cation exchange properties in micas, I, The relation between mica composition and potassium exchange in solutions of different pH, J. Soil Sci., 20, 357, 1969.

33. Hartley, J., Jarosite from carboniferous shales in Yorkshire, Trans. Leeds Geol. Assoc., 7, 19, 1957.

34. Nicholls, G.E. & Loring, D.H., The geochemistry of some British carboniferous sediments, Geochim. cosmochin. Acta, 26, 181, 1962.

35. Spears, D.A., The major element geochemistry of the Mansfield Marine Band in the Westphalian of Yorkshire, Geochim. cosmochim. Acta, 28, 1679, 1964.

36. Wandless, A.M., British coal seams: A review of their properties with suggestions for research, J. Inst. Fuel, 30, 541, 1957.

37. Nelson, J.B., Assessment of the mineral species associated with coal, Mon. Bull. Brit. Coal Utiliz. Res. Assoc., 27, 41, 1953.

38. Williams, J.D.H., Syers, J.K. & Walker, T.W., Fractionation of soil inorganic phosphate by a modification of Chang and Jackson's procedure, Soil Sci. Soc. Am. Proc., 31, 736, 1967.

39. Syers, J.K. & Walker, T.W., Phosphorus transformations in a chronosequence of soils developed on wind-blown sand in New Zealand, I, total and organic phosphorus, J. Soil Sci., 20, 57, 1969.

126

40. Lindsay, W.L. & Moreno, E.C., Phosphate phase equilibria in soils, Soil Sci. Soc. Am. Proc., 24, 177, 1960

41. Bromfield, S.M., The inadequacy of corrections for resorption of phosphate during the extraction of aluminium-bound soil phosphate, Soil Sci., 109, 1970.

42. Fitter, A.H., Mineral nutrition of grasses and the reclamation of colliery shale, Ph.D. thesis, Univ. Liverpool, 1972.

43. Pringle, W.J.S. & Bradburn, E., The mineral matter in coal, II, The composition of the carbonate minerals, Fuel, Lond., 37, 166, 1958.

44. Dixon, K. & Leighton, L.H., Ash composition of some British coals, J. Inst. Fuel, 32, 363, 1959.

45. Hem, J.D., Chemical equilibria and rates of manganese oxidation, U.S.D.I., Geol. Surv. Water-Supply Pap., 1667-A, 1963.

46. Berg, W.A. & Vogel, W.G., Manganese toxicity of legumes seeded in Kentucky strip mine spoils, U.S.D.A. For. Serv. Res. Pap., NE 119, 1968.

47. Moore, L.R., In Coal and Coal-bearing Strata, (ed. D. Murchison & T.S. Westoll), Oliver & Boyd, London, 1968.

48. Curtis, C.D., Trace element distribution in some British carboniferous sediments, Geochim. cosmochim. Acta, 33, 519, 1969.

49. Curtis, C.D., Diagenetic iron minerals in some British carboniferous sediments, Geochim. cosmochim. Acta, 31, 2109, 1964.

50. Mitchell, R.L., Trace elements in soil, In Chemistry of the Soil, (ed. F.E. Bear), 2nd Edition, Reinhold, New York, 1964.

51. Barshad, I., Chemistry of soil development, Chemistry of the Soil, (ed. F.E. Bear), 2nd Edition, Reinhold, New York, 1964.

DEPOSITIONAL ENVIRONMENT OF CARBONIFEROUS SEDIMENTS - A PREDICTOR
OF COAL MINE PROBLEMS

Frank T. Caruccio

Department of Geology, University of South Carolina,
Columbia, South Carolina, U.S.A.

THE ACID MINE DRAINAGE PROBLEM

When coal is mined the iron disulphides (FeS_2), occurring either
as marcasite or pyrite, in the coal and associated strata are exposed to the atmosphere and readily oxidize to a series of hydrous
iron sulphates. These compounds are soluble in water and produce
mine drainages that commonly have a pH values around 2, contain
acidities ranging from 4-20 mg/l (as H^+), have an abundance of iron
(usually 50-500 mg/l as ferrous iron), and a high sulphate content,
in the range of 500-12,000 mg/l (as SO_4^{2-}).

The reaction in its simplest form can be represented by
equation (1).

$$FeS_2 + 3\text{-}1/2O_2 + X\text{+}1H_2O = FeSO_4 \cdot XH_2O + H_2SO_4 \qquad (1)$$

The hydrous iron sulphate, in turn, can hydrolyse as in
equation (2).

$$FeSO_4 \cdot XH_2O + 2HOH \rightleftharpoons Fe(OH)_2 + SO_4^{2-} + 2H^+ + XH_2O \qquad (2)$$

Yellow and white crusts or coatings commonly appear along certain horizons on the exposed surfaces of the rocks and coal in
mines (Fig. 1). These minerals have been identified as melanterite
(white crystals of ferrous sulphate), copiapite (yellow crystals of
ferric sulphate), halotrichite (white crystals of iron or magnesium
aluminum sulphate) and alunogenite (white crystals of aluminum sulphate)[1]. These minerals contain water in the hydrated form the
amount of which varies with the mode of formation. The hydrated
iron sulphates are the minerals which react in accordance with

equation (2).

The ferrous iron generated in the reaction described by equation (2) can be oxidized to the ferric state with the generation of additional acidity, as in equation (3).

$$Fe(OH)_2 + HOH \rightleftharpoons Fe(OH)_3 + H^+ \tag{3}$$

Stumm & Lee[2] estimate that, "fifty percent of the acidity in acid mine drainage arises from the oxygenation of ferrous iron; the remainder arises from the oxygenation of sulphite or polysulphide." The ferrous and ferric hydroxides associated with equations (2) and (3) impart the red color that is characteristic of acid mine drainage. The precipitated iron hydroxide, popularly known as "yellowboy", is often observed in streams.

The reaction rate of equation (3) is generally very slow and the complete oxidation of iron from the ferrous to the ferric state takes place over a long section of a stream or drainage course. There are, however, iron bacteria (Thiobacillus and Ferrobacillus) that oxidize the ferrous ion to ferric and in essence catalyze the reaction in equation (3).

It has been shown by Caruccio[3] that the degree of acidity is a function of the calcium carbonate content of the strata (which produces a neutralizing medium), the pH of the ground waters before mining (which controls the occurrence of the various types of "iron bacteria" that catalyze the acid producing chemical reaction), and the mode of occurrence of the iron disulphide.

The iron disulphide occurs in coal strata as euhedral crystals (Fig. 2); coarse-grained (greater than 25 microns in diameter) masses which replaced original plant matter (Figs. 3 and 4); coarse-grained platy masses occupying joints in the strata (Fig. 5); framboidal pyrite - clusters of spheres of iron disulphide measuring about 0.25 microns in diameter (Fig. 6). Of these four basic types it is only the framboidal type that decomposes rapidly enough to produce severe acid mine drainage[4]. The other types or iron disulphide decompose at a relatively slow rate; the amount of acidity produced is generally low or easily neutralized by small amounts of alkalinity generated by associated calcium carbonate found in the rocks. The severity of acid mine drainage pollution from coal mines is, in part, a direct function of the amount of framboidal iron disulphide in the rock sequence disturbed and exposed to the atmosphere by the mining process.

GEOLOGICAL CONTROL

The occurrence of sulphurous materials in the stratigraphic

Figure 1.
Oxidation products
of pyrite coating
coal seam.

Figure 2.
Photomicrograph of
Euhedral pyrite.

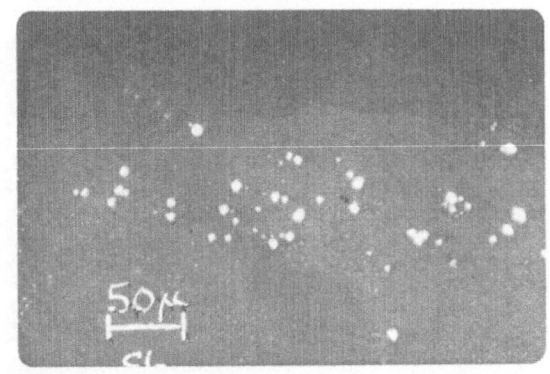

Figure 3.
Photomicrograph of
plant structures in
coal replaced by
pyrite.

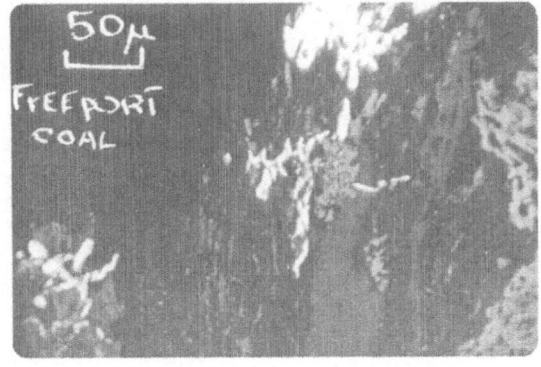

Figure 4.
Photomicrograph of
plant structures in
coal replaced by
pyrite.

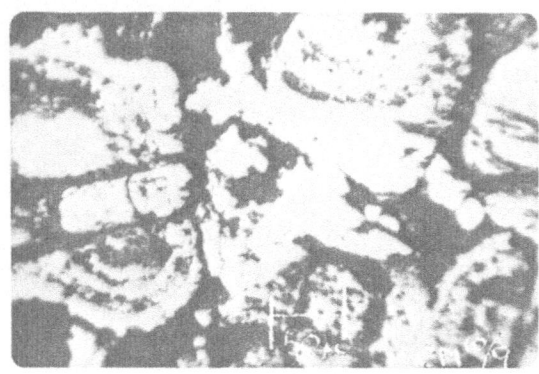

Figure 5.
Photomicrograph of
pyrite coating joints
in coal.

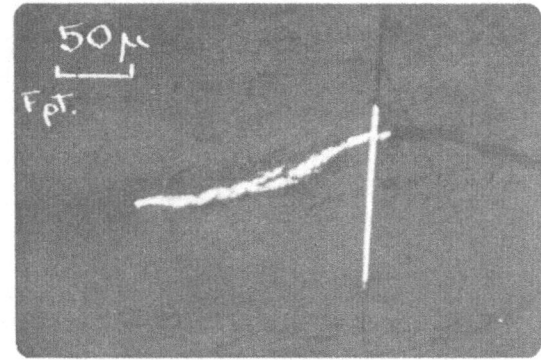

Figure 6.
Photomicrograph of
framboidal pyrite.

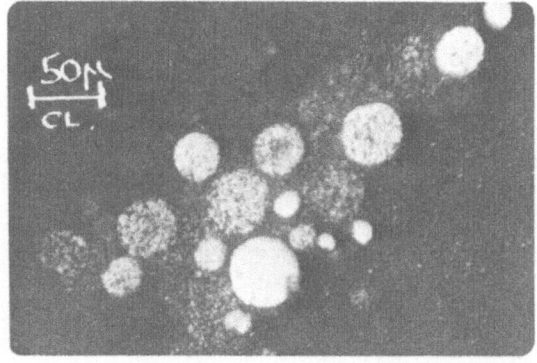

section of western Pennsylvania has been outlined in a general way by Williams[5] and Williams & Keith[6]. Based on fossil evidence, Williams[5] documented a transgression of paleoenvironments (the conditions under which the rock was deposited) in the Allegheny Group of the Pennsylvanian age rocks (Fig. 7). In the basal portion of the Allegheny of western Pennsylvania, the rocks contain fossils indigenous to a marine-brackish water paleoenvironment. In the upper portion of the Group, the rocks show evidence of having been deposited in a continental-fresh-water paleoenvironment. In a vertical succession the lower "marine" rocks, in general, grade upward into the upper "continental" rocks. Additional work by Ferm & Williams[7] has shown that the basal "brackish" rocks also exhibit a lateral transgression, becoming more "marine" in the western portion of Pennsylvania.

The rock established as having been deposited in a marine paleoenvironment in the Pennsylvanian (age) commonly contain bitumens which are preserved only under reducing conditions, whereas the rocks of continental origin show evidence of being oxidized. Because the formation and deposition of iron disulphide is favored by reducing environments, the lithologies of the basal portion of the Allegheny group in the eastern portion of the Bituminous coal field should be higher in sulphur (as pyrite) than the rocks in the top portion of the group. Further, the sulphur content in the basal section should increase as the paleoenvironment becomes progressively more marine toward the western part of the Bituminous area.

Mansfield & Spackman[8] traced the variations in the petrography and composition of the Lower Kittanning, Lower Clarion and Upper Freeport seams across the Bituminous coal field of Pennsylvania. They found this relation to hold true. With the exception of the Freeport coal, whose anomalous trend was attributed to chance, the coals became more sulphurous as the environment of deposition became more marine.

The occurrence of the framboidal iron disulphide within a particular rock unit is directly correlative with the paleoenvironment of the stratum; (the conditions under which the rock was deposited control the formation, deposition and occurrence of the framboidal iron disulphide). As shown above, Williams & Keith demonstrated that a coal seam in the bituminous coal field of Pennsylvania became increasingly more sulphurous as the paleoenvironment of the coal roof rock shifted from continental to marine. However, although the strata became more sulphurous (reflecting increases in iron disulphide), it was shown by Caruccio[9] that the type of iron disulphide and not the amount was the critical factor in the production of acid mine drainage. Strata containing large amounts of "sulphur" occurring as framboidal iron disulphide will produce more acid than the strata containing equal amounts of "sulphur" composed of coarse grained nonframboidal iron disulphide.

132

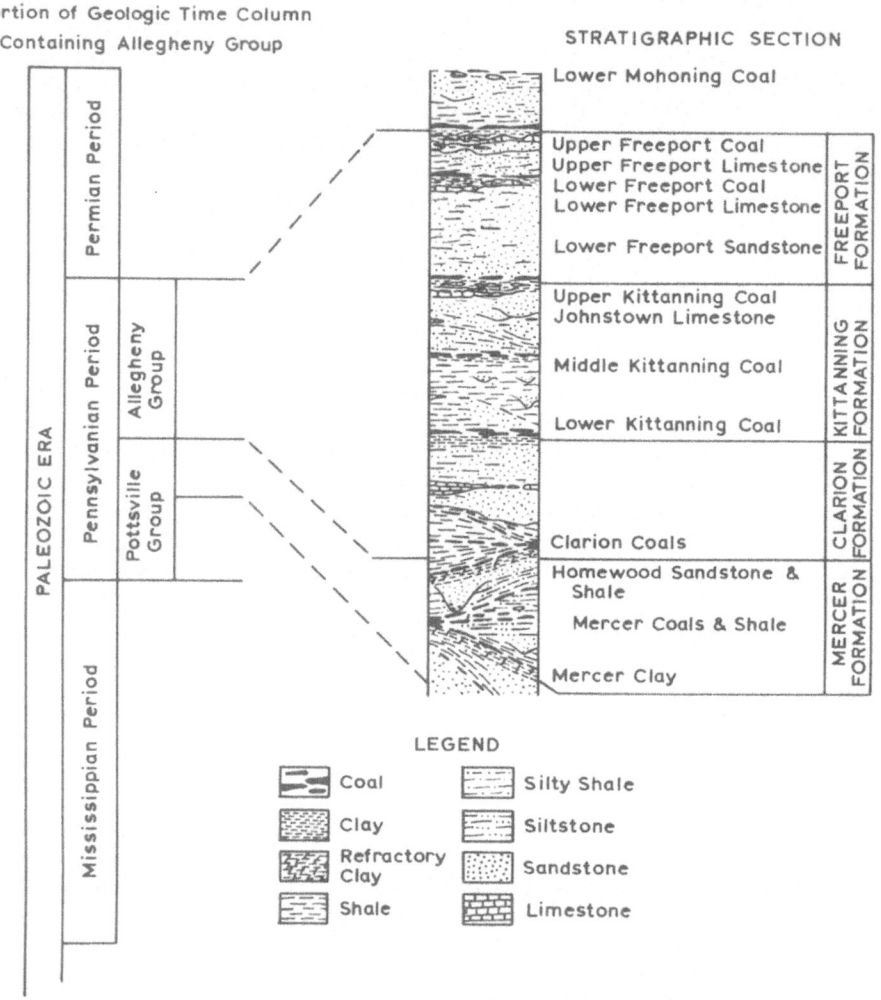

Figure 7. Stratigraphic nomenclature and generalized section of the Allegheny group in Western Pennsylvania[5].

Consequently, the association of framboidal pyrite with certain paleoenvironments is the key in identifying coal strata which when mined, will produce acid mine pollution.

On a regional scale Caruccio[10] postulated a model explaining the varying composition of mine drainages taking into account the mode of occurrence of the iron disulphide and the geo-chemistry of the ground water (Fig. 8). The model reasonably explains the acid mine drainage compositions in association with the paloenvironments of the strata of the Allegheny Group of the bituminous coal field of Pennsylvania. In Fig. 8 it will be seen that, at the upper part of the stratigraphic column, in the Freeport coal seams, the pyrite occurring in the strata is coarse-grained and is relatively inert. The pH of the ground water (before the mining phase) is above 6.4, which is not conducive to the support of iron bacteria, and there exists, within the natural system, a ground water with a high neutralizing capacity. In this environment, minor amounts of acidity produced by the pyrite are neutralized by the available alkalinity. The acid reaction is not catalyzed (due to the absence of the bacteria) and the mine drainages are chemically neutral and of low sulphate types.

On the other extreme, in the lower part of the stratigraphic column, in the Brookville-Clarion seams, most of the pyrite in the strata is fine-grained (less than 0.25 microns) and readily oxidizes to produce abundant acidity. The pH of the ground water is low and supports iron bacteria which catalyze the acid producing reactions and generate additional acidity. The buffering capacity of the ground water is extremely low and consequently, drainages from this environment are highly acidic and high in sulphate. The Kittanning seams represent an intermediate condition with correspondingly moderate amounts of acidity and sulphate.

Neutral, high sulphate drainages are generated when alkaline waters come in contact with acid waters and neutralization takes place. This situation occurred in North-Western Pennsylvania where coals mined from highly acid, high sulphate strata were capped by calcareous glacial drift. Here the highly alkaline waters from the glacial drift effectively neutralized the acidity created in the mining operation, producing a neutral-high sulphate type of mine drainage.

Emrich & Thompson[11] further confirmed the acid-producing model within western Pennsylvania by comparing the ratios of acid to neutral mine drainages to the ratios of minable "marine" to "continental" coal seams on a county by county basis.

Significantly, the total sulphur contents (reflecting pyrite content) of the coals and associated strata of acid and non-acid mine drainage areas are about the same. It was shown that a strip

134

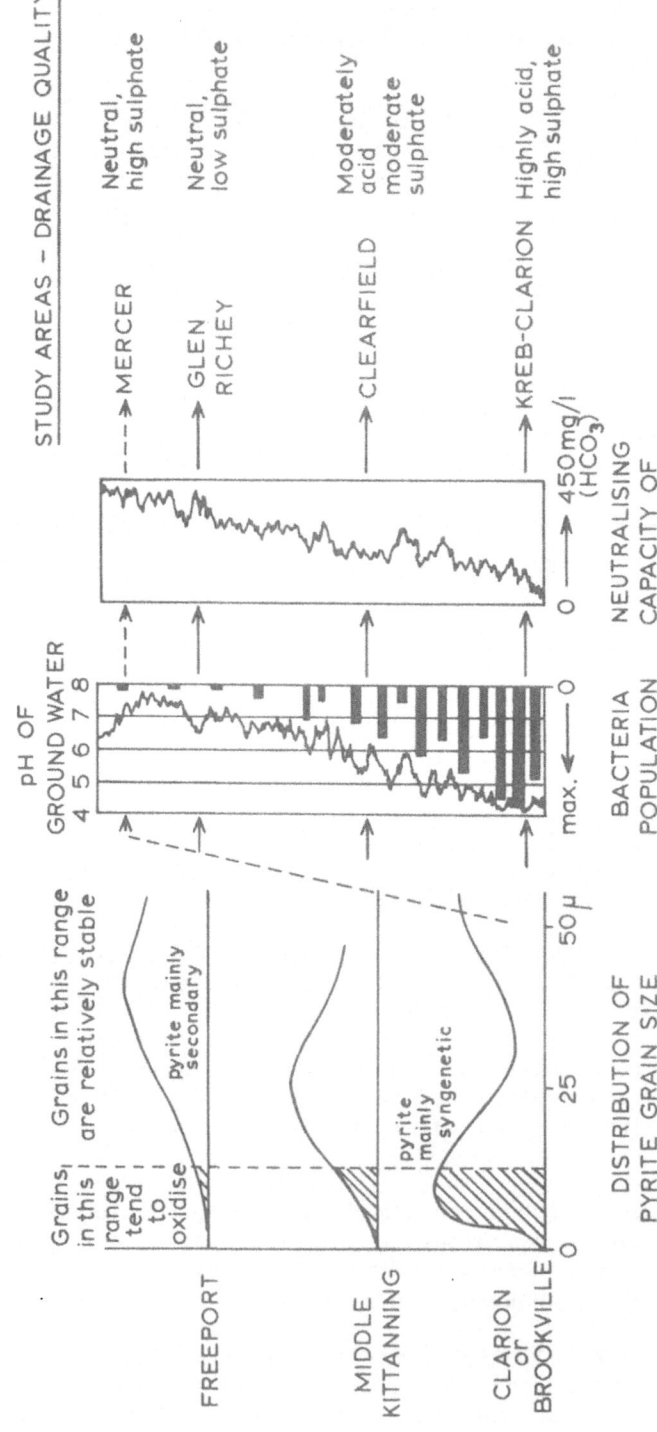

Figure 8. Model summarizing Hydro-Geochemical factors affecting Mine Drainage Qualities.

mine area underlain by continental strata, yielding neutral mine drainages had total sulphur contents about equal to the sulphur contents of another mined area underlain by marine strata and yielding moderately acid drainages (Fig. 9)[3]. From paleoenvironmental considerations the acid free area, underlain by continental strata should have contained less total sulphur than the acid producing area underlain by marine strata.

Another significant fact was that neutral drainages emanating from areas underlain by continental strata were low in sulphate anion concentrations. In fact, the concentrations were well within the limit set by the Public Health Service for drinking water standards (250 mg/l). The lack of sulphate anion in the drainage is taken to mean that the pyrite in this area is stable and not oxidizing.

The reason for the discrepancy in the results of the sulphur distribution between Williams & Keith[6] and Mansfield & Spackman[8] on the one hand, who showed the strata becoming less sulphurous with an increasing continental paleoenvironment, and Caruccio[3] on the other, who showed no sulphur variation between continental and marine beds can be understood if the sampling procedures used in both studies are reviewed.

The purpose of Williams & Keith's work[6] was to relate the paleoenvironment to the total sulphur content of a stratum. For this purpose, coal and rock samples were collected from beds which did not exhibit pyrite coatings on the joints or bedding plane surfaces. If a bed was coated with pyrite, the pyrite coating was scraped away in order to properly sample the rock and its intrinsic properties.

Caruccio[3, 4, 10] however, related the acid mine drainage potential to the pyrite content of the area (as measured by total sulphur content) and consequently the entire bed was sampled which included primary and secondary pyrite deposits.

Coincidentally, the primary pyrite deposits are finer grained, readily decompose and accordingly the preponderance of this type of pyrite in the Clarion and marine-like strata cause these strata to yield acid mine drainage. The Freeport, on the other hand, although containing the same amount of total pyritic sulphur as the Clarion seams, has a preponderance of secondary pyrite which is coarser grained, remains relatively stable and consequently does not produce significant amounts of acid mine drainage; as evidenced by the low sulphate contents of drainages from the Freeport mines.

Thus, coals from the lower part of the stratigraphic sequence are overlain by strata indicative of marine or brackish paleoenvironments and are found to be rich in framboidal pyrite. Coals in the

136

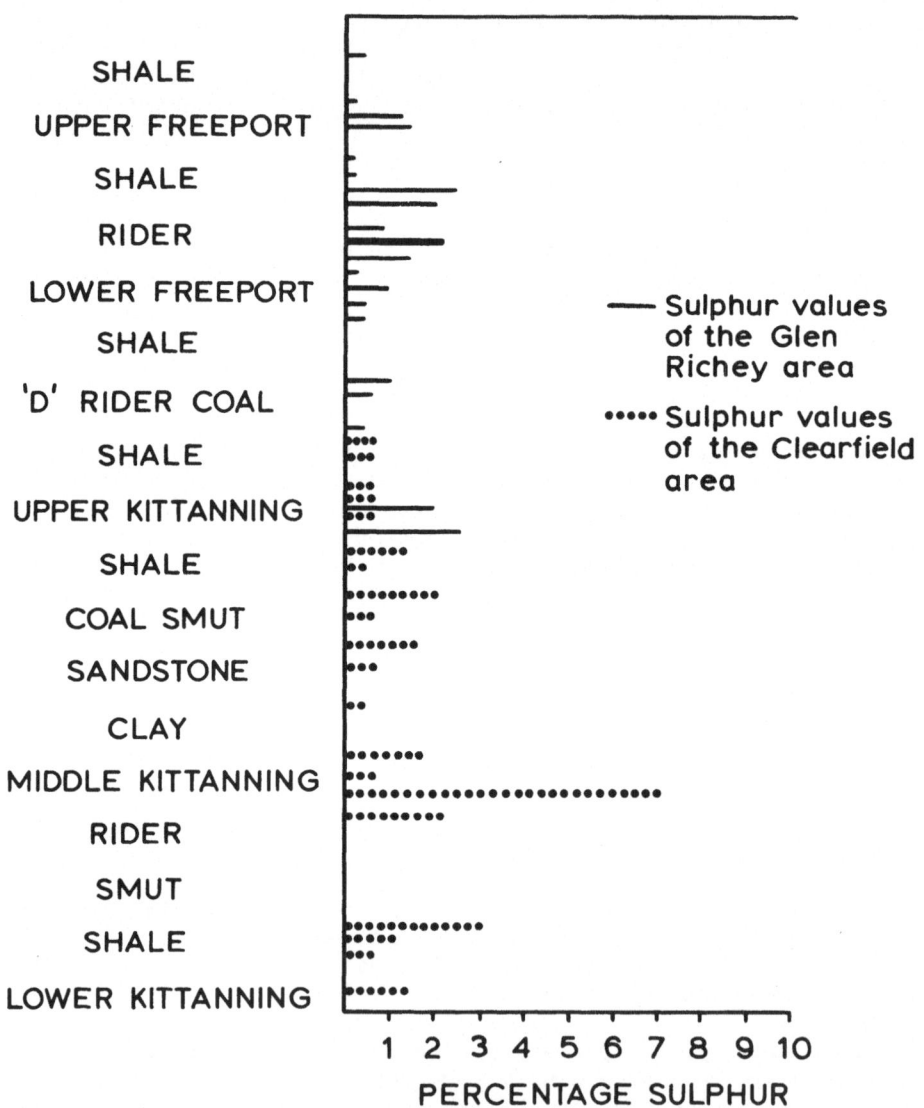

Figure 9. Distribution of Total Sulphur Content throughout a section of the Allegheny Group of Western Pennsylvania.

upper part of the sequence are overlain by strata of a continental
paleoenvironment and are found to be difficult in framboidal pyrite.
This is summarized on Fig. 8.

PREDICTION OF ACID MINE DRAINAGE POLLUTION PROBLEMS

Simultaneously with the work of Caruccio in western Pennsylvania[3, 4, 10], Ferm and his associates[12, 13, 14, 15, 16, 17, 18, 19]
have been engaged in an extensive program of recognizing and mapping
environments of coal deposition in the Appalachian coal field from
Pennsylvania to Alabama. By comparing ancient coal bearing rocks
to modern deposition analogues a set of criteria has been developed
for recognizing environments which range from alluvial plains through
upper delta plains and lower delta plains grading into a tidally influenced back barriers and beach barriers. Applying these criteria
to Caruccio's Pennsylvania data, the coals and associated rocks rich
in framboidal pyrite are recognizable as back barrier and lower delta
plain deposits and strata poor in framboidal pyrite can be associated
with upper delta and alluvial plain settings. In addition, where
Caruccio's samples represent different environments for the same
coal bed distributed across a broad geographic range, a significant
variation in the proportion of framboidal pyrite is observable.

Consequently, because many Appalachian coal beds are known to
represent several different environments in different places, and
because others can be recognized as belonging to at least two environments, a criterion appears to be available for predicting the
geographical distribution of framboidal pyrite without costly and
time consuming analytical procedures. If the different environments
of coal deposition can be readily recognized and shown on maps then
a predictive device for the occurrence of framboidal pyrite is
readily available. Strata of varying paleoenvironments could be
mapped (in both vertical and horizontal distributions) and zones
rich in framboidal pyrite delineated. Having constructed these
maps areas can be identified that would in advance of the active
mining stage:

(a) Delineate for each minable coal bed, areas of major and
    minor environmental impacts from acid drainage pollution
    in surface and deep mined areas.

(b) Outline surface mined areas amenable to successful restor-
    ation (in terms of reclaiming the backfilled surface) by
    surface revegetation. Strata can be identified for surface
    dressing suitability and for deep burial and hydrologic
    isolation.

(c) Identify coal seams by limits of high and low sulphur con-

138

tent and whether or not the high sulphur coals can be
cleansed to produce low sulphur fuels.

FUTURE STUDY

The hypothesis outlined above, i.e. that the occurrence of
framboidal pyrite in coal can be predicted by recognizing the depositional environment of the strata, is currently being tested by
Caruccio and Ferm at the University of South Carolina, under the
sponsorship of the Environmental Protection Agency. A coal seam
has been selected whose the paleodepositional environment varies
over a distance of 15 miles, has been the subject of many recent
studies and where abundant environmental data are available. Coal
and rock samples will be collected from various sections representative of different paleoenvironments and the mode of pyrite determined. Simultaneously, hydro-geochemical characteristics will be
determined for correlation coefficients. In turn, the depositional
history of the rocks will be related to both pyrite morphology and
the aqueous geochemistry of the mine drainages.

REFERENCES

1. Lorenz, Walter C., Progress in controlling acid mine water:
    A literature review: Information Circular 8080, Bureau of
    Mines, U.S. Dept. of the Interior, Washington, 1962.

2. Stumm, W. & Lee, G., Oxygenation of Ferrous iron, Ind. Eng.
    Chem., 53, 143, 1961.

3. Caruccio, F.T., An evaluation of factors affecting acid mine
    drainage production and the ground water interactions in
    selected areas of Western Pennsylvania, Second Symposium on
    Coal Mine Drainage Research, 107, Mellon Inst., Pittsburgh,
    Pennsylvania, 1968.

4. Caruccio, F.T., The quantification of reactive pyrite by grain
    size, Third Symposium on Coal Mine Drainage Research, 123,
    Mellon Inst., Pittsburgh, Pennsylvania, 1970.

5. Williams, E.G., Marine and fresh water fossiliferous beds in
    the Pottsville and Allegheny groups of Western Pennsylvania,
    J. Paleontol., 34, 1960.

6. Williams, E.G. & Keith, M.L., Relationship between sulphur in
    coals and the occurrence of marine reef beds, Econ. Geol. 58,
    720, 1963.

7. Ferm, J.C. & Williams, E.G., Stratigraphic variation in some Allegheny rocks of western Pennsylvania, Bull. Amer. Assoc. Petroleum Geologists, 44, 495, 1960.

8. Mansfield, S.P. & Spackman, W., Petrographic composition and sulphur content of selected Pennsylvania bituminous coal seams, Special Research Report No. SR-50, The Coal Research Section, Pennsylvania State Univ., 1965.

9. Caruccio F.T., Trace element distribution in reactive and inert pyrite, Fourth Symposium on Coal Mine Drainage Research, 48, Mellon Inst., Pittsburgh, Pennsylvania, 1972.

10. Caruccio, F.T., Characterization of strip mine drainage by pyrite grain size and chemical quality of existing ground water, in Ecology and Reclamation of Devastated Land, I, Hutnik, R.J. & Davis, G., Eds., Gordon & Breach, New York, 1975.

11. Emrich, G. & Thompson, D.R., Some characteristics of drainage from deep bituminous mines in Pennsylvania, Second Symposium on Coal Mine Drainage Research, 190, Mellon Inst., Pittsburgh, Pennsylvania, 1968.

12. Ferm, J.C., Dutcher, R.R., Flint, N.K. & Williams, E.G., The Pennsylvanian of Western Pennsylvania, Field trip no. 2, Geol. Soc. America Guidebook Series, Guidebook for the Pittsburgh Meeting, 61-114, G.S.A., 1959.

13. Ferm, J.C. & Williams, E.G., Allegheny paleogeography in the northern Appalachian plateau, Science, 137, 990, 1962.

14. Ferm, J.C. & Williams, E.G., Sedimentary facies in the Lower Allegheny rocks of western Pennsylvania, J. Sed. Pet., 34, 610, 1964.

15. Ferm, J.C. & Williams, E.G., Characteristics of a Carboniferous marine invasion in Western Pennsylvania, J. Sed. Pet., 35, 319, 1965.

16. Ferm, J.C., Ehrlich, R. & Neathery, T.L., A Field Guide to Carboniferous Detrital Rocks in Northern Alabama, Geological Society, Tuscaloosa, Ala., 1967.

17. Ferm, J.C. & Cavaroc, V.V., Siliceous spiculites as shoreline indicators in deltaic sequences, Bull. Geol. Soc. Am., 79, 263, 1968.

18. Ferm, J.C. & Cavaroc, V.V., A non-marine sedimentary model for the Allegheny of West Virginia: in Symposium on Continental Sedimentation, Klein, G.D., Ed., Geol. Soc. Am., Spec. Paper 106, 1, 1968.

ALUMINIUM AND MANGANESE TOXICITIES IN ACID COAL MINE WASTES

W. A. Berg

Department of Agronomy, Colorado State University,
Fort Collins, Colorado, U.S.A.

ABSTRACT

This review indicates that manganese toxicity will sometimes
be a problem on legumes but probably seldom a problem on perennial
grasses grown in acid coal mine wastes. Evidence for aluminium
toxicity to plants grown in acid coal mine wastes is indirect. However, information from soils and nutrient culture studies leads to
the conclusion that aluminium is a major cause of plant toxicity in
acid coal mine wastes.

INTRODUCTION

Plant toxicities resulting from acid conditions caused by sulphide oxidation are common in some coal mine wastes. The degree of
toxicity is often expressed in a general way in terms of pH. This
is understandable as the solubility of potentially toxic metallic
ions increases as acidity increases. A diagram of the solubility of
some of the ions of interest in soils is shown in Fig. 1 which follows the solubility diagrams of Lindsay[1] and Norvell[2].

It must be emphasized that in any soil system the solubilities
may vary considerably from that shown in Fig. 1. To construct such
a diagram certain assumptions and measurements have to be made on
the solubility of the most soluble solid phase of the ion in question, and in the case of iron and manganese on the oxidation state
of the system.

In this discussion the point of interest in Fig. 1 is the relationship between pH and solubility of the various ions. For each

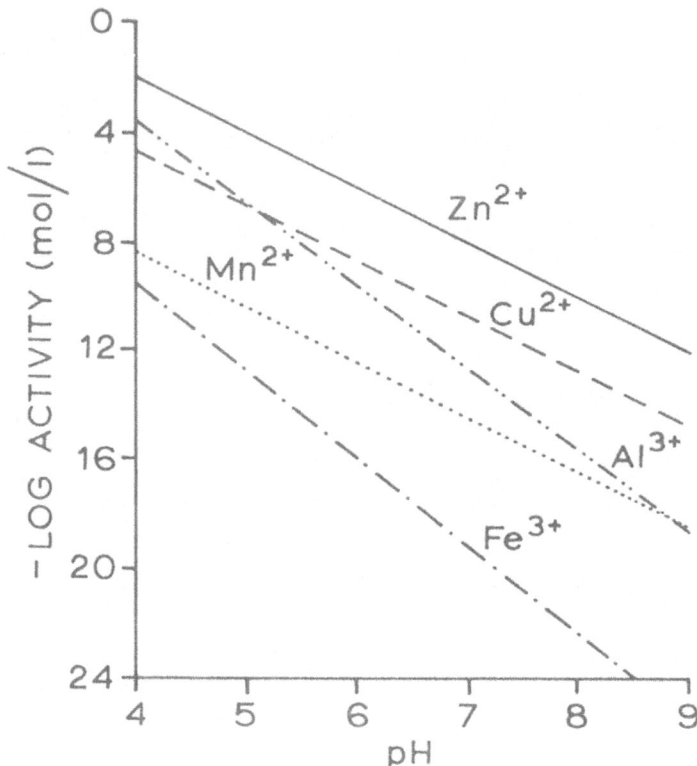

Figure 1. Activity relationships for H^+, Cu^{++}, Zn^{++}, Mn^{++}, Al^{+++}, and Fe^{+++} in soils.

unit increase in hydrogen ion activity, the activity of the divalent ions, zinc, copper and manganese increases ten-fold and the activity of the trivalent ions, aluminium and iron, increases 100-fold. Or in terms of a pH decrease of one unit (a 10-fold increase in hydrogen ion activity), the activity of the divalent ions increases 100-fold and the activity of the trivalent ions increases 1000-fold.

   In acid soils soluble aluminium and manganese are usually regarded as the major causes of plant toxicity. The research on this is extensive and has been comprehensively reviewed in a publication edited by Pearson & Adams[3]. Manganese toxicity symptoms on legumes grown in acid coal mine spoils have been described by Berg & Vogel[4]. Indirect evidence of aluminium toxicity to plants grown on acid coal mine wastes has also been demonstrated (Lorio[5]; Chadwick & Salt[6]; Beyer & Hutnik[7]; Berg & Vogel[8]).

Potentially phytotoxic levels of copper, nickel, zinc and iron, as well as aluminium and manganese, in extracts of extremely acid coal mine spoils were found by Massey & Barnhisel[9]. Rather large amounts of soluble copper as well as manganese and aluminium were found in acid deep mine waste by Chadwick[10]. It must also be recognized that under extremely acid conditions, toxicity due to hydrogen ion concentration per se may exist. Thus, the question is raised, are plant toxicities induced by ions other than aluminium and manganese of importance in acid coal mine wastes? So far this question has not been answered by plant growth studies.

The remainder of this discussion will focus on manganese and what are apparently aluminium toxicities on plants grown in acid coal mine wastes.

MANGANESE TOXICITY

Manganese toxicity symptoms on certain small-seeded legumes have been characterized by chlorotic margins on the leaflets (Morris & Pierre[11]; Lohnis[12]; Ouellette & Dessureaux[13]; Foy[14]; Andrew & Pieters[15]). Legumes grown in acid coal mine spoils and in nutrient solutions containing excess manganese were also found to have this pattern of chlorosis on leaflet margins (Berg & Vogel[4]).

In the latter study the chlorotic margins were distinct on Robinia pseudoacacia and on all Lespedeza species studied except L. striata (Fig. 2). Chlorosis caused by excess manganese on Lotus corniculatus was usually on the leaflet margins, but sometimes appeared as irregular chlorotic spots near the margins. When chlorosis caused by excess manganese occurred it was present on the first leaf or leaflets that emerged after the cotyledons opened. Growth of the legumes was sharply reduced as the degree of chlorosis increased (Fig. 3). With very severe chlorosis (70 to 100% of leaf surface) no further growth occurred.

Perennial grasses (Festuca arundinacea, Eragrostis curvula, Panicum virgatum, Holcus lanatus) grown in some of the same coal mine spoils that induced manganese toxicity symptoms on legumes had no visible symptoms of toxicity with one exception. The exception was Holcus lanatus which exhibited chlorosis similar to iron deficiency when grown in one spoil. When grown in nutrient solutions containing excess manganese this species showed a similar chlorotic pattern (Berg & Vogel, unpublished) but at manganese concentrations four to eight times greater than induced chlorosis in Lespedeza.

Rooting tests by Harding[16] indicated that manganese toxicity was not a problem in growth of Agrostis tenuis on acid colliery spoil.

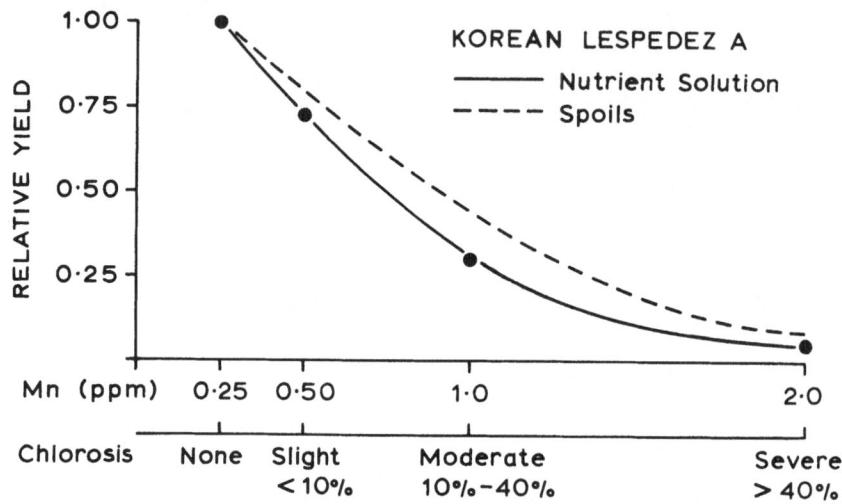

Figure 3. Yield of <u>Lespedeza</u> <u>stipulacea</u> as related to the degree
of marginal chlorosis induced by growth in coal-mine
spoils and nutrient solutions. (Mn ppm refers to Mn
added to nutrient solution.)

Thus, the very limited information indicates that manganese
toxicity will sometimes be a problem in legumes but probably seldom
a problem with perennial grasses grown on acid coal mine wastes.

Toxicity symptoms appear to be a reliable method of detecting
manganese toxicity, although the marginal chlorosis previously described for certain small-seeded legumes does not hold for all legumes (Morris & Pierre[11]; Andrew & Pieters[15]). Manganese toxicity
symptoms have been described for a number of species (Hewitt[17];
Lohnis[12]; Plant[18]; Labanauskus[19]). On grasses, Hewitt[17] reported
that excess manganese induced a chlorosis similar to iron deficiency
in <u>Avena</u> <u>sativa</u>, but on <u>Hordeum</u> <u>vulgare</u>, <u>Lolium</u> <u>perenne</u>, and <u>Festuca</u>
<u>elatior</u> the symptoms induced by excess manganese were small dark
spots on the leaves. Symptoms similar to this latter have been described and photographed on the grass <u>Chloris</u> <u>gayana</u> (Smith[20]). Manganese toxicity symptoms on <u>Betula</u> <u>alleghaniensis</u> grown in nutrient
solutions were described by <u>Hoyle</u>[21] as "marginal and/or interveinal
necrosis in the older leaves". Internal bark necrosis has been described as caused by excess manganese in apple (<u>Malus</u>) trees (Zeiger
& Shelton[22]).

Manganese content of plant tops may be a method of predicting
manganese toxicity in some plants. Labunauska[19] has reviewed toxic
manganese levels reported in various species. Recent studies

(Andrew & Hegartz[21]; Hoyle[23]) give additional information on manganese contents of various species when manganese toxicity has been
induced in nutrient solution cultures. However, Hoyle[21] indicates
that the effect of silicon in reducing manganese toxicity symptoms
as reported by Vlamis & Williams[24] should be investigated if toxicity
levels determined in nutrient solutions are to be used on plants
grown in soils.

Water soluble manganese in soils has been suggested as an indicator of manganese toxicity in soils (Morris[25]; Gupta[26]). Water
soluble manganese was not a useful indicator of manganese toxicity
to legumes grown in acid coal mine spoils by Berg & Vogel[4]. In the
latter study, spoil pH was found to be an indicator of potential
manganese toxicity if the relative manganese tolerance of the legume was known.

Methods to cope with manganese toxicity include treatment of
the coal mine waste by liming or other amendments to reduce the acidity, or to select species (Fig. 4, Berg & Vogel[4]) and varieties within species that are more tolerant to excess manganese. Literature
on manganese tolerance has recently been reviewed by Brown et al.[27].

ALUMINIUM TOXICITY

Stubby roots can be a symptom of aluminium toxicity (Rorison[28];
MacLeod & Jackson[29]; Clarkson[30]). However, toxicity caused by copper
and other metals may have similar effects on roots (Liebig et al.[31];
Bradshaw et al.[32]; Struckmeyer et al.[33]). Thus, no clear-cut symptoms are available that can be diagnosed as aluminium toxicity in
soil systems. Pratt[34] states, "Aluminium toxicity cannot be diagnosed either from visual symptoms or the aluminium content of the
tops of plants.... visual root symptoms as a basis for diagnosing
aluminium toxicity when the plant is grown in soils is a questionable
procedure".

As mentioned earlier, evidence for aluminium toxicity to plants
grown on coal mine wastes is indirect. Lorio[5] using a multiple regression approach found that growth of certain tree species on coal
mine spoils was inversely related to extractable aluminium. Agrostis
tenuis collected from an extremely acid deep coal mine waste by
Chadwick & Salt[6] produced more root growth into solutions containing
0.9 ppm aluminium than did A. tenuis collected from less acid waste
or soil. My interpretation of this is that A. tenuis population
growing on the more acid material was from a population naturally
selected for tolerance to acid conditions and probably specifically
tolerant of aluminium.

Somewhat similar results have been shown by Beyer & Hutnik[7]
working with Betula pendula, B. lenta, Pinus resinosa, and P. syl-

vestris. Their study compared growth of the birches and pines in
nutrient solutions having pH and aluminium concentrations as variables. The birches, which usually grow better than pines on extremely acid coal mine spoils, showed less sensitivity to aluminium
in nutrient solutions than did the pines.

Stunted stubby root growth on plants grown in acid strip mine
spoils have been shown by Berg & Vogel[8]. These symptoms probably
indicate aluminium toxicity, although as mentioned earlier, the
symptoms are not specific for aluminium. Restricted root growth of
Sorghum vulgare and Pinus taeda into acid coal mine spoils are shown
in Figs. 5 and 6.

Soluble aluminium concentrations in displaced soil solutions
or in saturation extracts have been suggested by Pratt[34] as a means
of diagnosing possible aluminium toxicity. He suggests a level of
less than 0.5 ppm as not likely to be harmful, 0.5-1.0 ppm will
probably be toxic, and more than 1.0 ppm has a high probability of
being toxic. Such criteria would vary among plant species and varieties. Medicago sativa grown in a soil from which a saturation
extract was taken that contained 0.45 ppm aluminium had restricted
growth (MacLeod & Jackson[29]). The latter researchers also reported
on the precipitation of aluminium from nutrient solutions and this
indicates that some of the older work citing concentrations of aluminium in nutrient solutions has to be used with caution as the
aluminium remaining in solution was not measured.

As mentioned in the introduction, pH has been and probably will
continue to be used as a general indicator of toxicity of coal mine
wastes to plants. In reality, the screening work carried out on
species suitability for growth in acid coal mine spoils (Limstrom[35];
Vogel & Berg[36]; Ruffner & Steiner[37]) may be in part a screening for
aluminium tolerance.

Selection of species and varieties within species that are more
tolerant to soluble aluminium have been made by a number of workers
(Brown et al.[27]). Methods to screen relatively large populations of
seedlings for tolerance to aluminium have been published by Kerridge
et al.[38] and Reid et al.[39]. Much of the screening work has been on
agronomically important species. Screening for aluminium tolerance
among and within species used for conservation and silviculture has
been rather limited. The technique would appear to hold promise on
a widely used species such as Robinia pseudoacacia which is indigenous over a wide range in the United States of America.

In some cases, surface exposure of toxic coal mine spoils can
be avoided by knowing the location of acid-producing strata in the
overburden (Berg & May[40]; Grube et al.[41]) and mining so this material is not left near the surface. However, in many cases extremely
acid coal-mine wastes will have to be treated with lime or other

amendments to increase the pH (Chadwick[42]; Sutton[43]). Here it is of interest that roots will not penetrate into extremely acid spoils (Sutton[43]). This is apparently explained by very restricted growth of roots into solutions containing soluble aluminium (Rios & Pearson [44]). Plants growing in limed coal mine wastes may thus tend to suffer water stress as the incorporation of lime or other ameliorants is usually restricted to the surface few inches.

REFERENCES

1. Lindsay, W.L., Inorganic equilibria of micronutrients in soils, in Micronutrients in Agriculture, Soil Sci. Soc. of America, Madison, 1972.

2. Norvell, W.A., Equilibria of metal chelates in soil solution, in Micronutrients in Agriculture, Soil Sci. Soc. of America, Madison, 1972.

3. Pearson, R.W. & Adams, F. Soil Acidity and Liming, Am. Soc. of Agron., Madison, 1967.

4. Berg, W.A. & May, R.F., Acidity and plant-available phosphorus in strata overlying coal seams, Min. Congr. J., 55, 31, 1969.

5. Lorio, P.L., Tree Survival and Growth on Iowa Coal-spoil Materials, Ph.D. thesis, Iowa State Univ., 1962.

6. Chadwick, M.J. & Salt, J.K., Population differentiation within Agrostis tenuis L. in response to colliery spoil substrate factors, Nature, Lond., 224, 186, 1969.

7. Beyer, L.E. & Hutnik, R.J., Acid and Aluminum Toxicity as Related to Strip-mine Spoil Banks in Western Pennsylvania, Res. Rep. SR-72, Off. Coal Res. Adm. Commonw. Penn., 1969.

8. Berg, W.A. & Vogel, W.G., Toxicity of acid coal-mine spoils to plants, in Ecology and Reclamation of Devastated Land, Gordon & Breach, London, 1973.

9. Massey, H.F. & Barnhisel, R.I., Copper, nickel and zinc released from acid coal mine spoil materials of eastern Kentucky, Soil Sci., 113, 207, 1972.

10. Chadwick, M.J., Methods of assessment of acid colliery spoil as a medium for plant growth, in Ecology and Reclamation of Devastated Land, Gordon & Breach, London, 1973.

11. Morris, H.D. & Pierre, W.H., Minimum concentrations of manganese necessary for injury to various legumes in culture solutions, Agron. J., 41, 107, 1949.

12. Lohnis, M.P., Manganese toxicity in field and market garden crops, Pl. Soil, 3, 193, 1951.

13. Ouellette, G.J. & Dessureaux, L., Chemical composition of alfalfa as related to degree of tolerance to manganese and aluminum, Can. J. Pl. Sci., 38, 206, 1958.

14. Foy, C.D., Toxic Factors in Acid Soils of the Southeastern United States as Related to the Response of Alfalfa to Lime, USDA, agric. Res. Serv. Prod. Res. Rep. 80, 1964.

15. Andrew, C.S. & Pieters, W.H.J., Manganese Toxicity Symptoms of One Temperate and Seven Tropical Pasture Legumes, Aust. Commonw. sci. ind. Res., Org., Div. of Tropical Pastures Tech. Paper 4, 1970.

16. Harding, C.P., Plant Available Nutrients in Colliery Spoil and their Relation to Ecotypic Differentiation within Populations of Agrostis tenuis L., Ph.D. thesis, Univ. York, 1970.

17. Hewitt, E.J., The Resolution of Factors in Soil Acidity: some Effects of Manganese Toxicity, in Long Ashton Res. Sta. Ann. Rep., 1946.

18. Plant, W., An analysis of the acid soil complex by the use of indicator plants, Pl. Soil, 5, 54, 1953.

19. Labanauskas, C.K., Manganese, in Diagnostic Criteria for Plants and Soils, Univ. California, Div. agric. Sciences, 1966.

20. Smith, F.W., Foliar Symptoms of Nutrient Disorders in Chloris gayana, Aust. Commonw. sci. ind. Res. Org., Div. Tropical Pastures Tech. Paper 13, 1973.

21. Hoyle, M.C., Manganese toxicity on yellow birch (Betula alleganiensis Britton) seedlings, Pl. Soil, 36, 229, 1972.

22. Zeiger, D.C. & Shelton, J.L., Leaf chlorosis and twig discoloration in Delicious apple trees having Mn-induced internal bark necrosis, Hort. Sci., 7, 494, 1972.

23. Andrew, C.S. & Hegartz, M.P., Comparative responses to manganese excess of eight tropical and four temperate pasture legume species, Aust. J. agric. Res., 20, 687, 1969.

24. Vlamis, J. & Williams, D.E., Manganese and silicon inter-action in the gramineae, Pl. Soil, 27, 131, 1967.

25. Morris, H.D., The soluble manganese content of acid soils and its relation to the growth and manganese content of sweet clover and lespedeza, Soil Sci. Soc. Amer. Proc., 13, 362, 1948.

26. Gupta, U.C., Effects of Mn and lime on yield and on the concentrations of Mn, Mo, B, Cu, and Fe in the boot stage tissue of barley, Soil Sci., 114, 131, 1972.

27. Brown, J.C., Ambler, J.E., Chaney, R.L. & Foy, C.D., Differential responses of plant genotypes to micronutrients, in Micronutrients in Agriculture, Soil Sci. Soc. of Am., Madison, 1972.

28. Rorison, I.H., The effect of aluminium on legume nutrition, in Nutrition of the Legumes, Butterworth, London, 1958.

29. MacLeod, L.B. & Jackson, L.P., Effect of concentration of the aluminum ion on root development and establishment of legume seedlings, Can. J. Pl. Sci., 45, 221, 1967.

30. Clarkson, D.T., Metabolic aspects of aluminium toxicity and some possible mechanisms for resistance, in Ecological Aspects of the Mineral Nutrition of Plants, Blackwell, Oxford, 1969.

31. Liebig, G.F., Vanselow, A.P. & Chapman, H.D., Effects of aluminum on copper toxicity, as revealed by solution culture and spectrographic studies of citrus, Soil Sci., 53, 341, 1942.

32. Bradshaw, A.D., McNeilly, T.S. & Gregory, R.P.G., Industrialization, evolution and the development of heavy metal tolerance in plants, in Ecology and the Industrial Society, Wiley, New York, 1965.

33. Struckmeyer, B.E., Peterson, L.A. & Hsi-Mei-Tai, F., Effects of copper on the composition and anatomy of tobacco, Agron. J., 61, 932, 1969.

34. Pratt, P.F., Aluminum, in Diagnostic Criteria for Plants and Soils, Univ. Calif., 1966.

35. Limstrom, G.A., Forestation of Strip Mined Land in the Central States, USDA Handbook 166, 1960.

36. Vogel, W.G. & Berg, W.A., Grasses and legumes for cover on acid strip-mine spoils, J. Soil & Wat. Cons., 23, 89, 1968.

37. Ruffner, J.D. & Steiner, W.W., Evaluation of plants for use on critical sites, in Ecology and Reclamation of Devastated Land, Gordon & Breach, London, 1973.

38. Kerridge, P.C., Dawson, M.D. & Moore, D.P., Separation of degrees of aluminum tolerance in wheat, Agron. J., 63, 586, 1971.

39. Reid, D.A., Fleming, A.L. & Foy, C.D., A method for determining aluminum response of barley in nutrient solution in comparison to response in Al-toxic soil, Agron. J., 63, 600, 1971.

40. Berg, W.A. & May, R.F., Acidity and plant-available phosphorus in strata overlying coal seams, Min. Congr. J., 55, 31, 1969.

41. Grube, W.E. Jr., Smith, P.M., Singh, R.N. & Sobek, A.A., Characterization of coal overburden materials and minesoils in advance of surface mining, in Research and Applied Technology Symposium on Mined-Land Reclamation, Bituminous Coal Research, Monroeville, 1973.

42. Chadwick, M.J., Amendment trials of coal spoil in the north of England, in Ecology and Reclamation of Devastated Land, Gordon & Breach, London, 1973.

43. Sutton, P., Establishment of vegetation on toxic coal mine spoils, in Research and Applied Technology Symposium on Mined-Land Reclamation, Bituminous Coal Research, Monroeville, 1973.

44. Rios, M.A. & Pearson, R.W., Some chemical factors in cotton root development, Soil Sci. Soc. Am. Proc., 28 232, 1964.

ON THE CHOICE OF SOWING DEPTH FOR BROOM (SAROTHAMNUS SCOPARIUS WIMM.)

Georg Schlützer

Danish Wildlife Planting Service

ABSTRACT

Although the best way to establish broom in spoil heaps is by direct seeding sowings are not always successful. In light sands, germination of shallowly sown seeds might be imperiled by droughts, or the seeds might simply be blown away.

Our experiments show that in sand the seeds, although quite small, may safely be sown as deeply as 3.5 cm and probably even deeper, and there is good evidence that for a wide range of sands, one has a free choice of sowing depth within the upper 3.5 cm.

INTRODUCTION

In Denmark, stripmined lignite fields cover some 3,200 hectares, 134 of which, including the 20 hectares of the Desert Arboretum, belong to the Ministry of Agriculture (under the Wildlife Fund) and are administered by the Wildlife Planting Service.

Climatic conditions and soil characteristics of the Ministry's 134 hectares have already been described by the author[5]. Practically all our spoil heaps consist of gravelly sand mixed with pyritic fossil-gyttja and lignite dust. On the surface, their pH largely ranges from 4.0 to 5.5, and most plant nutrients are present in extremely small amounts. Sand drift is a common feature of the naked spoil heaps, demanding special measures to ensure success of planting operations.

CHOICE OF SPECIES

If conifers are to be used, we prefer:- Pinus contorta Loud.;
Pinus mugo Turra var. rostrata Hoopes and var. rotundata Hoopes;
Larix leptolepis Gord.; or X L. eurolepis Henry; but we have found
that under these conditions deciduous broadleaved species are the
real pioneers. Of these, Alnus glutinosa Gaertn. is considered
the most important albeit shortlived species for the initial phase.
Normally it is planted in mixtures with longer living pioneer
species such as:- Betula pubescens Ehrh.; B. pendula Roth.;
X Populus tremula L. x tremuloides Michx.; Quercus robur L.;
Q. petraea Liebl.; Sorbus aucuparia L.; X S. latifolia Pers.;
and X S. intermedia Pers.

A number of Salixes are very effective, e.g. S.acutifolia
Willd.; S. daphnoides Vill.; X S. smithiana Willd.; and S. viminalis
L. The latter in not too acid sands. Alnus viridis DC is a fast
soil coverer if not very strongly exposed, and in exposed sites
Malus baccata Borkh., X Spiraea billiardii Herincq., and Prunus
virginiana L. have developed well, the latter at pH down to 4.4.
In a small plot in the Desert Arboretum, Amelanchier canadensis
Med. has competed favourably with Alnus glutinosa. It is now
being tested in large numbers against more common pioneer species,
in a bare spoil heap elsewhere.

Although strongly exposed, small groups of Pyrus communis L.,
Sorbus americana Marsh., and S. scopulina Greene have grown well
in the Desert Arboretum, as have three provenances of Betula
papyrifera Marsh. in more sheltered parts, whereas the roses used
have been rather unsatisfactory - perhaps because of the low
level of available copper. Up to now R. pendulina L. is considered
the best.

THE PROBLEM OF BROOM

Broom poses a particular problem. Being very difficult to
transplant out from the nursery, it is normally sown directly onto
the spoil heaps. If successful, such sowing results in a very
effective and beautiful cover. Often however, even direct sowing
leads to bad results.

Such lack of success might be due to seed coat dormancy, the
frequency and severity of which may vary between different seed
lots and at different times. Because the planting service has
to cater for non-professionals planting on their own, the author
advanced a practical method for overcoming this obstacle without
having to carry out soaking tests:[4] just before sowing, half of
the seed-lot is briefly flushed with (but not soaked in) boiling
water and immediately cooled under the tap. These pretreated

seeds are then mixed with the untreated half, and sown.

Another source of failures might flow from the choice of sowing depth. Since the seeds are small (C. 130,000/kg), one automatically tends to be guided by experience with small seeds and sow superficially, for instance at about 0.5 cm depth. However, in easily blown soils, germination of shallowly sown seeds might be imperiled. Which leads to the question, whether deeper sowing depths might result in more adequate germination.

In judging this, the available forestry literature does not offer much help. Bühler states from incomplete experiments with Robinia seeds (2-2.5 times the size of broom seeds) that 3%, 5% and 5% germinated from depths 5, 6, and 7 cm respectively. Rohmeder[3] found that broom seeds should be sown at shallow depths only, but his experimental technique was unrepresentative of spoil-heap conditions (moisture supply from the bottom of containers), and the duration of the experiment too short (21 days).

Recent observations from a sandy area where broom was removed and the ground ploughed and harrowed, reveal that some broom seedlings do emerge from depths down to at least 6 cm. This was unknown to us when the following experiments were carried out. It is also not known what percentage of seeds from such depths result in plants.

AN EXPERIMENT IN COARSE SAND

In 1965 an experiment was carried out with the aim of estimating to what extent one has a free choice as to sowing depth from a constant surface.

Sowings at 0.5, 1.5, 2.5, and 3.5 cm depth in a gravelly, coarse sand (Table 1) were carried out in a frame $1\frac{3}{4}$ x $\frac{1}{2}$ m, into which the sand was filled a fortnight before sowing and pressed so as to simulate a natural sand layer as much as possible. Depth of the sand layer was 25 cm. Each sowing depth was represented by 9 replications, distributed randomly within the frame, and each comprising 20 seeds, giving a grand total of 720 seeds for the whole trial.

Normal commercial stock seeds were used. Seed coat dormancy averaged 62.6% ± 3.4 (5 samples, each of 100 seeds, soaked for 10 days at room temperature). Before sowing, the seed lot was treated with boiling and cool water as described previously, and for each plot, 20 healthy looking, pretreated seeds were selected.

The seeds were sown May 17th, and the experiment stopped

	Particle size mm	1965 Sand pH 5.3		1970 8.5% Gravel excluded Clayey sand pH 5.2
		Gravel included %	Gravel excluded %	%
Loss on ignition:				
Clay	<0.002			7.4
Silt	0.002 – 0.2			5.4
Fine sand	0.02 – 0.2			25.8
	<0.06 – 0.25	6.3	7.6	
Coarse sand	0.2 – 2.0			59.3
	0.25 – 2.0	76.1	92.4	
Gravel	>2.0	17.6		

Table 1. Mechanical analyses.

Sowing Depth, cm	Number of seedlings per plot (Each plot: 20 seeds sown): Blocks									χ^2
	I	II	III	IV	V	VI	VII	VIII	IX	
0.5	10	10	11	10	13	15	14	8	9	9.12
1.5	15	10	11	7	8	11	16	11	15	16.47
2.5	5	16	11	12	7	7	12	9	10	17.80
3.5	11	9	10	9	14	11	11	13	10	0.46

Table 2. Results, by August 5th, 1965.

August 5th when heights of the plants were measured and the plants
lifted out of the frame. Counts of plants were carried out at
one to a few days' interval from the day when the first
germinating plants appeared until the end of the experiment.

The sums of these counts, including plants died after
germination, are shown in Fig. 1, which reveals that throughout
the period of observation, germination percentages did not vary
much among sowing depths.

The total numbers of plants emerged from the seeds by August
5th, when the experiment was stopped, are shown in Table 2. The
χ^2 test showed that there was no significant difference in the
numbers of seedlings emerging as between the various sowing
depths at $p<0.5$. As follows from the nature of seed shell hardness in leguminous seeds, the degree of which varies from seed
to seed, some germination took place throughout the period of
observation. Consequently, at the end of the experiment, plant
sizes from the single sowing depth varied much, as shown in Fig.2,
which also indicates that the output in terms of plant heights
from 1.5 and 2.5 cm depth is somewhat better than that from the
other depths, and in particular better than that from 0.5 cm
depth.

DISCUSSION OF THE RESULTS IN SAND

In dry weather the surface of a sand such as the one used
in this experiment will easily dry out. On the other hand,
evaporation will normally influence the deeper sand layers very
little, since the upper, dry layer, once formed, acts as an effective mulch. This is common knowledge which leads to the
opinion that over the years, germinations from 0.5 cm depth will
be more influenced by dry or excessively wet summers than those
from the other depths considered here. Also, seeds at 0.5 cm
depth are more exposed to the consequences of sand being blown
off the area than are those at the lower depths.

This focuses interest on the results from the three lower
depths. Expressed in plant height, results from 1.5 and 2.5 cm
seem better but not very much better than those from 3.5 cm depth.
And expressed in plant numbers, results did not differ significantly between sowing depths.

We have drawn the conclusion - and advise our clients
accordingly - that when handling broom seeds in an easily blown,
sandy soil, be it in a poor field or on a spoil heap, one has a
free choice of sowing depth within the range 1.5 - 3.5 cm.

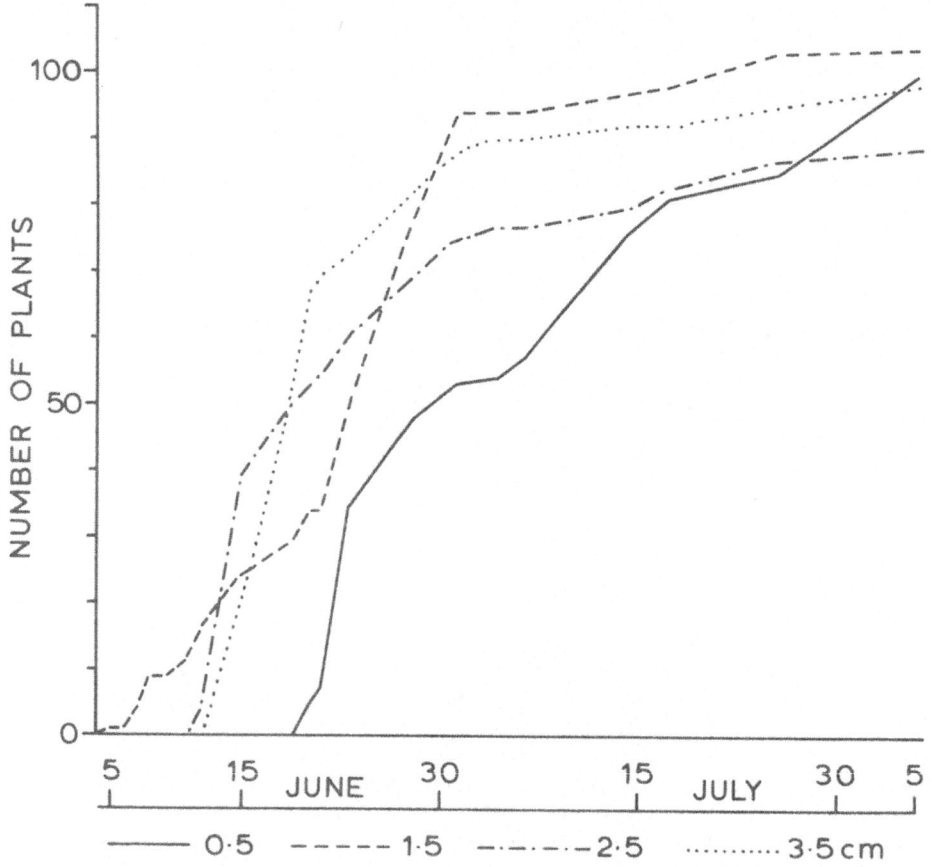

Figure 1. 1965 Numbers of broom seedlings emerging from 180 sown per treatment in gravelly coarse sand.

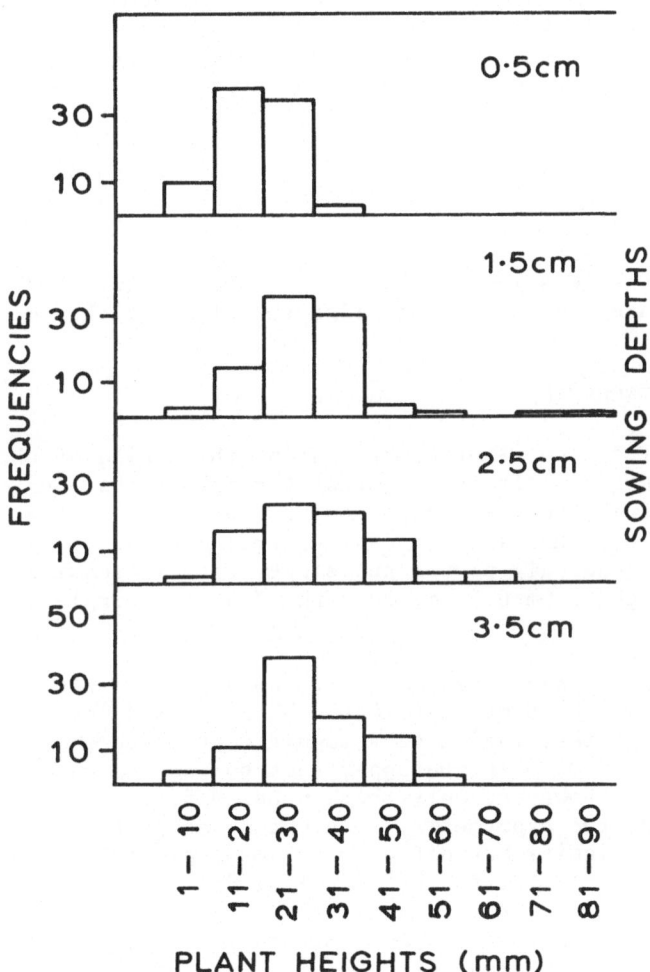

Figure 2. Heights of broom plants by August 5th, 1965, distributed to height classes. The plants measured at
this final day of the experiment cover various stages
of plant development.

AN EXPERIMENT IN CLAYEY SAND

In 1970 broom seeds were sown in clayey sand (Table 1) at
depths 0.5, 1.5, 2.5 3.5, 4.5 5.5, 6.5 and 7.5 cm. Pots of
thick, hard plastic were used as plots, and packed as closely as
possible in a frame (with sand filled between them). Into each
pot, 20 seeds were sown. Thus, with 5 randomized replications,
each sowing depth comprised 100 seeds. Seed pretreatment, observations, etc., followed the previous procedure. Commercial
stock seeds were used, with a seed coat dormancy for untreated
seeds of 7.18% ± 2.8, and for pretreated seeds of 25.8% ± 4.4
(test procedure in each case as in 1965). A one-month germination test in closed Petridishes (3 x 20 pretreated seeds) showed
germination 78.5% ± 3.3.

The seeds were sown June 2nd. On September 20th the last
counts were made, and on October 15th the plant heights measured.

RESULTS AND DISCUSSION

The results from plant counts during the period of observation
are shown in Fig. 3. It is seen that throughout that period,
germinations from the three uppermost depths do not differ much
from one another, whereas germinations from depths 3.5, 4.5 and
5.5 cm are increasingly poorer and slower with increasing depth.
No plants emerged from 6.5 cm, but from 7.5 cm a single plant
germinated.

For the six upper sowing depths, Table 3 gives total numbers
of plants emerged by September 20th. When all six depths are
considered, a χ^2 test indicated a significant difference between
treatments at $p<0.05$. If the upper four sowing depths only are
considered. If, however, only the depths, 0.5, 1.5, and 2.5 cm
are considered, the hypothesis is rejected, as is the hypothesis
that there are significant differences among plant figures from
these 3 depths (χ^2 2.16, f2, P95 level 5.99).

As in 1965, plant sizes from the singly sowing depth varied
very much (Fig. 4), whereas distribution to height classes shows
remarkably little variation among sowing depths.

The irregularity in germination figures from plot to plot
at the lower depths is very marked and might be attributed to
experimental error. However, this is not thought to be the case
as the degree of irregularity increases with depth from 1.5 cm
downwards. Moreover, from the lower depths is increasingly
slow with increasing depth. This might suggest some biological
explanation for the irregularity.

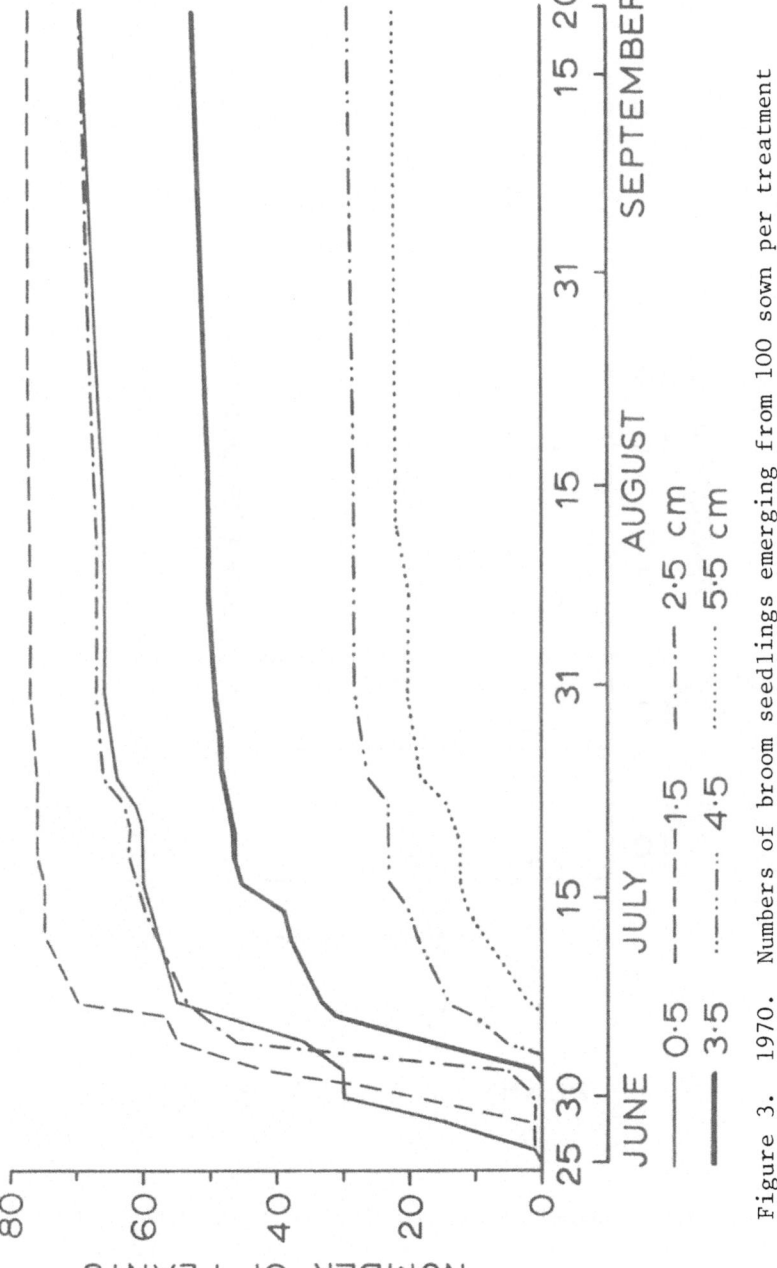

Figure 3. 1970. Numbers of broom seedlings emerging from 100 sown per treatment
in clayey sand.

160

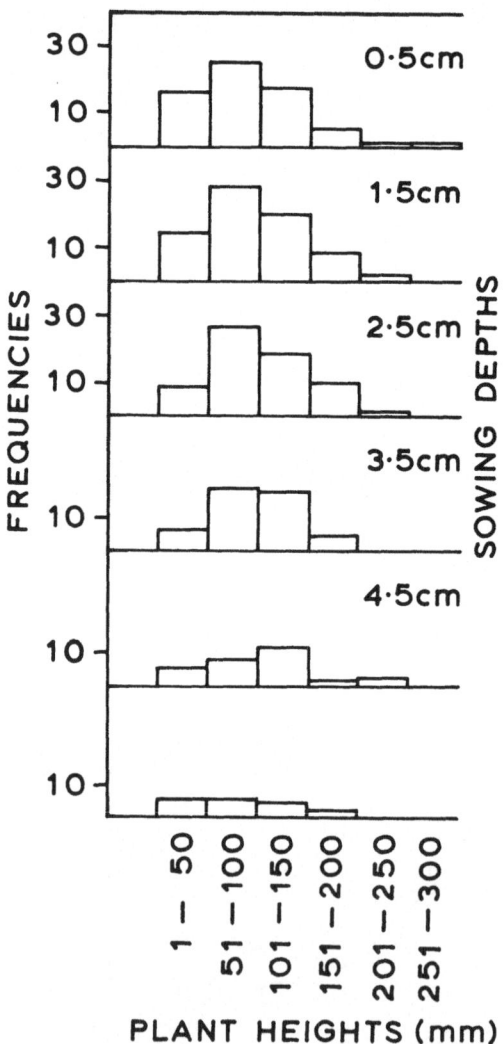

Figure 4.   Heights of broom plants by October 15th, 1970, dis-
            tributed to height classes. The plants measured at
            this final day cover various stages of plant develop-
            ment.

Sowing Depth, cm	Number of seedlings per plot, by September 20th (Each plot: 20 seeds sown): Blocks					χ^2
	I	II	III	IV	V	
0.5	17	14	10	16	12	7.67
1.5	14	18	15	15	15	2.60
2.5	11	12	15	15	16	4.39
3.5	15	4	5	17	11	27.08
4.5	6	4	5	13	1	19.14
5.5	2	3	1	11	5	18.41

Table 3.   Results from sowings 1970.

CONCLUSION

The experiment in clayey sand has revealed that even in such a relatively dense medium, broom seeds may germinate from depths completely out of proportion to the small size of the seeds.   The results corroborate those from the original experiment in coarse sand, and suggest that the conclusion may be extended to cover a wider range of sandy soils.   Additionally, they suggest that in pure sands the lower limit of the free choice may be extended to an even deeper depth than 3.5 cm.

ACKNOWLEDGEMENTS

The author is indebted to Statens Markforsag who has kindly carried out the mechanical soil analyses (Table 1), and to Messrs. P. Allerup and K. Frydendah, for valuable advice.

REFERENCES

1. Bühler, A., Saatversuche. Mittheil. der Schweiz. Centralanstalt f.d. forstliche Versuchswesen, 1892.

2. Lakon, G., Beiträge zur forstlichen Samenkunde 1. Der Keimverzug bei den Koniferen- un hartschaligen Leguminosensamen, 1911.

3. Rohmeder, E., Beiträge zur Keimungyphysiolgie der Forstpflanzen, München, 1951

4. Schlätzer, G., Forsog med forgehandling af gyvelfro. DST 49, pp. 379-393. (summary in English).

5. Schlätzer, G., Some Experiences with Species in Danish Reclamation Work, in Ecology and Reclamation of Devastated Land, Hutnik, R. and G. Davis, G. Eds., Gordon and Breech, London, 1969.

CHINA CLAY WASTE MANAGEMENT

CHINA CLAY WASTE MANAGEMENT - IDENTIFICATION OF PROBLEMS

One of the major problems associated with the mining of china clay is the long term use of pits. This means that they remain obtrusive on the landscape for many years, the conical tips contrasting to the smoother features of the surrounding land. Once the pits become disused they require reclamation techniques that allow an appropriate land-use pattern and integration into the landscape. This will involve the establishment of vegetation that is similar to the surrounding plant formations or allows agricultural use appropriate to the area. When areas involved are situated within regions of outstanding natural beauty or particular ecological value (like National Parks) then it is necessary to exercise great skill and sensitivity.

Clay waste may be particularly subject to mineral deficiencies particularly nitrogen, phosphorus and calcium. This leads to difficulties in establishing and maintaining vegetation and initial supplies of nutrients may be quickly leached out of the rooting zone. Under these conditions it is essential that ecologically suitable seeds mixtures are sown.

Physical problems are also encountered and where vegetation fails to establish satisfactorily erosion occurs. Organic additives, like peat, have possible uses in counteracting erosion and water and nutrient deficiencies. The more recent separation of micaceous material and sand has increased physical problems and the rapid loss of nutrients in the sandy material.

CHINA (KAOLIN) CLAYS: MINING AND RECLAMATION

Jack T. May

School of Forest Resources, University of Georgia,
Athens, Georgia, USA

INTRODUCTION

Clay has been the principal raw material in the field of ceramics since antiquity. A sort of porcelain was manufactured by the Egyptians during the Sixth Egyptian Dynasty, as early as 3,700 B.C. The Chinese claim that glazed pottery was invented by an Emperor, called Hwung-to, whose reign commenced about the year 2,697 B.C. From these early beginnings the use of clay gradually expanded until the introduction of porcelain into Europe; and enormously, since the art of making white porcelain was discovered by the Europeans, at the beginning of the eighteenth century (Mitchell[1]). At the present time, the uses of clay are so varied and its application so diverse, so universal, that this period has sometimes been called an Age of Clay.

The major types of commercial clay include kaolins, ball clay, fire and stonework clay, bentonite, fuller's earth and common clays and shales that are used in the manufacture of structural clay products, light weight aggregate and Portland cement (Ries[2]). The occurrence of clays is widespread over the world.

Kaolin is one of several types of clay and is commonly referred to as China Clay or Paper Clay. Strictly defined, kaolin is the name given to a group of hydrous aluminium silicates of which it is the most common. The kaolins are primarily indigenous clays which occur in situ or in the immediate place of origin where feldspar or other aluminium silicates have been decomposed, and in part chemically reconstituted into clays.

The term kaolin is derived from the Chinese word "kau - ling" meaning high ridge, referring to the locality from which the richest supplies were obtained (Falvey[3]). In the United States, kaolin was first mined in the Carolinas, Georgia and Florida during colonial days and shipped to England. In 1760 a Mr. Hodgson imported a cask of clay from North Carolina for an English china maker. In 1767-68 Josiah Wedgwood imported five tons of kaolin clay, known as Cherokee Earth from an Indian pit near the present town of Franklin, N. C. (Griffiths[4]; Goff[5]). The famous Wedgwood Pottery used considerable amounts of China Clay from the colonies until the English mines were opened. An American pottery and white ware industry gradually developed around two centres, Trenton, N. J. and East Liverpool, Ohio, using at first local clays and then domestic primary and imported English kaolins. The discovery and use of the English kaolins ended the mining of the sedimentary China Clays of Georgia for over a century.

Branner's[6] bibliography of clays and the ceramic arts of the United States and other countries indicate the range of clay uses prior to the twentieth century.

The China Clays were originally used primarily in pottery and china ware. The greatest use today is in paper manufacture. The more common uses are:

(a) Paper: as a filler to decrease porosity and as a coating to give a smooth glossy surface and printability;

(b) Ceramics: china, pottery, sanitary ware, flower pots, ash trays, porcelains, laboratory sinks and equipment, enamel coatings, abrasives, refractory brick, electric insulators, and missile nose cones;

(c) Rubber: filler in rubber goods, soles and heels of foot wear, floor tiles, tyres and tubes, conveyor belts, hose, gloves, insulated wire, adhesives, gaskets, neoprene compounds, toys, and novelties;

(d) Building materials: white brick, block and cement, paint, and adhesives;

(e) Plastics: phonograph records, floor covering, oilcloth, molded goods, dishes, toys, and novelties;

(f) Miscellaneous: food, medicine, insecticides, herbicides, and ink.

Kaolin imparts smoothness, gloss, brightness and printability to paper. It adds strength, abrasion resistance and rigidity to both natural and synthetic rubber goods. It is chemically inert, has a high covering power and gives desirable flow qualities to

paint. It is used in many process industries as a filler and sus-
pending medium and to provide color improvement, a more attractive
finish and good dimensional stability.

World production of China Clays by countries and continents
during 1968 to 1970 were as follows (Wells[7]; Gustavson[8]):

Area	Thousand Short Tons[†]	Percent of Total
United States	4,926	34
United Kingdom	3,509	24
USSR	2,000	14
Europe (14 countries)	2,534	17
Asia (12 countries)	1,077	7
Central and South America (6 countries)	346	2
Africa (10 countries)	186	>1
Australia	73	<1

These values included the highest production reported or
estimated during one of the three years. In addition to production
from the 46 nations included in these figures, about 10 other nations produce some China Clay, but data on production is not available. World production increased from 12,367 in 1967 to 14,390
short tons in 1970, or 16 percent. Georgia and South Carolina led
in 1970 production with outputs of 3,749,000 and 519,000 short tons,
respectively. A total production of 467,000 short tons was reported for nine states including Alabama, Arkansas, California, Florida,
Idaho, North Carolina, Pennsylvania, Texas, and Utah. Georgia
supplies approximately 27 percent of the World's production of China
Clays and 80 percent of the paper coating clays.

ORIGIN OF CHINA CLAYS

China Clays are of primary or secondary origin and have been
derived directly or indirectly from the decay and disintegration of
the original igneous rocks of the earth's crust. Feldspars of the
coarse granites and pegmatites are the chief source of residual

---

[†]1 Short Ton = 2,000 lbs

kaolins; but kaolins may also be derived from other sediments. Some
residual kaolins result from shales, schists, and arkosic sandstones
Some clastic kaolins occur as sedimentary deposits and are either of
estuarine origin or were deposited in offshore bodies of water.
Strictly speaking, the word kaolin is applicable only to primary or
residual kaolins. The so-called kaolins or China Clays of the
Coastal Plain of Georgia are sedimentary kaolins. The relationship
of kaolins or white clays to other clays based on origin is illustrated in the following classification (Ries[2]):

Classification of Clays

A. Residual clays. Formed in place by rock alteration due to
   various agents, of either surface or deep-seated origin.

   (i)   Those formed by surface weathering, the processes
          involving solution, disintegration, or decomposition
          of silicates.

          (a) Primary kaolins, white in color and usually white
               burning.

               Parent Rock   Shape

               Granite, Pegmatite,   Blankets; tabular
               Rhyolite, Limestone, Shale, steeply dipping masses;
               Feldspathic, Quartzite,   pockets or lenses.
               Gneiss, Schist, etc.

          (b) Ferruginous clays, derived from different kinds of
               rocks.

   (ii)   White residual clays formed by the action of ascending
          waters possibly of igneous origin.

          (a) Formed by rising carbonated waters.

          (b) Formed by sulphate solutions.

   (iii) Residual clays formed by action of downward moving
          sulphate solutions.

   (iv)   White residual clays formed by replacement, due to
          action of waters supposedly of meteoric origin.

B. Colluvial clays, representing deposits formed by wash from the
   foregoing and of either refractory or non-refractory character.

C. Transported clays.

   (i)   Deposited in water.

    (a)   Marine clays or shales. Deposits often of great extent. White-burning clays. (Sedimentary kaolins, ball clays). Fire-clays or shales. Buff burning. Impure clays or shales. Calcareous - Non-calcareous.

    (b)   Lacustrine clays. (Deposited in lakes or swamps.) Fire-clays or shales. Impure clays or shales, red burning. Calcareous clays, usually of surface character.

    (c)   Flood-plain clays. Usually impure and sandy. (Alluvial brick-clays of Georgia.)

    (d)   Estuarine clays. (Deposited in estuaries.) Mostly impure and finely laminated.

    (e)   Delta clays.

(ii)   Glacial clays, found in the drift, and often stony. May be either red- or cream-burning.

(iii) Wind-formed deposits. (Some loess.)

(iv)   Chemical deposits. (Some flint-clays.)

Grim's[9] classification of clays based on shape of the clay minerals and the expanding or non-expanding character of the 2:1 and 1:1 lattice is the more widely used system of classification. Grim[9] reports that kaolins occur in ancient and recent sediments; and are a product of alteration of primary mineral and also occur in sediments of marine, lacustrine and fluviatile origins.

The geologic origin of the kaolins in the Southeastern United States has been studied and described by Holmes[10], Ladd[11], Veatch[12], Bayley[13], Neumann[14], Smith[15], Kesler[16, 17, 18, 19], Bates[20] and Smith & Murray[21]. The sedimentary deposits in Central Georgia have been more throughly described by La Moreaux[22], LeGrand & Furcron[23], and LeGrand[24]. Clays occur in the Tuscaloosa Formation of the Upper Cretaceous Age and in the Tertiary sediments in an area that extends from North Carolina through South Carolina and Georgia, across Alabama, Tennessee, Kentucky, and into southern Illinois. In the Carolina's and Georgia, it appears in the transition zone between the Piedmont and the Coastal Plain known as the Fall Line, and varies from 8 to 40 km in width (Fig. 1). Features of the area have been summarized by Kesler[19].

The Cretaceous Sediments. A wedge-like series of unconsolidated sand, variably kaolinitic, micaceous, and cross-bedded, contains the kaolin deposits of commercial grade. The thickness of this series ranges from zero at the Fall Line to 150 meters or more along the

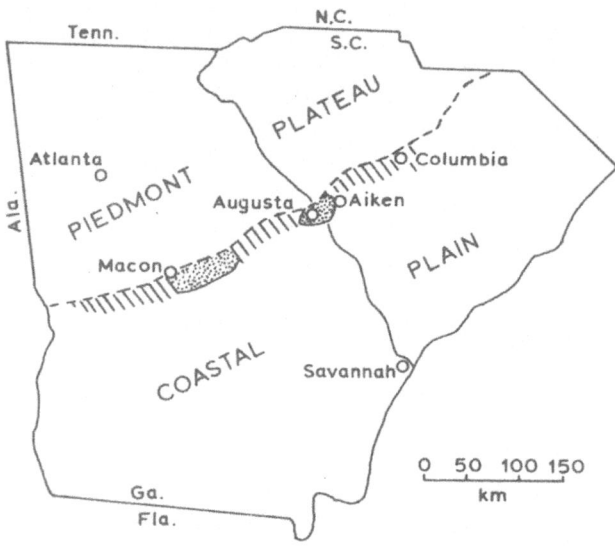

Fig. 1 Outline map showing location of the Cretaceous kaolin belt

southeastern edge of the mining areas, and increases southeastward
beneath a cover of Tertiary beds. The kaolin deposits occur as
lenses, particularly in the Tuscaloosa Formation (Fig. 2).

The sand is mostly coarse and sub-angular to angular. Sharp
particles of quartz up to 2.0 cm in length are known. Feldspar
particles of equal size, and kaolin pseudomorphs after feldspar,
also occur in some beds. Rounded gravel is very scarce and of small
size, the maximum dimensions rarely exceeding 2.5 cm. This series
reflects rapid deposition without sorting of the clastic material
according to size or density.

The Underlying Crystalline Rocks. The Piedmont erosion surface
slopes southeastward 10.5 to 15 meter per km beneath the Cretaceous
series. This surface is a weakly rejuvenated peneplain that has
been eroded to a secondary stage of maturity northwest of the Fall
Line, but probably has lower relief beneath the protecting Coastal
Plain sediments. Where the Piedmont rocks are exposed adjacent to
the Fall Line, they include granite, quartz monzonite, diorite,
gabbroic rocks, and a wide variety of felsic and mafic schists,
gneisses, and pyroclastics. These older rocks are cut by dikes of
Triassic diabase. Similar rocks probably underlie the Cretaceous
beds within the known limits of the kaolin belt, but the basement

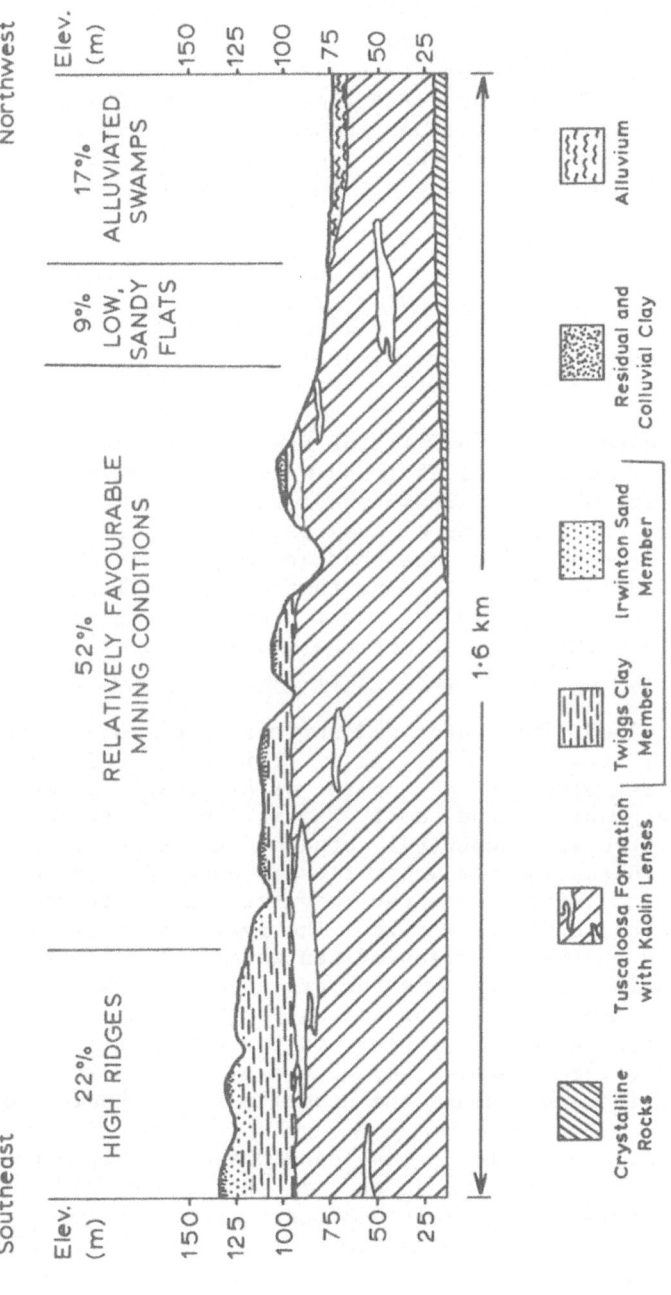

Fig. 2 Cross section illustrating stratigraphic and physiographic conditions in the kaolin producing area of central Georgia. (From Kesler, 1963.)

of the Coastal Plain formation consists of Palaeozoic rocks in southernmost Georgia and northern Florida.

Deposition of the Kaolin. The vast accumulation of coalescing delta deposits was built above sea level, as in major deltas known today, and chemical weathering attacked the feldspar of the exposed sands. The process was aided by minor erosional redistribution of the sands, and by rotting grasses whose presence is shown by thin lignite beds found above, below, and within a few of the kaolin lenses.

The erosion of the exposed sands was facilitated by the total lack of induration, and was gentle owing to the absence of appreciable gradient on the delta surface. The newly formed kaolinite was thereby separated from the coarser sediment, and was transported toward points of lowest elevation.

Such points, owing to the shifting of overloaded distributaries, contained cut-off stream segments as fresh-water ponds having irregular and commonly curved forms. These ponds received the inwash of suspended kaolinite. During times of heaviest rainfall and strong wind action, detrital sand, mica, and scarcer fine-grained heavy minerals were carried in. The two areas that contain relatively numerous deposits must have contained particularly strong distributaries, and therefore they may reflect the discharge areas of the stronger streams.

The ponds contained river water and surface water runoff, and hence, the water was fresh and mildly acid. This permitted slow settling of kaolinite, with the development of platy aggregates of comparatively large size - the "soft" kaolin. During times of storm at high tide, the sea undoubtedly breached the sands around some of the ponds. Alakaline sea water flushed out the fresh water, and newly inwashed kaolinite thereafter coagulated and settled rapidly forming the dense "hard" kaolin. The pyrite in such kaolin may have been formed by the reduction of sea water sulphates through decomposition of marine organisms, in the presence of soluble iron contained in the surface drainage furnishing the kaolinite. Thus, changes in pond environments were subject to the alternating influences of shifting stream channels and wave erosion, resulting in such unsystematic differences among the deposits as are known today.

As kaolin accumulated in the ponds, sand was washed from the margin into and over the edge of each deposit forming gradational and interfingering contacts. This lateral encroachment of the sands tended to complete the lens-like form of the deposits. The kaolin-filled ponds were covered by continuing sedimentation as delta building migrated northwestward by gradual subsidence of the source area. Thus, deposition continuously maintained the summit of the Cretaceous sediment above sea level, and subsidence contin-

uously permitted the outer part of this loose material to be planned
at wave base. The resulting truncated surface remained submerged
until the Tertiary beds were deposited.

Anomalous Features.

1. Kaolin deposits of economic purity are numerous only in two
   areas, but otherwise the Cretaceous series is nearly uniform in
   lithology and structure for at least 280 km along the strike.

2. Lenses of essentially pure kaolinite in coarse-grained, cross-
   bedded sands present an association of sediment reflecting
   almost perfect depositional classification with sediment
   reflecting almost none.

3. Very sharp grains of sand up to 0.6 cm in diameter occur separately embedded in otherwise homogeneous kaolin.

4. The same general source and environment have yielded beds of
   both "soft" and "hard" kaolin, which differ markedly in texture
   and utility.

5. Pyrite seems largely confined to a few deposits of "hard" kaolin,
   but these deposits display the effects of no special environmental features.

6. Ferromagnesian silicates and iron sulfides and oxides are common
   in the region from which the sediments were derived but iron-
   bearing minerals are very scarce in the Cretaceous sand, and
   the kaolin contains no syngenetic iron compound except the
   pyrite.

Stratigraphic Occurrence. The exposure, by erosion of the
Eocene beds, of kaolin lenses truncated by the unconformity above
the Cretaceous led to the early impression that the kaolin deposits
occur only in this position. Exploratory core drilling in recent
years, in places to depths 45 meters below the level of the unconformity, has found no vertical limit for the occurrence of the
deposits. It is not uncommon in a single drill hole to penetrate
two or more kaolin lenses in the Cretaceous sands. They show that
kaolin deposits occur to a depth of at least 53 meters below the
unconformity. Well logs indicate that deposits occur at random to
the base of the Cretaceous.

Knowledge of the random vertical succession of kaolin lenses
is bringing prospecting and mining to deeper parts of the Cretaceous
particularly in broad areas where the Tertiary and part of the
Cretaceous have been removed by erosion. Today, mining reaches
deposits at any economic stripping depth, whether the overburden
is Tertiary, Cretaceous, or both. Consequently, the deposits must

be described as in, not on, the Cretaceous beds.

The kaolin deposits are generally lenticular, but rarely symmetrical, and commonly of non-uniform thickness. In plan, the lenses are mostly elongate and curved or sinuous, although no distinct orientation is evident to date. In major dimension, they range from a meter to perhaps 1.5 km, but uniform quality is rarely found throughout a lens. Vertical thickness ranges from a few centimeters to about 15 m, and many of the mined deposits have averaged about 6 m. Contacts with the enclosing sands are mostly gradational, but a few are sharp.

Composition. Many of the kaolin deposits contain detrital sand and mica in quantity too great for refining, and even the purest contain them in small amounts. Extremely fine-grained heavy minerals are also present, and of these, titanium minerals have the widest distribution. These detrital minerals, together with fragments of vermicular kaolinite crystals, are collectively termed grit. With the exception of the grit, the kaolin consists almost entirely of kaolinite, although traces of montmorillonite are found in some deposits (Table 1).

Iron oxide is extremely low in relation to TiO_2. It is significant that the Cretaceous sands that enclose the kaolin lenses are deficient in iron-bearing minerals, a feature that must be considered in relation to the source of these sediments.

Occurrence of Pyrite. Nodules of pyrites are present in some of the deposits, but are largely restricted to "hard" kaolin, and are most abundant in the inner parts of the lenses. The nodules range in size from scarcely visible pellets to masses large enough to jolt a heavy truck-mounted drill. Deposits that contain the nodules in unusual abundance have a bluish cast derived also from the sulphide, which is disseminated in the kaolin as a pigment. A brown carbonaceous discoloration is commonly present as a halo around a pyrite nodule or group of nodules. In some deposits of very "hard" kaolin, the nodules are attached or arranged in close sequence forming somewhat sinuous and approximately vertical line groups.

The Overlying Eocene Deposits. Tertiary formations overlie the Upper Cretaceous series except near the Fall Line where, in most places, erosion has stripped them away and has also removed part of the Cretaceous beds. Thus, the Tertiary beds have been eroded in most of the mining areas near Aiken, South Carolina, leaving the Cretaceous beds exposed and partly dissected. In the central Georgia mining area Tertiary beds remain above the Cretaceous in the higher ridges even as far north as the Fall Line, but have been removed in the broad areas of lower relief. Southeastward (down dip) the lowland exposures of the Cretaceous are progressively

Table 1. Chemical analyses of white China clay from eight counties[1]

Analyses	Counties							
	Taylor	Crawford	Houston	Twiggs	Jones	Wilkinson	Washington	Richmond
Clay group	soft	soft	hard	soft	soft	soft	soft	soft
	----------------------------percent----------------------------							
Moisture @ 100°C [2]	.32	.82	.58	.32	.52	.98	.28	.28
Loss on ignition [3]	13.50	12.84	13.70	14.36	12.06	13.98	13.82	13.08
Soda (Na_2O)	Trace	.18	.12	.20	.12	.08	.04	.10
Potash (K_2O)	Trace	.20	.10	.08	.10	.06	Trace	.14
Lime (CaO)	.00	.00	.00	.00	.00	.00	.00	.00
Magnesia (MgO)	Trace	Trace	.14	Trace	.00	Trace	.07	Trace
Alumina (Al_2O_3)	35.82	34.80	37.90	37.00	32.20	37.60	37.36	37.67
Ferric oxide (Fe_2O_3)	1.70	1.10	1.30	1.41	1.95	.86	.92	1.18
Titanium dioxide (TiO_2)	.82	.90	1.08	.54	1.35	1.26	1.62	1.35
Sulphur trioxide (SO_3)	2.00	.00	.00	.69	.24	.32	Trace	Trace
Phosphorus pentoxide (P_2O_5)	Trace	Trace	.00	1.33	Trace	.00	Trace	.04
Silica (SiO_2)	46.08	49.14	45.24	43.96	47.54	45.50	46.04	46.24
	100.24	99.98	100.16	99.89	100.08	100.64	100.15	100.08

[1] Smith[15]

[2] Free water - held in the pores of the clay by capillarity

[3] Chemically combined water

narrower as only the deeper valleys have been cut through the Tertiary beds.

The two series are separated by a gently undulatory unconformity that truncates the Cretaceous beds at a very low angle, and is essentially parallel with those of the Tertiary. The unconformity between them has a southeastward slope of 2 to 2.8 m per km.

Mining and exploratory drilling have shown that the Tertiary in central Georgia consists of two fairly uniform units. The lower unit is 24 to 35 m thick and consists of fuller's earth containing limestone beds that are most prominent within 4.5 to 20 m of the base. The lowermost bed is very sandy and commonly about 1.2 m thick, and is leached to loose sand along the outcrops. These limestone beds are highly fossiliferous, and represent an interfingering of the Ocala limestone from the southwest. A dorsal vertebra of the "zeuglodon", _Basilosaurus_ _cetoides_, was found in these beds.

The upper unit consists of a weathered mantle of deep red sands and sandy clays at least 15 m thick, when not eroded. When last weathered the sand is weakly indurated. The mantle extends up the slopes and across the divides, and contains a few random thin beds of fuller's earth. Some colluvial material has migrated down slope and become mixed with sands and kaolin contemporaneously eroded from Cretaceous beds.

EXPLORATION AND MINING IN GEORGIA AND SOUTH CAROLINA

Since the clay deposits are discontinuous the only way to locate commercial bodies is to pattern drill where geologic and topographic conditions are favorable. In general, the drill pattern is based on a grid with location of drill holes determined by the size of the tract and geologic conditions. As drilling proceeds new information becomes available which in turn controls to a large extent the drilling program for that particular area. If the deposit is found to be usuable drilling and testing on 60 and then 30 m centers is completed prior to the preparation of stripping mining.

The mining techniques and the thickness of the overburden determine the characteristics of the spoil.

Early mining techniques were primitive by modern standards and mining was limited to relatively shallow deposits. In the colonial period, the overburden and the clay were worked by pick and shovel. The overburden was carried away in wheelbarrows and dumped on the slope of the hill or carried to a spot where it would not interfere with future mining. The clay was hauled from the mine in wagons to drying sheds or to a shipping point.

Since the renewal of mining in 1876, methods have changed. As earth moving equipment developed - highly sophisticated mechanization evolved. Hand and mule scrapers replaced wheelbarrows. At one time, the conventional method of overburden removal was by locomotive and dump cars (Falvey[3]). Another development was the belt conveyor and stacker. A series of 91.5 cm belt conveyors 152 m long feeding a stacker were placed parallel to the overburden wall. The conveyor was fed by feeders loaded by power shovels.

Motorized equipment is better adapted for the stripping of thicker overburden and longer haulage to processing plants (Scott[25]; Smith[26]; Smith & Murray[21]). Crawler tractors with bulldozer or K/G blades are used to strip the site of trees and brush. Rubber-tyred scrapers (Euclids) or draglines are utilized to remove the overburden.

One method of stripping uses the rubber tyre scraper - self loading or push-loader. Overburden can then be hauled considerable distances; however, 300 m is considered the maximum normal haul for economic reasons. A dump site is selected as close as possible to the cut where no commercial clay exists. The top soil may be stockpiled off the cut for later use.

Mining is accomplished by using small shovels and/or draglines to load trucks or to feed directly into large blungers which slurry the clay into a dispersed fluid. The slurry is then pumped to a primary beneficiation station and then to a processing plant.

As the clay is mined out, the pit can often be backfilled with overburden from adjacent cuts.

When the overburden is removed with draglines, long narrow parallel cuts are laid out with overburden placed in irregular conical steepsided piles on one side of the cut. One efficient dragline method is known as the "roll over method" - whereby there is continuous back filling of an area as mining progresses. This method is best suited to low overburden clay deposits that are homogeneous and uniform in quality. The final result is a long windrow of overburden at one end of the mining area and an open cut at the other end. The area in between is filled with overburden. With minimum effort, the area can be graded and planted for reclamation.

Rubber tyre scraper - self loader or push loaders are used to back fill previously mined pits with overburden from unmined clay. The overburden can be moved down hill, thus decreasing the cost of stripping as overburden is placed in the dump site. Every effort is made to keep the height of the spoil as low and even as practical. The overburden can then be covered with stockpiled top soil prior to reclamation.

Large irregular conical, steep-sided piles of overburden, left by dregline mining, are graded down with large tractor-dozers. Mining operations are gradually changing so that subsequent reclamation of spoil can be accomplished more easily and at less expense.

CHARACTERISTICS OF OVERBURDEN AND SPOIL MATERIAL ALONG THE GEORGIA-SOUTH CAROLINA FALL LINE

The sedimentary deposits of the Coastal Plain along the Fall Line attain a maximum thickness of possibly 90 to 150 m. The Barnwell and Tuscaloosa formations change in lithology from sand to alternating layers of variable thickness of clay, sand and gravel. Some layers are calcareous and change in part from a pale green hackly or blocky (fuller's earth) clay to a gray or olive green hackly fossiliferous clay or marl with scattered lime nodules (Tables 2 and 3). The channel sands consists of light-colored coarse crossbedded quartz sand interbedded with marl and clay (La Moreaux[22]).

In 1965, depth to operable clay ranged from about 3.5 to 14 m. In 1972, operational depths go about 30 m. Eventually, mining will reach all deposits, whether the overburden is Tertiary, Cretaceous or both (Murray[27]).

Physical properties of overburden

In 1966-67, the strata of 19 mine walls were delineated and described. Depth to operable clay ranged from 4.3 to 15 m (Johnson, May & Perkins[28]).

Distinguishable strata ranged in thickness from 0.3 to 6 m with a mean thickness of 1.5 m. Eight percent of the strata were over 2.5 m thick and 15 percent were 0.6 m thick or less.

Analysis for particle size distribution gave a range in clay from 16 to 77 percent and in sand from 3 to 76 percent (May et. al.[29]). Textural classes ranged from loamy sands to clay. Particle size distribution was not correlated with distance below the surface. Marl, fuller's earth and fossiliferous limestone occurred in some profiles.

Colors of the strata were quite variable. Reds, yellow and browns predominated. Calcareous strata ranged through grayish, greenish gray or olive gray to white. Fuller's earth consists of pale green blocky clay. Some sands ranged from white through pink, gray and tan. Mottles may occur in any strata below the surface.

Table 2. Well log in Washington County, near Deep Step[1]

Deposits	Depth (Meters)
Red clay and kaolin	0 - 0.9
Coarse red sand and kaolin	0.9 - 1.5
Light yellow sand	1.5 - 2.4
Medium white sand	2.4 - 3.4
Fine white sand and kaolin	3.4 - 6.1
Medium white sand	6.1 - 11.0
Coarse white sand	11.0 - 13.4
White kaolin	13.4 - 14.0
White sandy kaolin	14.0 - 14.9
Coarse white sand	14.9 - 17.7
White sandy kaolin	17.7 - 18.6
White kaolin and gravel	18.6 - 21.9
White kaolin	21.9 - 26.2
Sand and kaolin	26.2 - 29.9
Pink kaolin	29.9 - 35.4
Pink sandy kaolin	35.4 - 37.2
Gravel and kaolin	37.2 - 40.2
Coarse white sand and kaolin	40.2 - 44.2
Hard white kaolin	44.2 - 44.8
Sandy kaolin	44.8 - 46.6
Kaolin	46.6 - 52.7
Sandy kaolin	52.7 - 54.2
Gravelly kaolin	54.2 - 55.8
Red clay	55.8 - 57.9
Coarse white sand	57.9 - 60.6
Red clay	60.6 - 63.7
Red clay and sand	63.7 - 64.6
Red clay	64.6 - 72.2
Blue and white clay	72.2 - 73.4
Red and white clay	73.4 - 78.6
White clay and gravel	78.6 - 80.8
Blue clay	80.8 - 82.0

[1] La Moreaux[22]

Table 3. Section at Georgia Kaolin Company Pit[1]

Strata	Thickness (M)
Eocene (upper Eocene)	
Barnwell formation	
Upper sand member	
1. Sand, firm, massive, coarse, gritty, mottled gray and somewhat pebbly in lower half, brownish-red in upper half.	1.8
Irwinton sand member	
2. Clay, gray, purple, waxy, mottled red.	0.6-1.5
3. Sand, loose, white and yellow, fine-grained, interbedded with thin layers of purple clay.	6.1
4. Clay, light gray, bentonitic.	2.4
5. Sand, pink, buff and gray, in a fine clay matrix.	1.8
Twiggs clay member	
6. Marl, gray, fossiliferous, blocky.	1.8
7. Greenish gray nodular lime ledge.	0.15
8. Marl, buff and gray, medium hard, sandy fossiliferous.	1.2
9. Greenish-gray nodular lime ledge.	0.15
10. Marl, bluish-gray, massive, blocky, fossiliferous.	7.6
11. Clay, hackly pale green.	1.2
Ocala limestone	
12. Limestone, cream-colored, very fossiliferous, massive, becomes sandy in lower part. Contains _Periarchus pileus-sinensis_ (Ravenel), abundant in lower part; bryozoa abundant in upper part. _Pecten spillmani_ and _Ostrea_ sp.	5.5
13. Sand, buff, medium-grained.	3.6
Unconformity	
Upper Cretaceous	
Tuscaloosa formation	
14. Kaolin, white, massive, blocky.	4.3+

[1] La Moreaux[22]

Chemical Properties of Overburden

The pH ranges from 1.8 to 8.0 or extremely acid to moderately alkaline. Seventy percent of the strata have pH values between 5.0 to 5.9. Only two percent of the strata have values above 7.4 and 13 percent are at 6.0 or above. Seventeen percent of the strata have values below 5.0. High pH values were not associated with the highest levels of calcium, magnesium, sodium and potassium. One stratum with a pH of 7.9 has these levels: Ca - 390 ppm, Mg - 10 ppm, Na - 88 ppm, and K - 6 ppm.

The cations of most interest in soil fertility are calcium, magnesium, potassium and sodium. Each of these elements were extremely variable within highwalls and mines (Table 4). Exchangeable calcium ranges from a trace to about 3,660 ppm. In 15, 47 and 4 percent of the strata, calcium was below 11 ppm, 100 ppm and above 1000 ppm, respectively.

Magnesium follows the same pattern as calcium, ranging from a trace to 640 ppm. It was below 3 ppm in 41 percent of the strata; and above 100 ppm and 300 ppm in 21 and 3 percent of the strata, respecitvely. More exchangeable magnesium than calcium was found in 20 percent of the strata.

Sodium, although not an essential plant nutrient, is abundant in many of the strata. Concentrations ranged from a trace to 390 ppm. One sample point has 3,260 ppm. Concentrations were above 200 ppm for 23 percent of the strata and above 50 ppm for 77 percent. More exchangeable sodium than calcium was found in 51 percent of the strata.

Potassium is extremely low in nearly all strata, ranging from a trace to 200 ppm. Concentrations were above 100 ppm or 0.256 m.e./100 g for only three percent of the strata and above 39 ppm for only nine percent. Seventy-five percent of the samples had less than 21 ppm and 55 percent had 10 ppm or less.

Available phosphorus ranges from one to 160 ppm. Only eight percent of the strata have concentrations exceeding 21 ppm. Samples in strata of lime-rock material had concentrations of 160 ppm. Forty-one percent of the strata have concentrations of only 3 ppm or less.

Levels of available iron, manganese and zinc are extremely variable. Iron ranges from a trace to 25 ppm, except for one stratum with 80 ppm. Concentrations of 16 to 25 ppm are found at all depths but not in all strata. Fifty percent of the strata have only a trace of iron.

Manganese is extremely low, with only seven percent of the

Table 4A. Range of Chemical Characteristics of High Walls

Mine	Wall	No. of Stratagraphic Layers	Depth to Kaolin Meters	pH	P ppm	K
1	A	6	11.6	4.3-6.4	4-57	4-50
	B	6	14.9	5.2-7.9	1-160	6-70
2	A	5	8.2	5.4-6.7	3-6	2-16
	B	5	5.8	5.1-6.0	3-8	2-10
3	A	5	7.6	4.0-5.9	2-10	4-20
	B	3	4.3	4.2-4.7	1-5	12-18
4	A	5	7.6	5.0-5.5	1-10	10-38
	B	5	4.9	4.3-6.5	4-160	16-50
5	A	4	4.6	5.1-5.6	3-15	2-22
	B	5	9.4	5.0-5.2	2-5	1-4
6	A	7	7.6	4.8-5.1	1-7	2-12
	B	4	6.4	5.2-5.7	2-12	2-14
7	A	6	6.7	4.3-5.7	4-33	10-90
	B	5	7.0	5.0-5.9	1-10	4-14
8	A	6	8.2	5.0-5.7	1-12	6-20
	B	4	6.1	5.2-5.6	1-18	10-22
9	A	5	5.8	5.2-5.4	1-18	4-200
	B	7	8.2	5.3-5.8	6-27	4-178
10	A	7	11.3	4.6-6.0	6-23	4-104

T = Trace

Table 4B. Range of Chemical Characteristics of High Walls

Mine	Wall	Ca	Mg	Na	Fe	Mn	Zn
				ppm			
1	A	135-915	10-156	18-390	T-16.0	T-80.0	0.6-2.6
	B	390-1601	10-126	88-208	T-16.0	0.5-1.0	0.4-2.0
2	A	31-343	16-81	87-130	T-6.8	T-42.5	0.6-3.4
	B	10-530	4-68	88-204	7-16.0	T-43.0	0.4-4.7
3	A	16-1326	8-97	20-70	T-25.0	T-35.6	0.4-3.0
	B	31-391	87-170	19-22	16.0-25.0	1.0	0.4-0.8
4	A	208-572	54-253	10-32	T-25.0	T-67.6	0.5-1.5
	B	551-1830	36-260	15-60	T-16.0	1.5-117.0	0.6-4.3
5	A	10-52	10-23	104-143	T-15.0	T-17.5	0.5-1.4
	B	10-16	3-45	39-52	T-16.0	T-1.0	0.4-8.4
6	A	10-16	10-65	39-215	T	T	0.5-5.6
	B	10-62	8-44	78-104	T-80.0	T-43.0	0.5-0.5
7	A	62-676	9-274	156-229	T-6.0	1.0-8.4	0.6-2.3
	B	T-416	12-115	156-198	T-6.0	T	0.4-0.8
8	A	4-292	16-61	158-198	T-6.0	T-10.0	0.4-1.0
	B	83-218	12-68	158-177	T-12.0	T-20.4	0.4-1.0
9	A	T-3660	T-640	177-3260	T-13.0	T-13.0	0.6-3.3
	B	73-3203	27-560	250-338	T-8.0	T-9.0	0.4-3.0
10	A	52-385	24-429	8-228	T-9.0	T-6.0	0.4-4.0

T = Trace

strata with concentrations above 25 ppm. Forty-nine percent of the strata have only traces of manganese. All concentrations above 10 ppm are usually in the top strata.

Zinc is low ranging from 0.4 to 8.4 ppm. Eighty-two percent of the strata have concentrations of 2.0 ppm or less.

Copper was too low to measure.

Chemical properties were not correlated with distance below surface, except for manganese. High soil reaction values occur at the top and bottom of high walls. High concentrations of phosphorus, potassium, calcium, magnesium and sodium are associated with unusual geologic formations. There is no positive correlation of soil acidity with these cations. Some of the mildly alkaline strata have low concentrations of calcium, magnesium and sodium. Strata with the high concentration of these cations may be strongly acid. Much of the overburden is deficient in most of the nutrients essential for plant growth.

Characteristics of spoil material

Characteristics of the spoil are determined by the type of mining operation. Spoils are a mixture of all strata in the overburden. Sampling along line transects across spoils reveals a wide diversity in particle size distribution and elements within short distances and at all depths (Table 5). The percentage of sand ranges from 4 to 75 percent and clay from 17 to 85 percent. Generally the surface of the spoil contains a much higher percentage of clay than the undisturbed sites. Thirty-five percent of the spoil samples contain more than 40 percent clay as compared to 21 percent of the undisturbed top soil samples. Thirteen percent of the samples contain more than 60 percent clay; and five percent contain more than 80 percent clay (Johnson, May & Perkins[28]). In the mining operation, the rug material or low grade clay directly above the merchantable clay is frequently placed on top of the spoil and is mixed with the spoil in later movements.

The infiltration capacity and the erodability of spoil material is related to length and steepness of the slope, the sand-clay ratio, the diversity of the sand-clay structure, and the extent of crusting and compaction in the surface layer.

The soil moisture holding capacity of the spoil is extremely diverse for a single mining area (Table 6). For example, in the 0-15 cm layer, soil moisture content at 0.3 bar ranges from 6 to 41 percent. However, there is little difference in the moisture desorption curves for the 0-15 cm layer and 30-61 cm layer (Fig. 3).

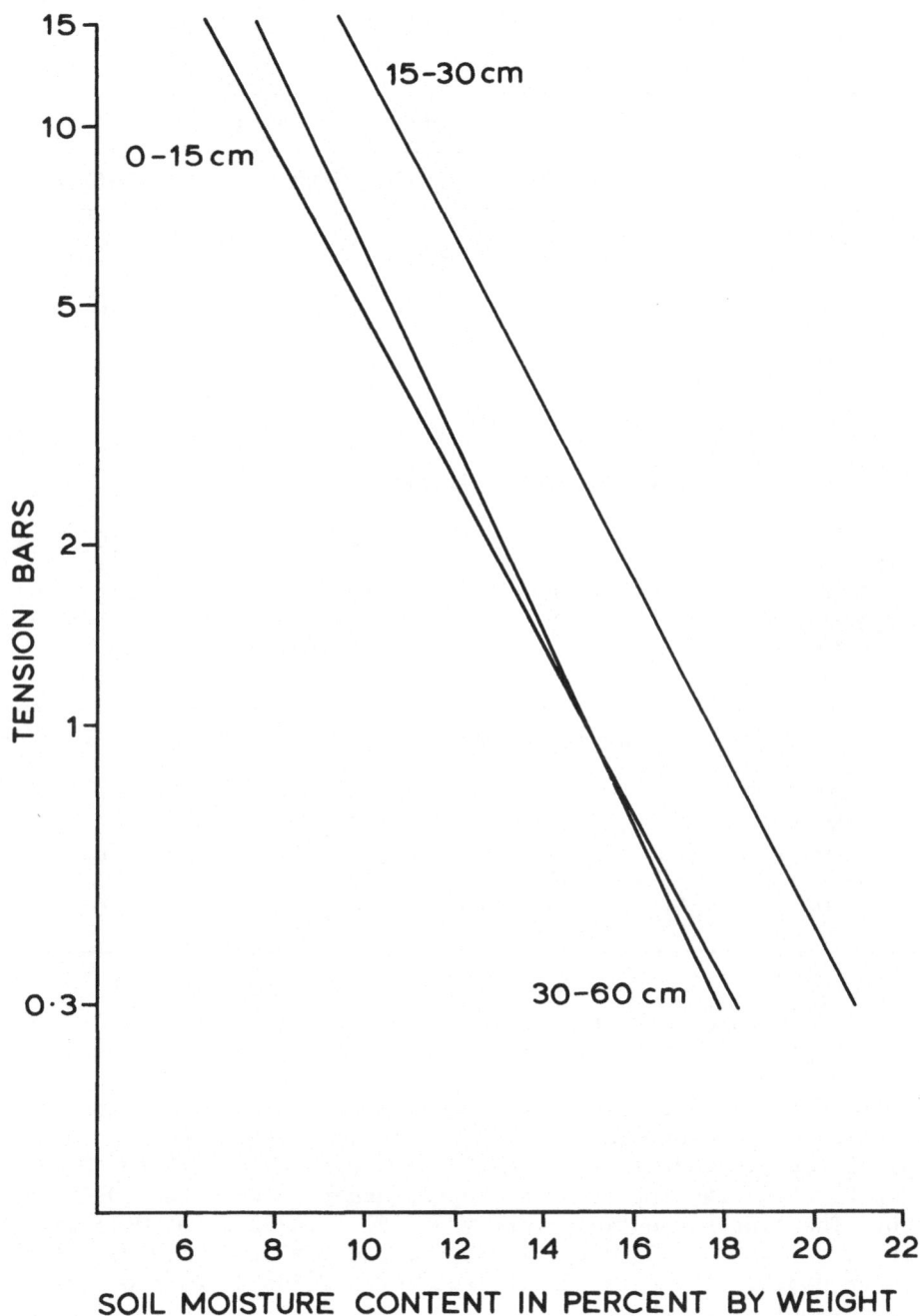

Fig. 3 Soil tension curve

Table 5. Range in particle size distribution of some spoil
material

Range Percent of Total	Percent of Samples		
	15 cm	15-30 cm	30-61 cm
Sand (2.0-0.5 mm)			
80	0	0	0
60-80	35	53	41
40-60	29	18	41
20-40	24	17	6
20	12	12	12
Silt (0.5-p.002 mm)			
25	18	0	0
20-25	18	23	18
15-20	18	18	18
10-15	23	23	17
5-10	23	18	29
5	0	18	18
Clay (0.002 mm)			
80	0	6	6
60-80	12	17	6
40-60	35	12	6
20-40	53	59	82
20	0	6	0

Soil moisture content at field capacity ranges from 6-40 percent. At wilting point moisture contents ranges from 2-25 percent.

Correlation coefficients for the 1/3 bar moisture tension as a function of percent clay plus fine silt was determined for each sampling depth. Correlation coefficients were r = .9439 for 0-15 cm depth, r = .6249 for 15-30 cm depth, and r = .6666 for 30-60 cm depth. The linear equations were y = 1.61 + .4792 x for 0-15 cm depth; y = 7.46 + .2753 x for 15-30 cm; and y = 4.40 + .2797 x for 30-60 cm depth. There is significant correlation at each of the various sampling depths.

In general the range in concentration of elements within the spoil is similar to the range within the combined strata of the overburden. The high concentrations of potassium evident in some

Table 6.   Range in soil moisture characteristics of the spoil
           material[1]

| Depth | | Bars | | |
cm		0.3	3.0	15.0
		(%)	(%)	(%)
0-15	Range	6.1-40.6	5.1-33.6	2.0-15.4
	Mean	18.2	13.4	6.7
15-30	Range	5.7-40.6	3.9-33.5	3.0-25.7
	Mean	21.0	14.0	9.7
30-61	Range	7.7-32.2	4.0-24.6	2.9-25.5
	Mean	17.8	9.6	7.8

[1] For one mining area

strata have been diluted due to mixing in the spoil. Sodium and
magnesium are lower in the spoils than in most stratum of overburden. Calcium levels and pH are generally low except in the presence of fuller's earth or calcareous spoil. The white clays are
extremely deficient in plant nutrients. The mixture of rug material
and other non-commercial kaolin clay with other components of the
overburden results in very infertile spoil. Major plant nutrients
such as P, K, Ca and Mg are deficient in about 80 percent of the
spoil (Table 7). For example, if we assume that 200 ppm or approximately 1.0 m.e./100 g represents a deficiency level for tree growth
60 percent or more of the spoils are calcium deficient for most
species.

     The soil reaction of the spoil material ranges from about 1.8
to 8.0. The major toxicity is associated with low acidity and high
iron concentrations. All spoil material is deficient in or devoid
of organic matter, nitrogen and micro- and macro-organisms.

Classification of spoils

     Strip mine spoils may be classified in many different ways.
However, for developing reclamation plans it is important that they
be classified according to factors or features that affect plant
growth and equipment operation. In the Georgia - South Carolina
China Clay mining area, the broad classification can be based on
slope, texture and color, and acidity.

Table 7. Ranges and means of chemical characteristics of spoil
          from three mining areas

Soil Characteristics	Mines 1	2	Mine 3 0-15cm	15-30cm	30-60cm
	----- 0-30cm -----				
pH	5.2-6.3	5.0-8.0	4.6-5.5	4.6-5.4	4.7-5.5
Mean	5.5	6.8	5.1	5.1	5.1
Avail. P ppm	5-6	4-160	4-25	2-31	2-29
Mean	7	37	7	8	10
Exch. K ppm	2-10	8-34	4-22	4-32	4-32
Mean	7	19	10	10	12
Exch. Ca ppm	15-395	9-5,960	7-63	8-88	6-82
Mean	65	1,865	29	37	34
Exch. Mg ppm	1-13	10-99	8-87	8-88	4-46
Mean	4	50	39	40	20
Exch. Na ppm	7-39	15-50	7-41	3-29	9-24
Mean	14	29	16	10	15
Mn ppm	2.0-19	0.5-88	0.5-14	0.5-22	0.5-28
Mean	8.5	11.2	27	4.8	7
Fe ppm	5-20	5-60	10-35	5-40	5-40
Mean	14	17	17	14	15
Zn ppm	1.2-4.2	1-18	0.7-2.4	1.0-4.5	1-15
Mean	2.9	6.3	1.7	2.6	3.5

A. - Slope Groups

1. 0-3 percent. Spoil is graded level to gently undulating.
   Erosion is slow or medium for most soils. Some broad terraces
   may be desirable.

2. 3-8 percent. Spoils may be graded to form gently rolling landscape. Runoff and erosion will be medium to rapid on spoils
   with long slopes.

3. 8-16 percent. Spoils may be graded to form strongly rolling
   landscape. Runoff is medium to rapid.

4. 16-35 percent. Slopes are moderately steep to hilly. Runoff
   is rapid. Sheet and gully erosion develop under normal rainfall.

5. 35 + percent. Spoils are developed by dragline mining. Runoff
   is rapid and slopes are highly erodable.

6. Highwalls. These are walls of mine or extremely steep back-
   slopes of some spoil.

   Slope classes 1, 2, 3 and 4 can be cultivated, seeded and fertilized with machinery. Slope class 5 presents serious problems.
Mulching and seeding have been effective in stabilizing spoil on
short slopes in excess of 38 percent. High walls are generally
stable; but present safety problems.

B. - Texture and Color Groups

1. Surface light colored - kaolinitic clay; the surface area consists of rug material or low grade clay.

2. Surface light colored - sands to sandy loam. The surface area
   consists of light textured material from the overburden; but
   does not include the original 0-0.6 m of surface soil.

3. Surface reddish to brown - sandy to clayey. The surface area
   consists of dark colored material from the overburden but does
   not include the original 0-1 m of surface soil.

4. Surface greenish to olive brown - fuller's earth. The surface
   area consists of a mixture of spoil containing more than 60
   percent fuller's earth.

5. Surface whitish to grayish or brown with sandy to clay marl or
   limestone containing marine fossils.

6. Surface dark gray to brown with either sands, clays or a mixture
   and contains a high percent of pyrite.

7. Surface gray to reddish or brown - sandy to sandy clay. This
   is a mixture of the A and B horizons that has been spread over
   spoil materials ; and is about 15-30 cm thick.

C. - Acidity Groups

1. Spoil pH is above 6.0. This material consists of fuller's earth
   marl or calcareous material. A few native pioneer plant species
   seed-in naturally. Establishment of natural vegetation is
   accelerated by application of nitrogen and sometimes phosphorus.

2. Spoil pH is between 4.5 and 6.0. This spoil group may contain a wide range of particle size distribution. Complete fertil-ization is required for establishment of all agronomic crops. All tree species require nitrogen and sometimes phosphorus.

3. Spoil pH below 4.5. Concentrations of pyritic materials res-ults in hot spots with pH as low as 1.8.

RECLAMATION

Site conditions on China Clay mined lands in the southeastern United States are so unique that a distinctive classification of factors influencing plant growth is needed. The two major elements that create problems in reclamation are (1) the rooting media (spoil material) and (2) the climate.

The rooting media is an unorganized mass of material consisting of a mixture of all strata above the mined seam. Some of its prop-erties have already been discussed. Basically, it is almost a sterile media; is highly erodible; has a variable infiltration rate and a low level of available water; and forms a hard surface crust. The nature of the spoil material can be modified by grading, redis-tribution of top soil, use of fertilizers and the use of surface mulches.

But major problems can still be caused by the one uncontrollable factor - the weather. The mean annual rainfall along the Fall Line is 109-145 cm per annum. Frequently devastating rains in the spring are followed by excessive periods of drought in summer and autumn. Land that has been graded, fertilized and seeded can be completely destroyed by sheet erosion resulting from torrential rainfall prior to establishment of a vegetative cover.

The Georgia Surface Mining Act of 1968 prescribes that mining affected lands must be reclaimed for specific purposes such as farming, grazing, forestry, recreation (includes wildlife), indus-trial or residential.

General Rules of the Georgia Surface Mining Act

Regulations regarding surface mining are issued under an act known as the Georgia Surface Mining Act of 1968 as amended in 1971. The Act provides for the licensing of operators of surface mines and for the reclamation of affected lands. Specific purposes are:

1. to assist in achieving and maintain an efficient and productive mining industry and increasing economic and other benefits attributable to mining;

2. to advance the protection of fish and wildlife and the protection and restoration of land, water and other resources affected by mining;

3. to assist in the reduction, elimination or counteracting of pollution or deterioration of land, water and air attributable to mining;

4. to encourage programs which will achieve comparable results in protecting, conserving and improving the usefulness of natural resources to the end that the most desirable conduct of mining and related operations may be universally facilitated;

5. to assist in efforts to facilitate the use of land and other resources affected by mining so that such use may be consistent with sound land use, public health and public safety, and to this end to study and recommend wherever desirable, techniques for the improvement, restoration or protection of such land and other resources.

The Land Reclamation Section of the Department of Natural Resources has the authority to prescribe standards of reclamation. In summation, the rules require or prescribe the following actions:

1. Operators of surface mines will obtain an annual license, which must be renewed annually.

2. An operator may substitute and reclaim an area mined in the past for lands currently being mined.

3. Each operator is required to submit a mined land use plan, including maps, plans and specifications.

4. The operator is to outline his particular method or technique of mining.

5. The operator is to clearly indicate what use will be made of affected lands following mining. He should indicate if lands will be developed for farming, pasture, forestry, recreation, industrial or residential purposes.

6. Operators will be required to establish approved permanent vegetative cover on all affected lands except sites specifically exempt on an approved mined land use plan.

7. Grading shall be carried out in such a manner as to leave affected lands so as to blend peaks, ridges and valleys to a rolling topography.

8. The operator is to show his plan for protecting the health and

194

welfare of a community, should mining be done in the immediate
vicinity of a town, village, heavily populated urban area or
public road system.

9. Development of lakes for recreation, fishing or wildlife habi-
    tats that will result from mining will be in keeping with good
    conservation practices for the intended purpose as recommended
    by appropriate state conservation agencies.

Early Reclamation Activities

Several mining companies initiated reclamation activities in
the early 1960s when forest tree seedlings were planted on a variety
of spoils. Although initial survival was nearly always high, subsequent growth was poor. The seedling vigor declined each year.
Most seedlings died from nutritional deficiency or they were eliminated by erosion. There was no site treatments, either before or
after plantings.

SCS Trials. In 1967, the Soil Conservation Service, United States
Department of Agriculture, established a study on a graded spoil
bank to determine species suitability (Haynsworth[30]). Some of the
plants were established from seed while others were established
from vegetative materials. All were planted in rows across the
plot. The seed were planted in a trench about 12 to 20 mm in depth
and were covered with loose soil. The plants were set slightly
deeper than they grew in the nursery and they were watered at the
time of planting. Fertilizer and lime was broadcast after the
plantings were made. There were four fertility variables. These
were:

a. 1681 kg of 6-12-12 fertilizer and 4483 kg of agricultural lime-
    stone per hectare.

b. 1681 kg of 6-12-12 fertilizer per hectare.

c. 4483 gk of agricultural limestone per hectare.

d. No fertilizer or agricultural limestone.

The fertility treatments were applied across the plant rows.
One of the sites received pine straw mulch at the rate of 4483 kg
per hectare. The other site was left unmulched.

The plants used in the evaluation were:

ALDER, HAZEL   Alnus rugosa (Du Roi) Spring.
AUTUMNOLIVE   Elaeagnus umbellata Thunb.
BAHIAGRASS, PENSACOLA   Paspalum notatum var. Saurae Parodi

BAHIAGRASS, WILMINGTON	Paspalum notatum
BERMUDAGRASS, COMMON	Cynodon dactylon (L.) Pers.
BLUESTEM KING RANCH	Andropogon ischaemum L.
BROOMSEDGE	Andropogon virginicus L.
BRUNSWICKGRASS, AMCORAE	Paspalum nicorae Parodi
BUTTONBUSH, COMMON	Cephalanthus occidentalis L.
CARPETGRASS	Axonopus affinis Chase
COTTONSEDGE, VIRGINIA	Eriophorum virginicum L.
GIANT REED	Arundo donax L.
INDIGO, FALSE ANIL	Indigofera pseudotinctoria
JOINTVETCH, SENSITIVE	Aeschynomene virginica (L.) BSP.
KUDZU 23	Pueraria thunbergiana (Willd.) Ohwi.
LESPEDEZA, KOBE	Lespedeza striata Thunb.
LESPEDEZA, SERICEA	Lespedeza cuneata (Dumont) G.Don
LESPEDEZA, THUNBERG	Lespedeza thunbergii
LESPEDEZA, VIRGATA	Lespedeza virgata
LOVEGRASS, WEEPING	Eragrostis curvula (Schrad.) Nees
MARIGOLD, TALL	Tagetes minuta
PEANUT, FORAGE	Arachis sp. L.
PEANUT, ANNUAL FORAGE	Arachis sp. L.
LOCUST, ARNOT BRISTLY	Robinia fertilis Ashe
SESBANIA	Sesbania sp. Scop.
SUMAC, SMOOTH	Rhus glabra L.
SWITCHGRASS, PANGBURN	Panicum virgatum L.
TRUMPETCREEPER	Campsis radicans L.
WILLOW, SANDBAR	Salix interior Rowlee

Established from seed or vegetative material
*Established vegetatively
(All others are from seed)

The quickest germination and overall best stand was secured under mulched conditions. Although good stands were obtained initially on the unmulched site, most of the plants lacked vigour and the stand weakened as the summer passed. The exposed surface of the spoil material tended to dry out readily although the moisture seemed fair at a depth of 12 to 20 mm. After about 10 days without rain, the surface cracked quite deeply and moisture became limited. Generally, only Arnot bristly locust from plants have given satisfactory results on the unmulched site. About half of the plants persisted, but they made little growth. The best plants, besides Arnot bristly locust, on the unmulched site, were autumnolive from plants and common bermudagrass, weeping lovegrass, and Pangburn switchgrass from seed.

Plant performance on the mulched site was dramatic. Stands were poor where the mulch was blown away by wind. The germination of most of the seeded plants was good. Differences in plant vigour according to fertility treatments became pronounced as the season progressed.

As the season progressed, it became increasingly evident that fertilization was necessary to maintain a stand of plants. Growth and development approaching normal conditions was obtained only where fertilizer and mulch was applied.

There was little or no growth in the unfertilized areas contrasted with fair to good growth of most of the plants in the fertilized area in the first year.

Nitrogen, at the rate of 56 kg/ha was applied to the grasses on the unmulched site in the second year, but there was no marked improvement in growth or cover. The application of agricultural limestone appeared to have no immediate effect.

A brief discussion of species that performed best on mulched sites follows:

Autumnolive was planted using seed. This was a failure with only one plant in the two sites surviving after two years. Autumnolive plants were also set out. Very little growth occurred in the first year. In the second year, growth was good. After four years these plants had grown to over 1.8 m tall and are producing food for wildlife. The ability of this plant to fix atmospheric nitrogen favors its success. Autumnolive is not expected to control erosion of the spoil banks in the early years. It is being looked upon as a wildlife food plant to be grown in association with other plants which can be depended upon for early erosion control.

Common bermudagrass came up to a good stand and spread rapidly the first year. It gave the quickest surface protection of any of the plants. It continued to spread and to provide land cover during the second year but provides very little new cover. It produced seed in both years.

Kobe lespedeza came up to a thick stand in the fertilized areas and made very good growth. A heavy seed crop produced an excellent volunteer stand in succeeding years. The leaves and stems of dead plants are providing a good cover of litter. Kobe lespedeza shows outstanding promise for quick establishment and cover on this site. It is also useful as a wildlife food plant.

Sericea lespedeza came up well and grew vigorously in the fertilized area. It produced seed and some volunteers have been observed. This planting further verified the usefulness of sericea lespedeza in planting. It is a standard plant for this use in the Southeast and is one of the best for establishing a quick cover.

Virgata lespedeza became well established with fair vigour in the fertilized area. It grew moderately well in the first year and better in the succeeding years. The leaf residue from this peren-

nial plant provides ground cover and mulch, It is a good species
for use in stabilizing mine spoil banks, is gaining acceptance for
roadbank stabilization, and is now considered an acceptable alter-
nate to sericea lespedeza.

Weeping lovegrass was quick to germinate and grew vigorously.
Lovegrass seed was transported by runoff water and germinated in
many of the plots where it was not planted. Excellent growth and
cover resulted everywhere it lodged in fertilized plots. Weeping
lovegrass, planted without seedbed preparation but with mulch and
fertilizer, is one of the best prospects for cover and erosion control on mine spoil banks. This further verifies the usefulness of
weeping lovegrass in critical area erosion control. It is a standard plant for this purpose in the Southeast.

Bristly locust was easier to establish from seed than from
plants. The seedling plants showed good initial vigour. They grew
to about 1.5 m in height where fertilized and only 15 cm when not
fertilized. The seedling plants attained about as much size in the
first year as the nursery grown plants. All plants in the fertilized area produced seed. The rhizomes spread out one meter as much
as in the first year and now reach out 9 m. It has excellent prospects for use on mine spoil banks.

Pangburn switchgrass produced a good stand and the plants were
vigorous the first year. On the fertilized area they remain fairly
vigorous and have slowly spread by tillers. Seeding has produced a
few volunteers. Pangburn switchgrass may have some value on these
sites.

Pensacola and Wilmington bahiagrass produced good stands. They
exhibited fair to good vigour in the first year, but have been slow
to cover the highly erodible spoils.

Thunberg lespedeza and bi-color lespedeza were grown from seed.
The plants were vigorous providing food and cover for wildlife.

ARS - University of Georgia. Three studies were initiated in 1966
by the Agricultural Research Service, USDA and the School of Forest
Resources, College Experiment Station, University of Georgia.

A greenhouse investigation using eight rates of phosphorus and
potassium with adequate lime was conducted on spoil material. The
mechanical composition of the spoil was 12 percent sand, 48 percent
silt and 40 percent clay. A chemical analysis showed a pH of 4.8
and only traces of phosphorus and potassium. Ladino clover (Trifolium spp.) was used as the test plant. Four thousand four hundred
eighty-three kg of lime per hectare raised the pH of the spoil from
4.8 to 7.2. Seventy-eight and five tenths kg per hectare of phosphorus was necessary for good plant growth and to raise the level

of extractable phosphorus in the spoil. If phosphorus was omitted, no plants became established. The spoil and seed contained ample potassium to support early plant growth; however, this was rapidly removed by the plants. Potassium at 112 kg/ha was needed for good plant growth (Parks, Perkins & May[31]).

Field plots were established on a graded spoil in 1966 to (1) determine the minimum fertilizer requirements for the establishment of forage plants on clay spoils; (2) develop a procedure for the establishment of forage grasses on spoils; and (3) determine the forage production potential of bahiagrass (seeded) and Coastal bermudagrass (sprigged). The spoil has a pH of 5.2; and a moisture percent at 0.5 bar of 29.8 and at 15 bars of 14.6. After grading, the plots received a broadcast application of 4,480 kg/ha of dolomitic limestone. Twelve fertilizer treatments included four levels of nitrogen, phosphorus and potassium. The best yields of dry matter during the second and third growing seasons were 10,960 and 8,855 kg/ha for Coastal bermudagrass and bahiagrass, respectively. These yields were obtained by using 4,480 kg of lime, 448 kg of nitrogen, 90 kg of phosphorus and 186 kg of potassium/ha. Yields exceeding 5,000 kg/ha were obtained only on plots receiving levels of nitrogen, phosphorus and potassium of 224, 45 and 93 kg/ha respectively. Nitrogen influenced the growth more than phosphorus and potassium. The yield increased with each addition of nitrogen and phosphorus, except at the higher rates when potassium was omitted. Minor elements had no effect on yield.

Plots on a graded slope were fertilized with lime and 6-12-12 at a rate of 4,480 and 662 kg/ha, respectively, and seeded to rye (Secale cereale L.), millet (Pennisetum glaucum L.), fescuegrass (Festuca arundinacea Schreb.), ryegrass (Lolium spp.), and Lespedezas (Lespedeza striata Thunb., L. cuneata (Dumont) G. Don and L. virgata). One half of each plot was mulched with straw. A supplemental fertilization with 10-10-10 at 662 kg/ha was applied in the spring of the second growing season. The millet, ryegrass, and rye provided good cover for one year, with the heaviest cover on the mulched area. Where the rye was mowed, a heavy cover of rye was obtained the following year. After the second year, the rye, ryegrass and millet began to disappear from the stand. The lespedezas and fescuegrass tests provided excellent ground cover and stabilization of the spoil.

A third area of spoil was graded to a rolling topography, fertilized and seeded to rye and fescuegrass in the fall of 1967. Fertilizers were 4,483 kg of lime, 560 kg of 6-12-12, and 560 kg of 0-20-0/ha. Seven species of trees were planted on part of the area in the winter of 1967-68. After five years the rye and fescue almost completely disappeared, and gully erosion is occurring on the slopes. However, the loblolly pine (Pinus taeda L.), slash pine (P. elliottii), sycamore (Platanus occidentalis L.) and black alder

(Alnus glutinosa (L.) Gaertn) have grown vigorously (May, Parks & Perkins[32]).

Hydroseeding and Helicopter Seeding

Hydroseeding or hydromulching, designed to apply a slurry mix of seed, fertilizer and a cellulose mulch in one operation, has been used successfully on golf courses, lawns and some coal mining spoils. This technique was tested extensively on non-prepared and graded China Clay mining spoil between 1967 and 1969. Plant species included Lespedezas, bahiagrass, fescuegrass, rye, common bermudagrass, millet, oats (Avena sativa L.), sudangrass (Sorghum sudanensis (Piper) Stapf.) and loblolly pine. Wood fibre mulch was applied at rates up to 1793 kg/ha. Organic compost from sewage sludge and lespedeza hay were also tested as mulches.

Most of the hydroseeding treatments were considered failures after the second years. Causes of failures included:

1. Frost heaving of plants.

2. Lack of adequate moisture to insure germination.

3. Extended droughts after germination of seed.

4. Extensive erosion associated with rainstorms.

Seeding and fertilizing with helicopters was tested on prepared sites in 1969. Materials were very evenly applied to the site. But vegetative covers were not successfully established.

Grading for Mechanical Seeding and Planting

Requirements of the Georgia Surface Mining Act coupled with the almost impossibility of getting vegetation established on non-treated spoil has resulted in (1) modifications of mining techniques and (2) the intensive treatment of spoils (Gronow[33]).

The roll over method of mining and other techniques that fill the completed mine cuts with overburden from adjacent cuts are being utilized when compatible with the available equipment and the stratigraphy of the mining area. The filled cuts are graded to a level or gently rolling landscape.

Special problems resulting from the mining are:

1. Spoils that consist of high irregular conical, steep-sided piles. These can be made plantable by pushing the top of the pile down

hill and creating a concave or saucer-shaped depression at the highest elevation. The depression must be deep enough to hold all precipitation from the heaviest rains.

Water moves through the spoil by gravity. Long slopes can be broken by wide benches or terraces that will hold all water from the slope above. Generally the slopes should be on less than 35 percent grade and less than 15 m long. The landform must provide for water control and the use of agricultural equipment for fertilizing, seeding and mulching.

2. Spoils that are rich in kaolin. Vegetation is more difficult to establish in heavy clay spoils than the more sandy material or the red clay-sand mixes of the Barnwell formation. Where mining conditions permit, fuller's earth or top soil from adjacent cuts or from stockpiles should be spread over the kaolin clay spoil.

3. Acidic sands and clays are toxic to vegetation. Heavy applications of lime have not reduced the acidity. Fuller's earth or top soil should be spread over these acidic spoils. The occurrence of acidic conditions is generally limited to very small areas.

Establishment of a Vegetative Cover

The Georgia Surface Mining Act of 1968 requires that stripped mined lands be restored to productive uses compatible with adjacent land use. About 75 percent of the land in the clay mining area is in forests. Experience has shown that tree planting alone is not the complete solution, in a planned program of reclamation involving a wide variety of sites.

The Use of Forest Tree Species

Selection and use of trees in reclamation depends primarily on four considerations, namely, (1) characteristics of the spoil material; (2) characteristics of the vegetative cover; (3) silvical requirements of the species and (4) ultimate objectives of the owner.

Some native tree species have become established on the older spoils (Troth [34]). These include cottonwood (Populus deltoides Bartr.), red maple (Acer rubrum L.), dogwood (Cornus florida L.), sweetgum (Liquidambar styraciflua L.), persimmon (Diospyros virginiana L.), hawthorn (Crataegus spp.), river birch (Betula nigra C.), waxmyrtle (Myrica cerifera L.), black cherry (Prunus serotina Ehrh.), sycamore, black willow (Salix nigra Marsh), sassafrass

(Sassafras albidum (Nutt.) Nees), yellow poplar (Liriodendron tulipfera L.), loblolly pine, and shortleaf pine (P. echinata Mill). Most of the trees seeded in on wet or moist sites, after other vegetation had provided a protective cover. Pines were established slowly on some of the drier sites, where the spoil contained a high percent of top soil.

Tree planting by mining companies was initiated in the late 1950's and several plantations were established on spoils between 1960 and 1965. These were almost complete failures due to nutrient deficiencies and erosion. In 1967, some of the early plantings were fertilized and/or mulched. Response within 2 to 3 months was very evident. During the next five year period fifteen species were planted on a variety of sites. Eighteen fertilizer treatments were tested on one or more species.

Species with the best survival and growth over a range of sites were the pines (P. taeda L. and P. elliottii), sycamore and European black alder. Other species that show suitability to some sites are Virginia pine (P. virginiana Mill.) and sawtooth oak (Quercus acutissima Carruthers). Three oaks, (chestnut, northern red and white) (Q. prinus L., Q. rubra L., and Q. alba L.) survive well on some sites but growth is slow (Khemnark[35]; Thomas[36]).

Blackgum (Nyssa sylvatica Marsh, cottonwood, shortleaf pine, sweetgum, black walnut (Juglans nigra L.) and yellow poplar are not suited to most spoil sites.

Most of the plantings on kaolin mining spoils have been on graded sites - many of which are highly erodible. The only successful plantings were on level or gently rolling sites where erosion was not a problem; or on sites that have been stabilized with a vegetative cover.

Fertilization is essential for the successful establishment of tree seedlings. On bare sites, and considering both survival and growth, a complete fertilizer with about one ounce (28 gms) of nitrogen per seedling was optimum. Mortality increased with an increase of ammonium nitrate from three ounces (85 gms) to six ounces (170 gms) per seedling. Superphosphate at rates up to 10 ounces (283 gms) did not adversely affect survival or growth of seedlings. Applications of calcium and potassium have not significantly affected survival or growth of most of the species tested on clay spoils.

Fertilizers have been applied in a circle or a spot on the ground surface near the seedlings, in slits adjacent to the seedlings, and as pellets in the planting slot and in adjacent slits. In general, surface fertilization and slit fertilization provide comparable growth response. The application of pellets in the

planting slits has an adverse effect on survival of pines but not
hardwoods. Growth response is slow from pellets in adjacent slits.

The common mycorrhizae, Thelephora terrestris, Ehr., that
develops on seedling roots in the nursery is not ecologically adapted to adverse sites. During the first year after pine seedlings
are planted, there is a proliferation of Pisolithus tinctorius
(Pers.) Cok. & Couch on the seedling roots. This symbiotic relationship must contribute to the vigour and growth of planted seedlings

Where spoils are predominately mixtures of fuller's earth or
of a highly calcareous nature, European black alder, sycamore and
bi-color lespedeza have given better survival and growth than other
species. Pine survival and growth has not been good on these sites.
Fertilization of these sites stimulates the invasion of native
grasses and lespedezas which crowd out planted seedlings.

Machine and hand planting have given comparable survival and
growth rates. Observations that indicate that root growth is more
extensive in the trenches of machine planting than in the slit of
bar planting. No studies have been made on root growth in mining
spoil.

Use of Herbaceous Vegetation

A protective cover of a fast growing annual or perennial should
be established as quickly as possible on steep, rolling and erodable
sites.

Annuals that have been successful on clay mining spoil are rye,
sudan grass-sorghum hybrid, lovegrass, common lespedeza, and brown
top millet (Panicum ramosum L.).

Perennials that are good to excellent are Lespedeza sericea,
virgata lespedeza, bi-color lespedeza, bermudagrass, bahiagrass,
white clover (Trifolium repens L.) and fescuegrass (May, Parks &
Perkins [32]).

Fertilizer requirements vary with soil reaction and nutrient
levels of the spoil. Where pH is between 4.5 and 6.0, 2,244 to
4,483 kg/ha of lime are required for many of the species. Where
soil reactions are below 4.5 requirements are 4,483 to 6,725 kg/ha
of lime; or a cover of top soil.

Some fertilizer rates that have resulted in a good vegetative
cover over a wide range of spoils are:

Nitrogen   56 - 112 kg/ha   N

Phosphorus   112 – 196 kg/ha P_2O_5

Potassium   56 – 202 kg/ha K_2O

Plant mixtures that have produced good vegetative covers are:

1. Lespedeza cuneata (Dumont) G. Don

2. Secale cereale L.

    Lespedeza cuneata (Dumont) G. Don

    Cynodon dactylon (L.) Pers

3. Secale cereale L.

    Festuca arundinaceae Schreb.

    Cynodon dactylon (L.) Pers

    Lespedeza cuneata (Dumont) G. Don

4. Paspalum notatum var. Saurae Parodi.

    Cynodon dactylon (L.) Pers.

    Trifolium repens L.

5. Lespedeza cuneata (Dumont) G. Don

    Eragrostis curvula (Schrad.) Nees

6. Paspalum notatum var. Saurae Parodi

    Cynodon dactylon (L.) Pers.

    Panicum ramosum L.

A fertilizer supplement, especially nitrogen, is needed during the summer of the first year or early in the second year (Troth[37]).

Seeding trials have been made throughout the year. Fall seeding is used when rye or fescuegrass are desired as the initial cover.

The most successful treatments have been fall seeding with rye, with and without mulch after seeding. Fescuegrass, bahiagrass, bermudagrass and lespedeza sericea may be sowed with the rye. The fescuegrass seed will germinate in the fall and the other species in the spring. The rye can be left standing or knocked down in the spring. A successful variation is to knock the rye down in the spring and then overseed or sow lespedeza sericea and other species.

Spring sowing is preferable for the lespedeza's, millet, love-grass, bahiagrass and bermudagrass. If seeds are drilled into the soil (spoil) with a grain drill followed by a cultipacker, germination and establishment may occur before heavy spring rains create erosion problems or the surface develops a hard crust. Mulching spring-sowed sites is desirable and is essential for summer sowing.

The maintenance requirements for a vegetative cover are still to be determined. The lespedezas have provided good cover for six years, and appear to be permanently established on many level and slightly rolling sites. Bahiagrass and common bermudagrass covers have disintegrated after four years, without supplemental fertilization.

The vegetative covers established since 1967 have functioned primarily to (1) provide a protective cover to control erosion, (2) provide a habitat for wildlife, and (3) initiate a process of soil building through recycling and the accumulation of a soil biota consisting of organic matter and soil organisms. The areas have not been used for production of hay or forage crops. However, some of the rye and sericea lespedeza stands compare favorably with similar crops on non-mined lands, and could possibly compete in the production of hay or seed.

Mix Vegetative - Tree Cover

A permanent vegetative cover, except for pasture, will normally include a mixture of grasses, herbaceous plants and woody plants (shrubs and trees). Four possible establishment techniques are:

1. Establish tree vegetation on relatively level sites.

2. Establish herbaceous vegetation on graded rolling sites.

3. Plant tree seedlings on sites with established vegetation.

4. Establish herbaceous vegetation and trees during the same year on graded sites.

The first technique is relatively simple on all level sites except where the soil pH is above 6.5 or below 4.5. Trees can be machine or hand-planted. Fertilization is required at time of planting or before the active growing season. Maintenance fertilization may be required about the third year for most sites and for most species. It is possible that a third or fourth treatment may be required before nutrient recycling becomes effective.

The second procedure has been successful on a wide range of sites. One or more maintenance fertilizations are required for most vegetation except lespedezas. Bermudagrass and bahiagrass covers began to disappear after four years and only one maintenance fertilization. A third fertilization after six growing seasons resulted in vigorous growth of the existing cover.

The third technique presents special problems, namely:

1. The established vegetation provides too much competition for space, moisture and nutrients.

2. Fertilization of tree seedlings within a complete cover or a disintegrating cover stimulates growth of herbaceous vegetation, which subsequently crowds out the tree seedlings.

The procedure of establishing tree vegetation simultaneously with a herbaceous cover crop has been tested on few sites. It shows promise - when a forest cover is the ultimate objective (Dipert[38]).

SOIL DEVELOPMENT

Soil is defined as the collection of natural bodies on the earth's surface, containing living matter, and supporting or capable of supporting plants. Each of the natural bodies have a unique morphology resulting from a unique combination of climate, living matter, parent material, relief and time.

Mining spoil is the parent material that will at some future time become soil. The results of soil formation will be stabilization of sites through aggregation of mineral and organic matter, increased nutrient status, initiation of nutrient cycling and reduced erosion.

Chemical and physical weathering are continuous processes in the warm, humid climates of the Southeast United States. Nitrogen is supplied to the soil through precipitation. As the process of weathering continues plant and animal life gets established in favorable niches. Airborne soil organisms invade the spoil with the establishment of available sources of energy or a supply of food. Soil formation on a mined spoil is directly influenced by any factor that influences the establishment of vegetation.

When the original top soil is a component of the surface spoil, adequate nutrients and soil micro-organisms may provide a favorable niche for successful invasion of any one of several native plant species. Thus, many of the early mining spoils are vegetated by natural invasion of pioneer species and the subsequent succession of more complex communities.

When sites are fertilized, nutrient levels become adequate for the successful establishment of many species.

The combination of natural physical and chemical weathering, establishment of plant communities and subsequent decomposition of organic residues creates a nutrient cycle that in turn accelerates the process of soil genesis. Studies of natural and artificially revegetated sites reveal that (Troth[37]; Troth[34]):

1. Structure development is occurring at the surface, but decreases with depth.

2. Clay content increased and sand content decreased with depth.

3. The amount of clay in the subsurface layer increased with time.

4. Organic matter was incorporated to the greatest depths at those sites undisturbed for the longest period of time.

5. Cation exchange capacity values are related to amount and kind of clay as well as organic matter content.

6. There is a recycling of K, Ca, Hg, Zn, and Mg.

7. Most extractable P was found in the upper 5 cm of spoil.

8. The amount of extractable K in the upper 0.5 cm layer generally increased with an increase in the rate of K fertilization.

9. Five years after fertilization, Ca and Mg were well distributed in the upper 15 cm of spoil; but levels decreased sharply in the 15-30 cm depth.

10. Organic matter accumulation in the 0.5 cm layer beneath bahia-grass was closely associated with N application rates.

SUMMARY

Mining of China Clays is an extensive operation along the Fall Line of the Southeastern United States and in the United Kingdom and occurs on a limited scale in numerous other parts of the world. The spoils from mining are diverse mixtures of all strata including non-merchantable and viscous clays. The spoil is generally comp- letely devoid of nitrogen and soil organisms. Erosion is a major problem in areas subject to frequent floods. Herbaceous and woody plant vegetation can be established on graded and fertilized sites; but maintenance fertilization will be required until nutrient cycling can create a favorable nutrient ratio. The actual cost of reclamation per hectare is a factor which varies depending upon the individual site conditions and the mining methods used by the operator.

Table 8.   Common and scientific names of plants mentioned

Scientific Name	Common Name
Acer rubrum L.	Maple, red
Aeschynomene virginica (L.) BSP	Jointvetch, sensitive
Alnus glutinosa (L.) Gaertn	Alder, European black
Alnus rugosa (Du Roi) Spring	Alder, Hazel
Andropogon ischeamum L.	Bluestem, King Ranch
Andropogon virginicus L.	Broomsedge
Arachis sp. L.	Peanut, annual forage
Arachis sp. L.	Peanut, forage
Arundo donax L.	Giant reed
Avena sativa L.	Oats
Axonopus affinis Chase	Carpetgrass
Betula nigra L.	Birch, river
Campsis radicans L.	Trumpetcreeper
Cephalanthus occidentalis L.	Buttonbush, common
Cornus spp. L.	Dogwood
Crataegus spp.	Hawthorn
Cynodon dactylon	Bermudagrass, coastal
Cynodon dactylon (L.) Pers.	Bermudagrass, common
Diospyros virginiana L.	Persimmon
Elaeagnus umbellata Thunb.	Autumnolive
Eragrostis curvula (Schrad.) Nees	Lovegrass, weeping
Eriophorum virginicum L.	Cottonsedge, Virginia
Festuca arundinacea Schreb.	Fescuegrass

Scientific Name	Common Name
Indigofera pseudotinctoria	Indigo, false anil
Juglans nigra L.	Walnut, black
Lespedeza cuneata (Dumont) G. Don	Lespedeza, sericea
Lespedeza striata Thunb.	Lespedeza, kobe
Lespedeza thunbergii	Lespedeza, thunberg
Lespedeza virgata	Lespedeza, virgata
Liquidambar styraciflua L.	Sweetgum
Liriodendron tulipifera L	Yellow-poplar
Lolium spp.	Ryegrass
Myrica cerifera L	Waxmyrtle
Nyssa sylvatica Marsh.	Blackgum
Panicum ramosum L.	Millet, browntop
Panicum virgatum L.	Switchgrass, Pangburn
Paspalum nicorae	Brunswickgrass, Amcorae
Paspalum notatum	Bahiagrass, Wilmington
Paspalum notatum var. Saurae Parodi	Bahiagrass, Pensacola
Pennisetum glancum L.	Millet
Pinus echinata Mill	Pine, shortleaf
Pinus elliottii Engelm.	Pine, slash
Pinus taeda L.	Pine, loblolly
Pinus virginiana Mill	Pine, Virginia
Platanus occidentalis L.	Sycamore, American
Populus deltoides Bartr.	Cottonwood, eastern
Prunus serotina Ehrh.	Cherry, black

Scientific Name	Common Name
Pueraria thunbergiana (Willd.) Ohwi.	Kudzu 23
Quercus acutissima Carruthers	Oak, sawtooth
Quercus alba L.	Oak, white
Quercus prinus L.	Oak, chestnut
Quercus rubra L.	Oak, northern red
Rhus glabra L.	Sumac, smooth
Robinia fertilis Ashe.	Locust, Arnot Bristly
Salix interior Rowlee.	Willow, sandbar
Salix nigra Marsh.	Willow, black
Sassafrass albidum (Nutt.) Nees	Sassafrass
Secale cereale	Rye
Sesbania sp. Scop.	Sesbania
Sorghum sudanensis (Piper) Stapf.	Sudangrass
Tagetes minuta	Marigold, tall
Trifolium repens L.	Clover, white
Trifolium spp.	Clover, ladino

REFERENCES

1. Mitchell, L., Ceramics: Stone Age to Space Age, McGraw-Hill, New York, 1963.

2. Ries, H., Clays; their Occurrence, Properties and Uses, Wiley, New York, 1927.

3. Falvey, A.E., Introduction to kaolin mining, Ga. Min. Soc. Newsl., 6, 82, 1953.

4. Griffiths, T., A journal of the voyage to South Carolina in the year 1767. Ga. Min. Soc. Newsl., 3, 130, 1950.

5. Goff, J.H., Thomas Griffiths's A Journal of the Voyage to South Carolina, 1767 to obtain Cherokee Clay for Josiah Wedgewood, with annotations. Ga. Min. Soc. Newsl., 12, 113, 1959.

6. Branner, J.C., Bibliography of clays and the Ceramic Arts, Bull. U.S. Geol. Surv. No. 143, Washington, 1896.

7. Wells, J., Clays, in Minerals Year Book, 1969, U.S. Dept. Interior, Washington, 1969.

8. Gustavson, S.A., Clays, in Minerals Year Book, 1970, U.S. Dept. Interior, Washington, 1970.

9. Grim, R.A., Clay Minerology, McGraw-Hill, New York, 1968.

10. Holmes, J.A., Notes on the Kaolin and Clay deposits of North Carolina, Trans. Am. Inst. Min. Eng., 25, 929, 1896.

11. Ladd, G.E., A preliminary report on a part of the Clays of Georgia, Geol. Surv. Ga. Bull. 6-A, 1898.

12. Veatch, O., Second report on the clay deposits of Georgia. Geol. Surv. Ga. Bull., No. 18, 1909.

13. Bayley, W.S., Kaolin in North Carolina, with a brief note on hydromica, Econ. Geol., 15, 236, 1920.

14. Neumann, F.R., Origin of the Cretaceous white clays of South Carolina, Econ. Geol., 22, 374, 1927.

15. Smith, R.W., Sedimentary kaolins of the coastal plain of Georgia, Geol. Surv. Ga. Bull., No. 44, 1929.

16. Kesler, T.L., Occurrence and exploration of Georgia's kaolin deposits, Min. Engng, 3, 879, 1951.

17. Kesler, T.L., Environment and origin of the cretaceous kaolin deposits of Georgia and South Carolina, Econ. Geol., 51, 1956.

18. Kesler, T.L., Environment and origin of the cretaceous kaolin deposits of Georgia and South Carolina, Ga. Min. Newsl., 10 1957.

19. Kesler, T.L., Environment and origin of the cretaceous kaolin deposits of Georgia and South Carolina, Ga. Min. Newsl., 16, 1963.

20. Bates, T.F., Geology amd minerology of the sedimentary of kaolins of the Southeastern United States - A review, in Clays and Clay Minerals, Proc. 12th Nat. Conf., Pergamon, New York, 1964.

21. Smith, J.M. & Murray, H.H., Kaolins of the Southeastern U.S., Proc. Symp. Am. Inst. mech. Engr., Birmingham, Alabama, 1972.

22. La Moreaux, P.E., Geology and ground-water resources of ·the Coastal Plain of East-Central Georgia, Geol. Surv. Bull., 52, Ga. Dept. Mines, Mining and Geology, 1946.

23. Le Grand, H.E. & Furcron, A.S., Geology and ground-water resources of Central-East Georgia, Geol. Surv. Bull., 64, Ga. Dept. Mines, Mining and Geology, 1956.

24. Le Grand, H.E., Geology and ground-water resources of the Macon area, Georgia, Geol. Surv. Bull., 72, Ga. Dept. Mines, Mining and Geology, 1962.

25. Scott, J., Some Clay's good, Ga. Min. Newsl., 7, 93, 1954.

26. Smith, J.M., Spoil placement and use to meet reclamation requirements, in Rehabilitation of Drastically Disturbed Surface Mined Lands, Ga. Surface Mined Land Use Board, Macon, 1971.

27. Murray, H.M., Mining and processing industrial kaolins, Ga. Min. Newsl., 15, 12, 1963.

28. Johnson, H.H., May, J.T. & Perkins, H.F., Some characteristics of overburden on kaolin clays in central Georgia, Ann. Meeting Am. Soil Sci. Soc., New Orleans, 1968.

29. May, J.T., Johnson, H.H., Perkins, H.F. & McCreery, R.F., Some characteristics of spoil material from kaolin clay strip mining, in Ecology and Reclamation of Devastated Land, Gordon & Breach, New York, 1973.

30. Haynsworth, H.J., Experience of Soil Conservation Service in revegetating drastically disturbed sites, in Rehabilitation of Drastically Disturbed Surface Mined Lands, Ga. Surface Mined Land Use Board, Macon, 1971.

31. Parks, C.L., Perkins, H.F. & May, J.T., A greenhouse study of P and K requirements for Ladino clover establishment on kaolin strip mine spoil, Ga. agric. Res., 9, 8, 1967.

32. May, J.T., Parks, C.L. & Perkins, H.F., Establishment of grasses and tree vegetation on spoil from kaolin clay strip-mining, in Ecology and Reclamation of Devastated Land, Gordon & Breach, New York, 1973.

33. Gronow, C.W., Reclamation problems and needs of the surface mining industry, in Rehabilitation of Drastically Disturbed Surface Mined Lands, Ga. Surface Mined Land Use Board, Macon, 1971.

34. Troth, M., Soil Formation on Kaolin Spoil, Unpublished M.S. thesis, Univ. of Georgia, Athens, 1972.

35. Khemnark, C., Establishment of Trees on Kaolin Mining Spoil Material, M.S. thesis, Univ. of Georgia, Athens, 1970.

36. Thomas, J.P., Survival and Growth of Selected Tree Species on Kaolin Mining Spoil, Unpublished report, Univ. of Georgia, Athens, 1971.

37. Troth, J.L., Evaluation of Movement and Effectiveness of Lime and Fertilizer Materials on Kaolin Spoil Sites, M.S. thesis, Univ. of Georgia, Athens, 1971.

38. Dipert, D.D., Establishment of Trees in Herbaceous Vegetation on Kaolin Clay Mining Spoil, M.S. thesis, Univ. of Georgia, Athens, 1972.

THE NITROGEN PROBLEM IN DERELICT LAND RECLAMATION WITH SPECIAL
REFERENCE TO THE BRITISH CHINA CLAY INDUSTRY

J.F. Handley, W.S. Dancer, J.C. Sheldon and A.D. Bradshaw

Department of Botany, University of Liverpool,
United Kingdom

ABSTRACT

Nitrogen is extremely deficient in china clay waste. The
relative merits of two alternative strategies of supplying
nitrogen are discussed: as a grass/legume sward or by frequently
adding nitrogen fertilizer to a predominently grass sward.
Immobolisation of nitrogen fertilizer may conserve it against
leaching but rates of mineralization are subsequently slow
compared to 'legume-N'.

INTRODUCTION

The wide range of waste materials encountered in derelict land
reclamation present a formidable array of problems for plant
growth. However one feature is common to most - a low organic
matter content and a very limited ability to supply nitrogen for
plant growth. The development of a functional nitrogen economy
within the ecosystem must be regarded as a central objective in
derelict land reclamation. Where reclamation involves
afforestation the problem may be masked as a large soil volume
is exploited and rapid growth rates are not usually anticipated.
By contrast the establishment of a vigorous grass sward cannot
be achieved without a sustained nitrogen input as fertilizer or
by nitrogen fixing nodule bacteria. Even in cases where the
final objective is a semi-natural vegetation with high species
diversity it may still be necessary to enrich the soil system
and establish a labile pool of organic matter. In this paper
we shall describe the development of a reclamation technology
for a coarse sand waste produced in large quantities by the China

Clay industry of Cornwall and Devon. On this material, in the absence of soil toxicity and other complicating factors, the nitrogen problem can be isolated and examined in some detail.

THE BASIC SITUATION AND APPROACH

The China Clay industry in Britain is concentrated on three granite bosses in the South West of England - Bodmin Moor, Dartmoor and the St. Austell region. In the mining areas the granite has undergone a late magmatic transformation in which plagioclase feldspar was hydrolysed to give kaolin, secondary mica and quartz (Exley[1]). Cone shaped deposits of clay bearing rock are worked in quarries where a mixture of sand, mica and kaolin is washed from the quarry face by a powerful jet of water. The sand settles out rapidly and is carried away on conveyors to form large heaps; clay and mica are pumped out of the pits in suspension and after flocculation of the clay the mica is carried to large settlement lagoons walled in by sand material.

The two residuals, sand and mica, are finding increasing use in the construction industry, but these outlets are much too small to cope with the current rate of production, around 15,000,000 tons of sand per year, so that the waste materials continue to accumulate in vast heaps and lagoons. It is essential therefore that techniques are developed to enable clay winning operations to be harmonised as far as possible with the moorland landscapes of Bodmin Moor and Dartmoor. In the St. Austell region mining has created its own spectacular landscape and here the primary objectives must be rapid stabilisation of the sand dumps to prevent sand blow and gulley erosion, together with a general softening of the landscape at the margins of the working area. In the longer term it is hoped that a functional landscape can be produced for agriculture and recreation. Experiments in affor-estation have been started but in this paper we shall concentrate on the development of a low ground cover.

In the past the sand was dumped in conical heaps but today large stepped pyramids are produced. The lifts should eventually be less than 100 feet with a gradient of 30 degrees but at present the slopes are frequently much longer and steeper than this. Conventional methods of seed and fertilizer application are ruled out on these unstable and inaccessible sand slopes. Hydro-seeding, a technique in which sand and fertilizer in suspension with water is sprayed onto the slopes, is appropriate in this situation. On the moorland sites the grass swards will be maintained by sheep grazing but this may not be possible in the St. Austell region where a natural progression from a grass/clover sward to a dwarf shrub heath will be encouraged. The micaceous waste has proved very satisfactory as a growing medium and

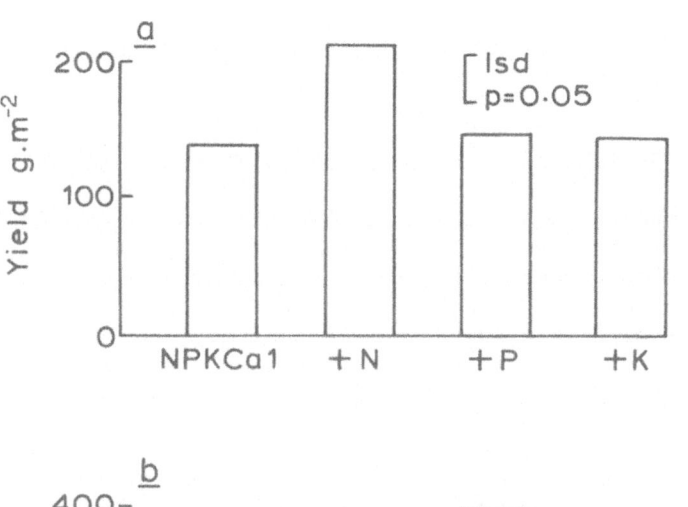

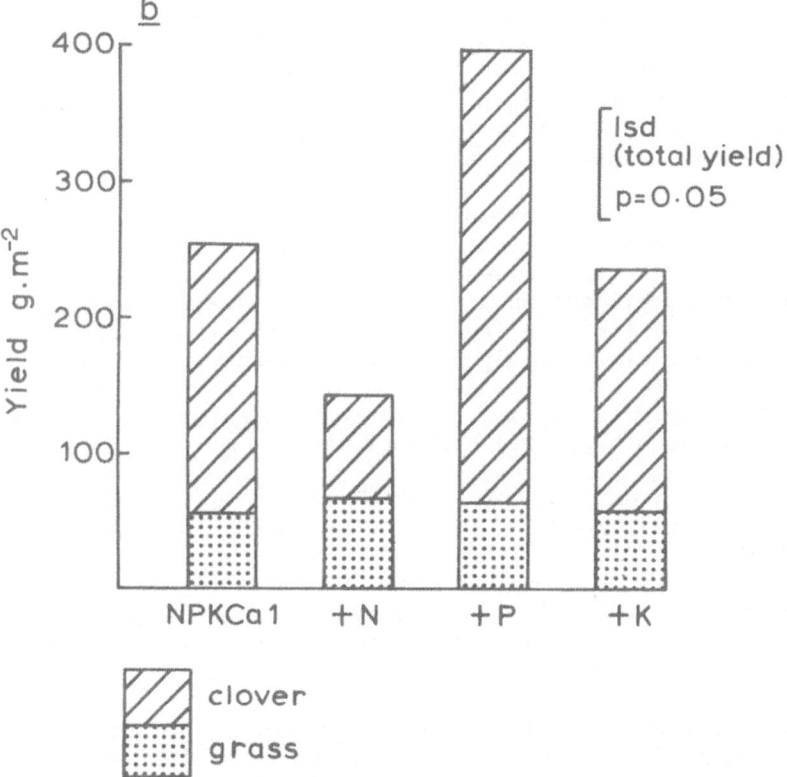

clover

grass

Figure 1.   Regrowth of a moribund grass sward on sand following
            application of fertilizers.   Application rate in kg.ha^{-1},
            50, N; 22, P; 42, K; 25, Mg; 1000 Lime; micronutrients
            in glass frit, 100.

various horticultural and agricultural uses are now being explored - this aspect will not be considered here.

PROPERTIES OF THE SAND MATERIAL IN RELATION TO PLANT GROWTH

The sand materials are non-toxic but the starting capital of all the major plant nutrients is extremely low. Nitrogen in particular is almost absent; the total N content of raw sand waste is in the range 10 - 20 ppm. The sand is extremely coarse (up to 40% particles in >2 mm diameter) so that in dry conditions water deficits develop rapidly whilst under wet conditions mineral nutrients especially nitrogen are leached away in percolating rainfall. Mean annual rainfall is 150 cm at St. Austell and well distributed through the year so that a leaching regime predominates even in the summer (mean excess rainfall over potential transpiration 16 cm from April to September). The pH of the raw sand waste in water is 5.0-5.5 but the lime requirement is unusually low, some 100 kg.ha^{-1} ground limestone are needed to raise soil pH to 6.5. This low lime requirement reflects the low cation exchange capacity of the materials, less than 2 meq. 100g^{-1}.

In preliminary hydroseeding trials grass swards were established without difficulty on the sand slopes but rapid deterioration occurred unless fertilizer was applied regularly. A fertilizer experiment on a moribund grass sward demonstrates clearly that the limiting factor was availability of soil nitrogen, Fig. 1. Before discussing practical approaches to the nitrogen problem the process of nitrogen accumulation in natural succession will be briefly reviewed and the pattern of natural colonisation on the sand waste heaps will be outlined.

THE PROCESSES LEADING TO NITROGEN ACCUMULATION IN PRIMARY SUCCESSION

Nitrogen accumulation in primary succession has been reviewed by Stevenson[2] and the following processes identified as potentially important:-

1. Input of combined nitrogen compounds from the atmosphere in dust, aerosols and precipitation.

2. Sorption of ammonia from the atmosphere.

3. Fixation of atmospheric nitrogen by free living microorganisms.

4. Fixation of atmospheric nitrogen by symbiotic microorganisms.

Eriksson[3] summarised data on combined nitrogen (NH_4^+ and NO_3^-) in precipitation for North America and N.W.Europe and gave the range as 0.74-21 kgN.ha^{-1}. annum^{-1} with most values between 4 and 7 kg. ha^{-1}.annum^{-1}. Allen et al[4] measured input in precipitation at five localities in Britain where total N input ranged from 8.7-19 kg.ha^{-1}.annum^{-1} of which 28-50% was organic. Nitrogen inputs much in excess of 10 kg.ha^{-1}.annum^{-1} are considered to be influenced by atmospheric pollution. The contribution of combined nitrogen compounds from the atmosphere may be further increased by dry deposition of dust and aerosol particles especially in woodland canopies (White & Turner[5]). Nitrogen supply in precipitation has been considered important in the nitrogen balance of woodland ecosystems (Miller[6]; Ovington[7]) and Williams[8] has suggested that nitrogen in precipitation may help to maintain grass productivity on reclaimed colliery shale in the heavily industrialised Yorkshire coalfield.

Gaseous ammonia may be absorbed from the atmosphere by foliage (Wehrmann[9]) or soil (Malo & Purvis[10]). Eriksson[11] constructed a map of ammonia concentrations over Western Europe for 1958 from which the dry deposition of ammonia can be predicted. In the South West peninsula atmospheric pollution is slight (H.M.S.O.,[12]) and a total input of combined nitrogen from this and other sources in excess of 10kgN.ha^{-1}.annum^{-1} would seem unlikely.

Free living nitrogen fixing microorganisms include blue green algae and a number of bacteria particularly the aerobic saprophyte Azotobacter and the anaerobic saprophyte Clostridium. The activity of blue green algae in rice paddies is well known e.g. De & Mandal[13] who reported rates of N fixation between 13 and 70 kg.ha^{-1}.annum^{-1}. But these organisms are also present in natural ecosystems and Stewart[14] has shown that N15 fixed by blue green algae in a dune slack environment was later transferred to bryophytes and vascular plants in close proximity. An algal mat is frequently present on naturally colonised sand waste heaps and may contain appreciable quantities of nitrogen (300 kg.ha^{-1}) but the presence of blue green algae has not so far been demonstrated.

The energy requirements of N fixing saprophytic bacteria are considerable. Alexander[15] has calculated that 1,000 kg of organic matter would be required by Azotobacter to fix 5-10 kgN.ha^{-1}.annum^{-1} under field conditions. The potential contribution of these organisms has therefore been treated with caution. However, recent progress with the acetylene reduction technique has demonstrated a capacity for rapid nitrogen fixation in the rhizosphere of various tropical grasses (Dobreiner, Day & Dart[16]) and temperate dicotyledonous plants (Harris & Dart[17]). The ecological implications of the work of Harris & Dart[17] are

considerable for their plants (<u>Stachys sylvatica</u>, <u>Mercurialis</u>
<u>perennis</u>, <u>Heracleum sphondylium</u> etc) were taken from the
Broadbalk Wilderness at Rothamsted Experimental Station. Arable
cultivation was abandoned on this site in 1881 and the accumulation
of organic matter and associated nitrogen has been well documented.
Jenkinson[18] has presented a thorough analysis of the situation.
In one section where woodland had developed the net annual
accumulation of nitrogen was 65 $kg.ha^{-1}.annum^{-1}$; 12 $kg.ha^{-1}.annum^{-1}$
in the standing vegetation and 53 $kg.ha^{-1}.annum^{-1}$ in the soil.
In an adjacent area where trees and shrubs have been stubbed
regularly since 1900 the net gain of N was 55 $kg.ha^{-1}.annum^{-1}$.
Legumes have been more or less absent since the turn of the
century and a combined estimate for nitrogen input (organic N,
NH_3, NH_4^+, NO_3^-) from the atmosphere gave a value of 20 $kg.ha^{-1}.$
$annum^{-1}$. Harris & Dart[17] measured nitrogenase activity in soil
cores and litter samples and calculated a contribution by free
living nitrogen fixers of 4-5 $kg.ha^{-1}.annum^{-1}$. These various
sources of nitrogen input account for less than half the observed
rate of N accumulation and nitrogen fixation in the rhizosphere
of the woodland herbs could be involved in making up this
discrepancy.

It is of interest that Hassouna & Wareing[19] reported a
net gain in the soil plant system of 35 $kg.ha^{-1}.annum^{-1}$ when
marram grass (<u>Ammophila arenaria</u>) was grown on dune sand. Large
numbers of nitrogen fixing bacteria were found on and adjacent to
the root surfaces but relatively small numbers in the bulk soil.
Marram is a pioneer plant in sand dune succession on a substrate
which is severely deficient in nitrogen.

The importance of nitrogen accumulation by symbiotic N
fixing bacteria is well established on virgin soils and derelict
land substrates (Stevenson[2]). Legume species with their
associated <u>Rhizobia</u> often play the major role (e.g. <u>Trifolium</u>
<u>repens</u> and <u>Melilotus</u> <u>alba</u> on strip mine spoil banks; Leisman[20])
but non-legumes especially alder may be important (e.g. primary
succession on glacial moraines in Alaska; Crocker & Major[21]).
Crocker & Major[21] reported a mean rate of nitrogen accumulation
in litter and soil under alder of 49 $kg.ha^{-1}.annum^{-1}$. The rate
of nitrogen accumulation under legumes is variable but values in
excess of 100 $kg.ha^{-1}.annum^{-1}$ are not uncommon in agricultural
situations (Stewart[14]).

NATURAL COLONISATION ON THE SAND WASTE HEAPS

Left to itself China Clay waste is gradually colonised by
plant material. Examination of aerial photographs spanning a
period of thirty years has shown that the rate of succession
varies considerably and may be a function of the seed available.

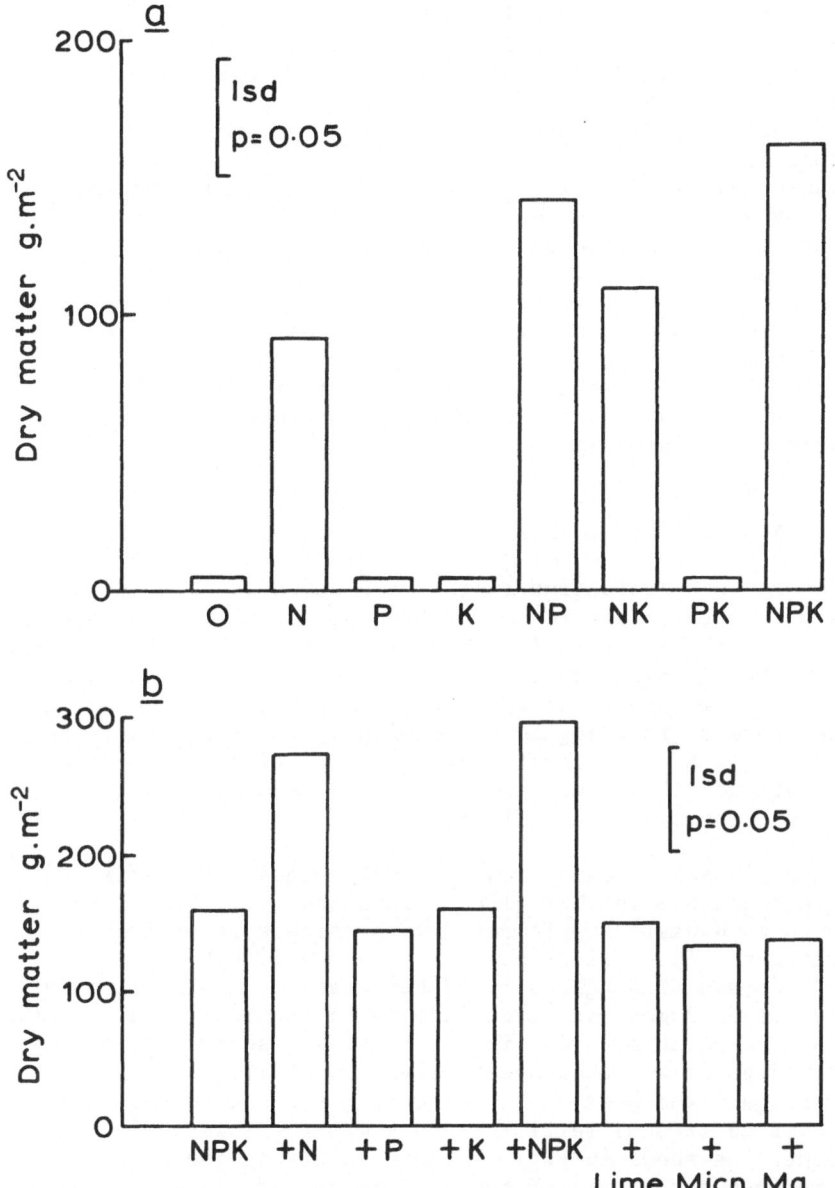

Figure 2. Mean yields of different fertilizer treatments applied to grass/clover swards established on a range of china clay wastes in a preliminary pot experiment. Application rate in kg.ha^{-1},50, N; 22, P; 42, K;. All treatment combination supplemented by ground limestone at 1000 kg.ha^{-1}. a) First season (grass + clover); b) second season.

In some cases little change is visible over the entire thirty year period (e.g. stands of Calluna vulgaris and Erica cinerea with a surface mat of algae and bryophytes) whilst in other situations large heaps may be covered within a decade by shrubby legumes including Ulex europaeus, Sarothamnus scoparius and particularly Lupinus arboreus. The larger legumes are frequently accompanied by willow Salix cinerea ssp. atrocinerea and Salix aurita. Eventually the heaps may become blanketed by Rhododendron ponticum or pass on through to climax oak woodland. However, burning usually occurs in the shrub stage and a drastic redistribution of nitrogen in the soil/plant system must follow this. It appears that after a hot burn from which the shrubs are unable to regenerate a relatively permanent herbaceous vegetation is established. The process of nitrogen accumulation and the seral relationships of these plant communities are currently being investigated.

NITROGEN ACCUMULATION IN PRACTICE - ALTERNATIVE STRATEGIES

The rapid colonisation of the sand dumps by shrubby legumes suggests that these species might have potential in reclamation work. However the vegetation which is produced has little amenity or agricultural value and forms a serious fire hazard in the spring. The two practical approaches to achieving a ground cover are therefore a grass/legume sward with agricultural varieties of clover etc. or a predominantly grass sward maintained by nitrogen fertilization. These two approaches have been evaluated in glasshouse experiments and field trials.

In a glasshouse experiment a simple grass clover mixture (Festuca rubra S59 and Trifolium repens Kent Wild White) was sown onto sand waste supplemented with various fertilizer combinations. The basic treatment received 50 kgN.ha^{-1}, 22 kgP.ha^{-1} and 41.5 kgK.ha^{-1} and this rate of application was doubled for each macronutrient in turn in the additional treatments. The yield of plant material in the first and second seasons is shown in Figs 2a and 2b respectively. Initially the high N treatment gave the best yield but during the second season in absence of additional fertilizer this was the lowest yielding treatment. As shown in Fig. 2b regrowth during the second year was very much a function of the performance of clover. In the high N treatment clover establishment was suppressed but in the high P regime establishment and growth of clover was encouraged.

Using a more complex grass/clover mixture a similar experiment was carried out under field conditions. The experiment involved a factorial design with four lime levels and five fertilizer treatments:-

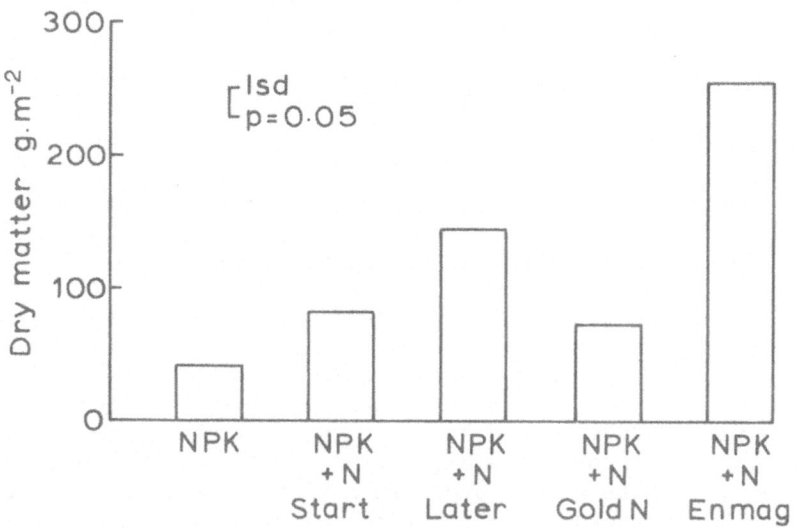

Figure 3.   Field trial on sand waste to investigate the effect of
various fertilizer treatments on the establishment of
a grass/clover sward.   Data presented as mean herbage
yield 16 months after application of seed and fertilizer.

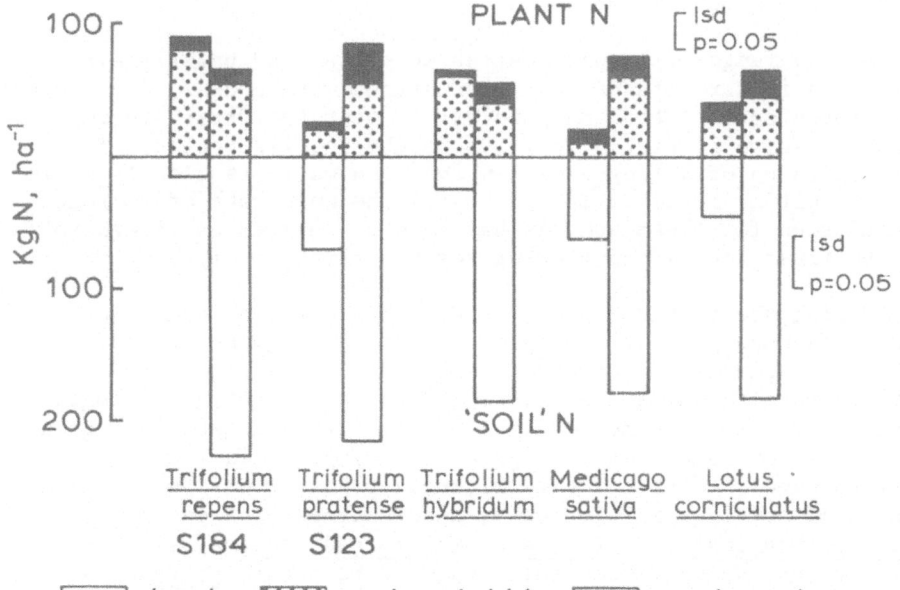

Figure 4.   Nitrogen accumulation in plant material and sand waste
within pure stands of various legumes.   The field
plots were sampled 7 months (H1) and 17 months (H2)
after application of seed and fertilizer.

1. Basic NPK - 50 kgN.ha^{-1}; 22 kgP.ha^{-1}; 41.5 kgK.ha^{-1}.

2. Basic NPK + 50 kgN.ha^{-1} initially as NH_4NO_3.

3. Basic NPK + 50 kgN.ha^{-1} after 3 months as NH_4NO_3.

4. Basic NPK + 50 kgN.ha^{-1} as sulphur coated urea (Gold N)

5. Basic NPK + 22 kgP.ha^{-1} as NH_4MgPO_4 (Enmag)

The experiment was set up in April 1972 and harvested in August 1973. As shown in Fig. 3 yield was doubled by increasing the initial nitrogen input and doubled again by applying the extra nitrogen 3 months after establishment. However the legume component was heavily suppressed in these treatments and prospects for regrowth were poor. The highest yield was obtained with additional phosphate which supported a vigorous growth of clover during the second season. Sulphur coated urea, a controlled release fertilizer formulated to give nitrogen release over a 180 day period, had no special advantage over ammonium nitrate at this rate of application.

MAXIMISING THE LEGUME COMPONENT

A wide range of legume species were assessed under field conditions in pure stands and mixed stands with grass. The process of nitrogen accumulation in pure stands over the first growing season is shown in Fig. 4 for five of the species tested. The rapid build up of nitrogen within the sand waste is clearly shown, from an initial level of 18 kgN.ha^{-1}. The mean rate of nitrogen accumulation for these species has been calculated as 76 kgN.ha^{-1} for the first and second growing seasons respectively.

During the second season the dry matter yield from mixed stands was much greater than yields from the legumes alone (Table 1). The grass component was particularly vigorous in combination with white clover, suggesting an effective nitrogen transfer from the legume. Dilz & Mulder[22] investigated nitrogen transfer from legume roots to perennial ryegrass in a carefully designed experimental system. Nitrogen transfer was poor in the absence of cutting (N released as a percentage of N in tops Trifolium pratense 1%, T.repens 3%, Medicago sativa 8%) but improved considerably when the legumes were cut (T.pratense 6%, T.repens 32%, M.sativa 16%). The precise mechanism of nitrogen transfer in the absence of grazing is not well understood. Dilz & Mulder[22] reported greatest N transfer when the legume component was killed out and suggested that under field conditions root death over the winter period would be important. Similar effects may follow short periods of summer drought in Cornwall when severe

Species	Pure	Mixed	P÷Mx100
Trifolium repens S184	53	219	24
Trifolium pratense S123	109	305	38
Trifolium hybridum	78	206	41
Lotus corniculatus	124	251	50
Medicago sativa	88	246	36

Table 1. Herbage yield (g dry matter m^{-2}) of legumes in pure and mixed stands on sand waste during second season.

	NH_4^+	K^+	Na^+	Mean
NO_3^-	65	100	80	82
Ce^-	4	90	90	61
$H_2PO_4^-$	0	75	30	35
Mean	23	88	67	59

(Seedling establishment was 75% over 20 mM mannitol)

Table 2. Number of white clover (Trifolium repens S100) seedlings established in salt treatments at 10 mM as a percentage of seedlings established over deionised water.

water deficits will develop in the sand material.

Chestnutt & Lowe[23] emphasised the importance of a soil pH above 5.5 with adequate levels of soil calcium and phosphate for successful growth of clover. In order to select optimum conditions for clover growth a mixed sward containing grasses and three clovers (Trifolium pratense, T.repens and T.hybridum) was sown at four lime levels with five phosphate sources. Productivity was strongly influenced by lime/phosphate interaction as shown in Fig.5. Superphosphate, an acidic fertilizer, gave poor results without lime and at the lower lime rate of 1000 kg.ha^{-1}. Seedling establishment may have been inhibited by phosphate toxicity under these conditions (Table 2.). Basic slag, by contrast, was satisfactory in the absence of lime but phosphate availability declined as the lime rate increased. This effect was very marked with rock phosphate and bone meal which could provide a long term supply of phosphate, as dissolution rate increases due to natural acidification. Magnesium ammonium phosphate (Enmag) an expensive material with slow release properties gave satisfactory results at all lime levels, with an optimum at 2,500 kg.ha^{-1}.

The grass/clover ratio in a mixed sward increases with the level of fertilizer nitrogen applied. Clover does not persist at fertilizer rates in excess of 150 kgN.ha^{-1}.annum^{-1} in British grasslands (Chestnutt & Lowe[23]) and is excluded at lower rates on china clay sand (<100 kgN.ha^{-1}). Suppression of clover is usually attributed to increased competition from the grass component rather than direct effects of inorganic N on the legume. Stern & Donald[24] demonstrated competition for light in grass/ clover swards but competition for water and soil nutrients is likely to be more important in the rather open swards on china clay waste.

ESTABLISHMENT PROBLEMS WITH CLOVER

In the hydroseeding operation seed and fertilizer are placed in close proximity on the sand surface and this has led to serious difficulties with seedling establishment. A modest fertilizer treatment (NPK at 60, 26 and 49 kg.ha^{-1} respectively) produced a 60% inhibition of white clover establishment when applied to a sand surface kept moist under greenhouse conditions. Experiments in solution culture (Table 2) demonstrate the susceptibility of clover seedlings to toxicity of NH_4^+ and $H_2PO_4^-$ during the establishment phase. The dominant influence in practice will be the ammonium ion which may exceed 10mM in the soil solution following modest applications of a compound fertilizer to sand waste. Under field conditions the toxicity effect is intensified by evaporation of water and crust formation at the sand surface. Clover mortality may be reduced by the use

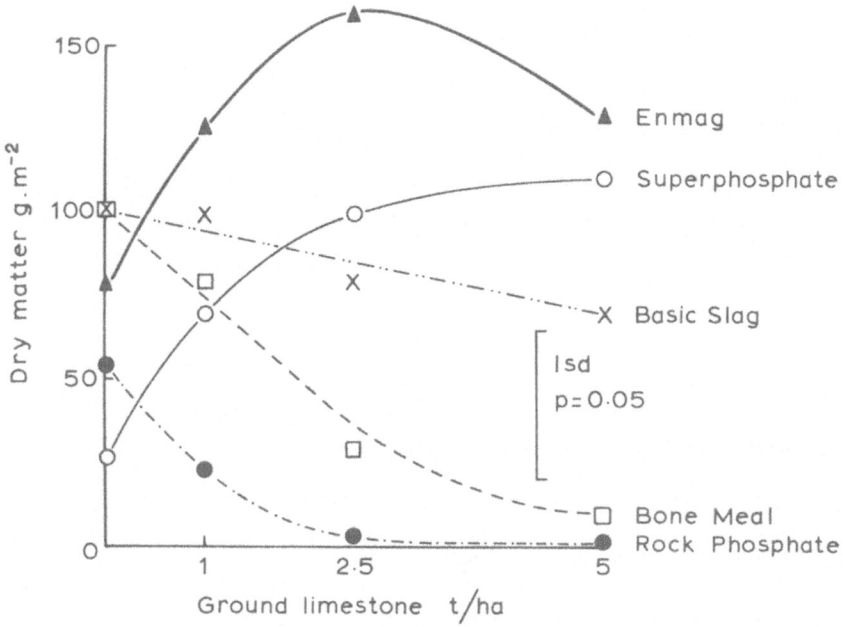

Figure 5. Mean herbage yield of a grass/clover mixture 16 months
after establishment on sand waste with five phosphate
sources at four lime levels.

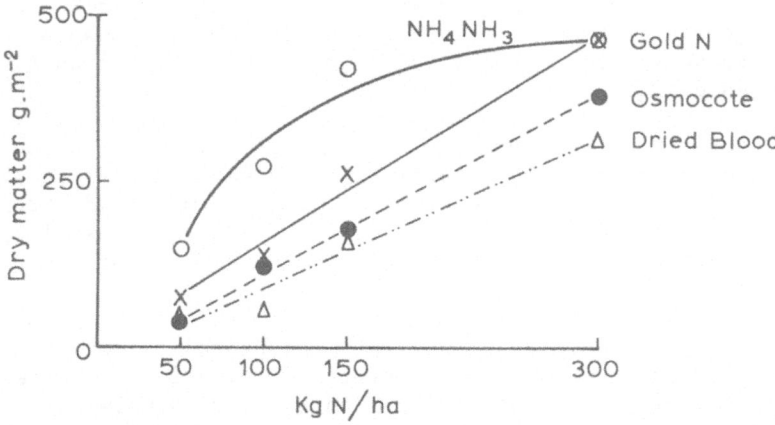

Figure 6. Response of a grass sward on sand waste to a maintenance application of four nitrogen fertilizers over the
range 50 – 300 kg N. ha^{-1}. Data presented as mean
herbage yield during 5 months following fertilizer
application in May.

of a protective peat mulch together with an appropriate slow
release nitrogen fertilizer.

One further complication is the inoculation of legume seeds
with the appropriate Rhizobium. Seeds were inoculated from slope
cultures in the field trials but peat cultures which should be
suitable for larger scale work (Vincent[25]) are difficult to
obtain commercially in Britain. Pelleting seed with lime and
Rhizobia has proved successful on acidic sands in Australia
(Loneragan et al.[26]) and may prove a useful future approach.

MAINTENANCE OF GRASS SWARDS BY FERTILIZER NITROGEN

Fertilizer nitrogen may be used to bring about a rapid build
up of nitrogen in the soil/plant system. The nitrogen may be
supplied in various forms:-

1. Conventional fertilizers.

2. Synthetic slow release materials.

3. Natural slow release materials.

4. Bulky organic manures.

The response of a grass mixture (Lolium perenne, Agrostis tenuis,
Festuca rubra) to fertilizer nitrogen supplied at 100 kg.ha^{-1}
from various sources is shown in Fig. 6. Dry matter yield over
the first growing season may be compared with the response of
one year old swards to a repeat addition of 100 kgN.ha^{-1} on a
rerandomised design.

The contrast between ammonium and nitrate nitrogen in the
first season probably reflects differential leaching. Bates &
Tisdale[27] demonstrated that leaching of nitrate increased with
the percentage of coarse sand and large pore space in a soil.
Dancer[28] investigated the leaching of nitrate and ammonium ions
under field conditions on sand waste. The mean depth of inorganic
N was found to be highly correlated with rainfall which accounted
for 89 and 92 per cent of variance for mean depth of inorganic
N for ammonium sulphate and sodium nitrate respectively. Linear
regression demonstrated that one centimetre of rain moves the
front of NH_4^+ downward 2.5 cm compared with 4.1 cm for NO_3^-.
This leaching rate for nitrate is comparable with published data
but the ammonium leaching rate is unusually rapid. However the
lag of NH_4^+ behind the wetting front indicates that the ammonium
ion is resistant to leaching, despite the low cation exchange
capacity of the sand waste.

Nitrogen loss by leaching can be reduced by the use of slow release fertilizer (Lunt[29]). The fertilizers available include synthetic materials (coated fertilizers, metal ammonium phosphates, urea formaldehydes etc.) and naturally occuring materials (hoof and horn, dried blood, fish manure etc.). At low rates of application these materials were less successful than ammonium sulphate (Table 4). Analysis of the nitrogen content of the plant material has shown that dry matter production is broadly correlated with per cent nitrogen recovery. Holmes[30] published curves for nitrogen recovery which also show a high N recovery from ammonium sulphate compared to slow release formulations. Nevertheless, slow release nitrogen fertilizer, for example sulphur coated urea with a heavy coating weight, may reduce toxicity effects during the establishment phase. Osmocote a plastic coated material with a controlled dissolution rate, as shown in Fig. 6, is very effective in this respect.

Unpublished data from this Department indicates that the development of a satisfactory rye-grass sward (L.A.I. 2, bare ground <10%) requires a dry matter yield of 2,000 $kg.ha^{-1}$ during the first growing season. Longer term productivities are less easily defined but Hanley et al[31] reported a mean dry matter yield of 3,500 $kg.ha^{-1}.annum^{-1}$ for a range of unfertilised British grasslands, so that 2,000 $kg.ha^{-1}.annum^{-1}$ may also be appropriate in the long term. A good quality rye-grass pasture with heavy nitrogen fertilization should yield 10,000 $kg.ha^{-1}.annum^{-1}$ dry matter in British conditions (Lazenby & Rogers[32]). Nitrogen recovery by plant material is low on derelict land substrates and applications in excess of 100 $kgN.ha^{-1}.annum^{-1}$ may be required to give an acceptable yield. Mean dry matter yield from maintenance applications up to 300 $kgN.ha^{-1}$ is shown in Fig. 6. Response to fertilizer N was linear to 300 $kg.ha^{-1}$ for the slow release materials, in contrast to the asymptotic curve for ammonium nitrate. Davies[33] has shown that yield and N recovery by established rye grass from one application of Gold N, a sulphur coated urea at 375 $kgN.ha^{-1}$, compared well with 5 applications of NH_4NO_3 at 75 $kgN.ha^{-1}$ on good pasture.

Berryman[34] presented data on the mean nitrogen content of a range of bulky organic manures. The nitrogen content is generally less than 2 per cent and large quantities of material must be applied (at least 10,000 $kg.ha^{-1}$). This is feasible in situations where large supplies of a material are available locally and gradients on site permit spreading and incorporation of the manure. Where this has proved possible bulk organics have been used with great success e.g. chicken manure on colliery shale (Bradshaw, Fitter & Handley[35]) and sewage sludge on toxic metalliferous waste materials (Weston[36], Gemmell[37]). On the sand waste heaps very large quantities of material would be required, application would be very difficult and incorporation would be generally out

Site	Sward Type	Age in years	Total N	Inorganic N	ΔN Sand	ΔN Sand+ plant debris	ΔN Sand+ plants + 45 ppmN as NH_4^+
Maggie Pie	Untreated sand	Control	7	2	0	−	+44
Maggie Pie	Hydroseeded; no clover	2	262	12	+6	−7	−59
Park	Hydroseeded; no clover.	2	244	12	+4	−7	−27
Zee Moor	Sown;sparse clover.	3	177	16	+39	−9	+4
Stannon	Sown; good clover and grazing	4	291	53	+63	+66	+4

Table 3. Net release of inorganic nitrogen ($\Delta NH_4^+ + \Delta NO_3^-$) during 14 day incubation of samples from reclaimed sand waste. Data presented as ppmN.

Treatment	First year		Second year	
Ammonium sulphate	94	a	262	a
Ammonium nitrate	63	b	168	ab
Sodium nitrate	<1		216	a
Sulphur coated urea	21	c	68	bc
Ureaformaldehyde	2	d	13	c
Dried Blood	7	d	33	c
Hoof and Horn	11	d	43	c

(Treatment means sharing the same letter are not significantly different at 5% probability).

Table 4. Response of a grass sward on sand waste to nitrogen fertilizers applied at 100 kgN.ha^{-1}. Data presented as shoot dry matter in g.m^{-2}.

of the question.

NITROGEN TURNOVER

The addition of organic matter to a soils system will
stimulate microbial activity which may lead to a net increase
or a net decrease in inorganic nitrogen – these processes are
known as mineralisation and immobilisation respectively. The
capacity for mineralisation can be predicted to some extent by
the C/N ratio in the organic matter (Tinsley[38]) and it might be
expected that immobilisation would predominate beneath the grass
swards whilst mineralisation would occur where legumes are
present. A study of the behaviour of two year old grass swards
and grass/legume mixtures using the incubation technique of
Keeney & Bremner[39] demonstrates a striking difference in this
respect (Table 3).

The organic matter and total nitrogen content was high in the
soil from the hydroseeding where heavy maintenance dressings of
nitrogen fertilizer had been applied, but immobilisation predominated
in this situation in contrast with the clover rich sward from
Stannon where mineralisation occured. Immobilisation could be
seen as a method of conserving fertilizer nitrogen against
leaching but the rate of turnover of nitrogen immobilisation in
this way is rather low at 1-4 per cent/annum (Bartholomew[40]).
Interpretation of Table 3 is complicated by the effect of grazing
on the legume rich sward and grazing would certainly improve
nitrogen turnover on the hydroseeded areas. Nevertheless nitrogen
release from legumes to the grass component is very apparent in
field trials where grazing animals are excluded (Table 1).

AFTERMANAGEMENT AND THE DEVELOPMENT OF A LOW MAINTENANCE GROUND
COVER

On moorland sites grass/legume swards will be maintained
by grazing and a number of reclaimed areas are already managed in
this way. It is important that these areas should harmonise with
surrounding moorland and rather low rates of lime and phosphate
application are recommended so that soil fertility corresponds
approximately to improved upland pastures.

In the St. Austell area grazing may not be feasible and in
this situation an attractive low maintenance ground cover is
required. The dwarf shrub heath containing Calluna vulgaris,
Erica cinerea and Ulex gallii which develops naturally on certain
heaps provides a suitable landscape alternative to grassland or
woodland. It is intended that the hydroseeding operation should
produce rapid stabilisation of the sand surface followed by a

period of nitrogen accumulation by clovers. Seed of the dwarf
shrubs will be included in the initial treatment in the expect-
ation that they will gradually form a closed sward as soil pH
falls and nitrogen content increases.

ACKNOWLEDGMENTS

We would like to acknowledge the considerable help of
Mr L.D.C. Owen and his staff of English China Clays Ltd. through-
out this work and Mr W. Newnes for his technical assistance.
Mr A. Pattison of Rothamsted Experimental Station kindly provided
all the Rhizobium cultures. The Natural Environment Research
Council have supported some of the work described in this paper.

REFERENCES

1. Exley, C.S., Magmatic differentiation and alteration in the
St. Austell Granite, Quart. J. Geol. Soc., 114, 197, 1959.

2. Stevenson, F.J., Origin and distribution of nitrogen in soil,
In Soil Nitrogen, (ed. W.V. Bartholomew & F.E. Clark),
Amer.Soc. Agrom., Madison, 1965.

3. Eriksson, E., Composition of atmospheric precipitation I
Nitrogen compounds, Tellus, 4, 215, 1952.

4. Allen, S.E., Carlisle, A., White, E.J. & Evans, C.C., The
plant nutrient content of rainwater, J. Ecol., 56, 497,
1968.

5. White, E.J. & Turner, F., A method of estimating income of
nutrients in a catch of airborne particles by a woodland
canopy, J. appl. Ecol., 7, 441, 1970.

6. Miller, R.B., Plant nutrients in hard beech, N.Z. J. Sci.,
6, 365, 1963.

7. Ovington, J.D., Quantitative ecology and the woodland
ecosystem concept, Adv.ecol. Res., 1, 103, 1962.

8. Williams, P.J., Investigations into the nitrogen cycle in
colliery spoil, In The Ecology of Resource Degradation and
Renewal, (ed. M.J. Chadwick & G.T. Goodman), Blackwell,
Oxford, 1975.

9. Wehrmann, J., Nitrogen requirements of forests, Span, 5, 55,
1962.

10. Malo, B.A. & Purvis, E.R., Soil absorption of atmospheric ammonia, Soil Sci., 97, 242, 1964.

11. Eriksson, E., Air and precipitation as sources of nutrients, In Handbuch der Pflanzenernahrung und Dungung, Vol. III, Part I, (ed. H. Linser & K Scharrer), 774, Springer-Verlag, Berlin, 1968.

12. H.M.S.O., National Survey of Air Pollution 1961-1971, Vol. 2, South West, Wales & North West, 1972.

13. De, P.K. & Mandal, L.N., Fixation of nitrogen by algae in rice soils, Soil Sci., 81, 453, 1956.

14. Stewart, W.D.P., Nitrogen Fixation in Plants, Athlone Press, London, 1966.

15. Alexander, M., Introduction to Soil Microbiology, Wiley, New York, 1961.

16. Dobereiner, J., Day, J.M. & Dart, P.J., Nitrogenase activity in the rhizosphere of sugar cane and some other tropical grasses, Pl. Soil, 37, 191, 1972.

17. Harris, D. & Dart, P.J., Nitrogenase activity in the rhizosphere of Stachys sylvatica and some other dicotyledonous plants, Soil Biol. Biochem., 5, 277, 1973.

18. Jenkinson, D.S., The accumulation of organic matter in soil left uncultivated, Rep. Rothamsted Experimental Station 1970, 113, 1971.

19. Hassouna, M.G. & Wareing, P.F., Possible role of rhizosphere bacteria in the nitrogen nutrition of Ammophila aernaria, Nature, Lond., 202, 467, 1964.

20. Leisman, G.A., A vegetation soil chronosequence on the Mesabi iron range spoil banks, Minnesota, Ecol.Mon., 27, 221, 1957.

21. Crocker, R.L. & Major, J., Soil development in relation to vegetation and surface age at Glacier Bay, Alaska, J. Ecol., 43, 427, 1955.

22. Dilz, K. & Mulder, E.G., Effect of associated growth on yield and nitrogen content of legume and grass plants, Pl. Soil, 16, 229, 1962.

23. Chestnutt, D.B.M. & Lowe, J., Agronomy of white clover/grass swards, In White Clover Research, (ed. J. Lowe), Brit. Grass Soc., 191, 1970.

234

24. Stern, W.R. & Donald C.M., Light relationships in grass – clover swards, Aust. J. agric. Res., 13, 599, 1962.

25. Vincent, J.M., A Manual for the Practical Study of Root-nodule Bacteria, I.B.P. Handbook No. 15, Blackwell, Oxford, 1970.

26. Loneragan, J.F., Meyer, D., Fawcett, R.G. & Anderson, A.J., Lime pelleted clover seeds for nodulation on acid soils, J. Aust. Inst. Agric. Sci., 21, 264, 1955.

27. Bates, T.E. & Tisdale, S.L., The movement of nitrate nitrogen through columns of coarse-textured soil materials, Soil Sci. Soc. Amer. Proc., 21, 525, 1957.

28. Dancer, W.S., Leaching losses of ammonium and nitrate in the reclamation of sand spoils in Cornwall, J. Environ. Qual., 4, 499, 1975.

29. Lunt, O.R., Controlled-release fertilizers: achievements and potential, J. Agric. Fd Chem., 19, 797, 1971

30. Holmes, M.R.J., Evaluation of nitrogenous fertilizers, In Nitrogen and Soil Organic Matter, Min. Ag. Fish & Fd Tech. Bull., 15, 129, 1969.

31. Hanley, J.A., Improvement on Pastures, H.M.S.O., London, 1932.

32. Lazenby, A & Rogers, H.H., Selection criteria in grass breeding – effect on Lolium perenne of differences in population density, variety and available moisture, J. agric. Sci., 62, 285, 1964.

33. Davies, L.H., Two grass field trials with a sulphur coated urea to examine its potential as a slow release nitrogen fertilizer in the U.K., J. Sci. Fd Agric., 24, 63, 1973.

34. Berryman, C., Composition of organic manures and waste products used in agriculture, N.A.A.S., Paper No. 2, 1965.

35. Bradshaw, A.D., Fitter, A.H. & Handley, J.F., Why use top soil in land reclamation?, Surveyor, 39, 1973.

36. Weston, R.L., Gadgil, P.D., Salter, B.R. & Goodman, G.T., Problems of revegetation in the Lower Swansea Valley, an area of extensive industrial dereliction, In Ecology and the Industrial Society, (ed. G.T. Goodman, R.W. Edwards & J. Lambert), Blackwell, Oxford, 1965.

37. Gemmell, R.P., Use of waste materials for revegetation of chromate smelter waste, Nature, Lond., 240, 569, 1972

38. Tinsley, J., Nitrogen releasing properties of various types of organic matter, In Nitrogen and Soil Organic Matter, Min. Ag. Fish & Fd Tech. Bull., 15, 30, H.M.S.O., London, 1969.

39. Keeney, D.R & Bremner, J.M., Determination of isotope-ratio of different forms of nitrogen in soils, VI, Mineralizable nitrogen, Soil Sci. Soc. Amer. Proc., 31, 34, 1967.

40. Bartholomew, W.V., Mineralisation and immobilisation of nitrogen in the decomposition of plant and animal residues, In Soil Nitrogen, (ed. W.V. Bartholomew & F.E. Clark), Amer. Soc. Agrom., Madison, 1965.

RIVER POLLUTION BY CHINA CLAY WASTE AND OTHER SOLIDS IN SUSPENSION

J.S. Alabaster

Water Pollution Research Laboratory, Department of the Environment, Stevenage, United Kingdom.

Most natural waters contain solid matter in suspension, sometimes at very high concentrations, particularly when natural processes of erosion are exacerbated through the activities of man, for example by deforestation and the logging operations that attend it, bad agricultural practices, general construction operations involving the disturbance of large volumes of earth, the removal of gravel and sand direct from streams, and perhaps above all mining operations. Such conditions in rivers can destroy or transform freshwater fisheries in a number of ways, including direct effects of the solids on the survival, growth, and resistance to poisoning and to disease of the fish themselves or their eggs and larvae, effects on fish behaviour, particularly on their feeding, natural movements, and migrations, and indirect effects on fish through reduction in the food supply either from direct effects on fish organisms or from reduced productivity of algae caused by light attenuation.

The effects of china clay wastes on trout streams in Cornwall have been fairly fully described by Herbert et al.[1]. These were considered together with data for other types of inert solids in suspension by the European Inland Fisheries Advisory Commission (EIFAC[2]) which put forward tentative water quality criteria for such solids; it concluded that in the absence of other pollution freshwater fisheries were unlikely to be harmed at concentrations less than 25 mg.1^{-1}, there should be good or moderate fisheries at 25-80 mg.1^{-1}, good fisheries were unlikely to be found at 80-400 mg.1^{-1}, while at best only poor fisheries would exist at 400 mg.1^{-1}.

Solids of domestic and industrial origin may also be dis-

238

charged to rivers in sewage, sewage effluent, storm run-off from
urban areas, and in direct discharges of trade and industrial
wastes. These, unlike those of geological origin that are
essentially chemically inert, exert their effects in different
ways; those consisting largely of oxidizable organic material may
be capable of reducing the dissolved oxygen concentration of the
water to levels that are asphyxial to aquatic life, while others,
for example basic salts of zinc (Lloyd[3]) and precipitated ferric
hydroxide (Sykora, Smith & Synak[4]) may have toxic properties. At
the same time organic solids of sewage origin may remove a proportion of heavy metals and other poisons from solution and thus
reduce the effective toxicity of streams containing these wastes.

Nevertheless, where there are severe effects of pollution in
rivers in the United Kingdom (approximately 4 per cent of fishless
river length) the concentration of suspended solids is generally
low (e.g. annual median monohydril value in the River Trent, less
than 40 mg.1^{-1}) and the predicted toxicity relatively high based
upon the commonly occurring poisons ammonia, phenol, cyanide,
copper and zinc (e.g. annual median 48 h LC_{50} to rainbow trout in
the River Trent, greater than 0.3). These topics have recently
been reviewed briefly by Alabaster[5].

REFERENCES

1. Herbert, D.W.M., Alabaster, J.S. Dart, M.C. & Lloyd, R., The
    effect of china-clay wastes on trout streams, Int.J.Air
    Wat. Poll., 5, 56, 1961.

2. E.I.F.A.C. Report on finely divided solids and inland
    fisheries, European Inland Fisheries Advisory Commission,
    Tech. Paper No. 1, F.A.O., Rome, 1964.

3. Lloyd, R., The toxicity of zinc sulphate to rainbow trout,
    Ann.appl.Biol., 48, 84, 1960.

4. Sykora, J.L., Smith, E.J. & Synak, M., Effect of lime
    neutralised iron hydroxide suspensions on juvenile brook
    trout (Salvelinus fontinalis, Mitchell), Wat. Res., 6, 935,
    1972.

5. Alabaster, J.S., Suspended solids and fisheries, Proc. R. Soc.
    Lond. B, 180, 395, 1972.

DISPOSAL OF SOLID WASTE IN THE MARINE ENVIRONMENT WITH
PARTICULAR REFERENCE TO THE CHINA CLAY INDUSTRY

N.A. Holme and P.K. Probert

The Marine Laboratory, Plymouth, U.K.

ABSTRACT

China clay deposits in the South West of England, derived
from the kaolinisation of feldspar contained in the granite
intrusions of Cornwall, have been extensively mined in open pits
since the 18th century. Jets of water break up the friable
granite and the china clay is separated from micaceous residues
which have been discharged to the sea, particularly Mevagissey
Bay, via various rivers.

Although this practice is now being discontinued, an
estimated 17,000,000 tons of fine micaceous waste material has
been discharged into Mevagissey Bay by the White River since
mining began. This gives a dense 'white-water' plume at the sea
surface, often with an abrupt boundary.

The material is non-toxic and although experiments have
shown that high concentrations of suspended solids may cause gill
damage in freshwater fish, no gill damage has been observed in
fish taken from Mevagissey Bay. Deposition of fine material on a
former rocky surface suitable for crab or lobster fishing,
produces a smooth soft surface which may eventually become
suitable for trawl-fishing.

Experiments to determine whether pipline discharge of the
waste to the sea bed would successfully disperse the deposited
material were inconclusive as to whether the method would
completely avoid beach pollution and so the pipeline scheme was
discontinued.

ORIGIN AND PRODUCTION OF CHINA CLAY

One of the more obvious geological features of south-west England is the presence of a number of granite intrusions, such as Dartmoor, Bodmin Moor, Hensbarrow or St. Austell Moor, Carnmenellis, Lands End and the Scilly Isles, which were pushed up during a period of intense mountain forming activity in Europe about 250 million years ago. During its final stage of cooling the granite gave off various gases, including fluorine, sulphur dioxide, superheated steam and carbon dioxide, emissions which effected certain alterations in the granites. One of these processes, which has particularly affected the Hensbarrow intrusion, (Fig. 1), is Kaolinisation. This means that the granite, which is largely composed of the minerals quartz, mica and feldspar, has had the feldspar converted into the clay mineral kaolinite, of which kaolin or china clay is mainly composed.

China clay is mined in open pits by employing jets of high-pressure water which break up the friable kaolinised granite, and wash the clay and coarse-grained material to the bottom of the pit. Here the coarser fraction is separated off and transported by conveyor belt to a tip. The fine components flow or are pumped to a series of settling tanks in which the clay is separated, leaving a fine-grade waste known as micaceous residue. A typical micaceous residue has a median particle diameter of 36 μm, and the greater part is finer than 100 mesh (152 μm). Its mineralogical composition is approximately 29 per cent quartz, 27 per cent muscovite, 25 per cent kaolinite, 15 per cent feldspar and 4 per cent tourmaline (Fig. 2). The exact composition of the waste varies according to which pits are being worked and on the type of clay being extracted in the settling tanks.

WASTE DISPOSAL

China clay workings in Cornwall produce large quantities of waste. For every ton of clay extracted there are 4 to 5 tons of sand, 1 ton of rock, 1 ton of overburden, and 0.9 tons of micaceous residue. The micaceous residues, being non-stackable, have until recently been mainly discharged into large storage lagoons (mica dams) or into two small rivers which flow into Mevagissey and St. Austell Bays respectively. Residues may also be routed to pits which are not being worked. The latter method was not favoured in the past because the resources of each pit had not been fully exploited, the kaolin deposits occurring to a greater depth than the workings, but has now become necessary because of the decision to stop discharging waste into the rivers.

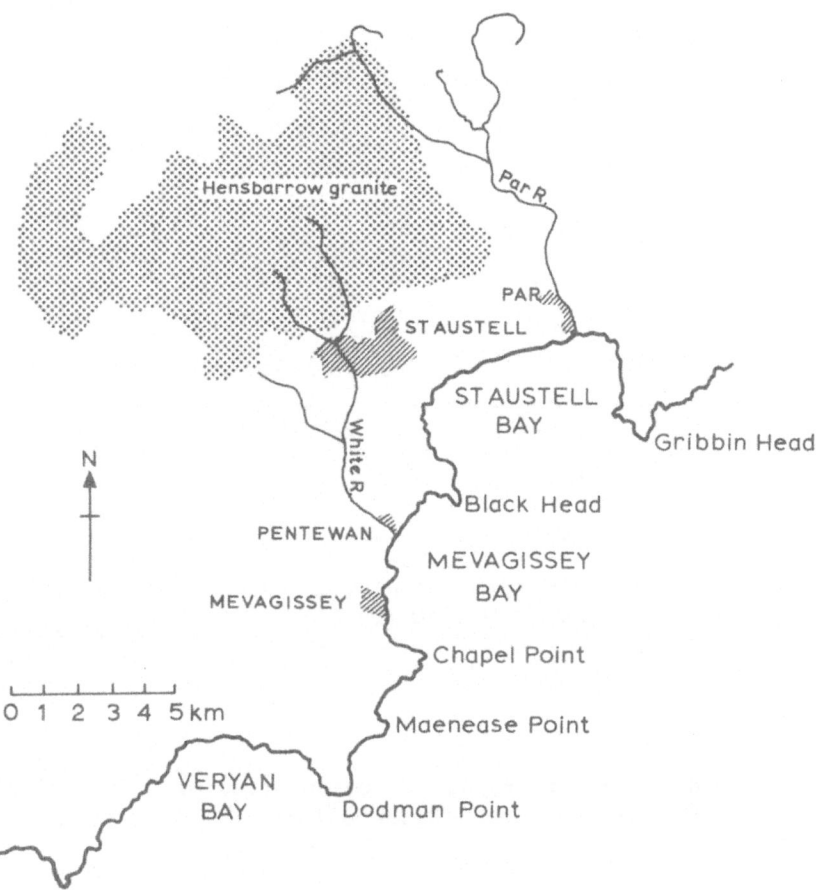

Figure 1.   Map of the St. Austell area, showing the White River
and the Par River which formerly discharged waste from
the china clay workings into the sea.

242

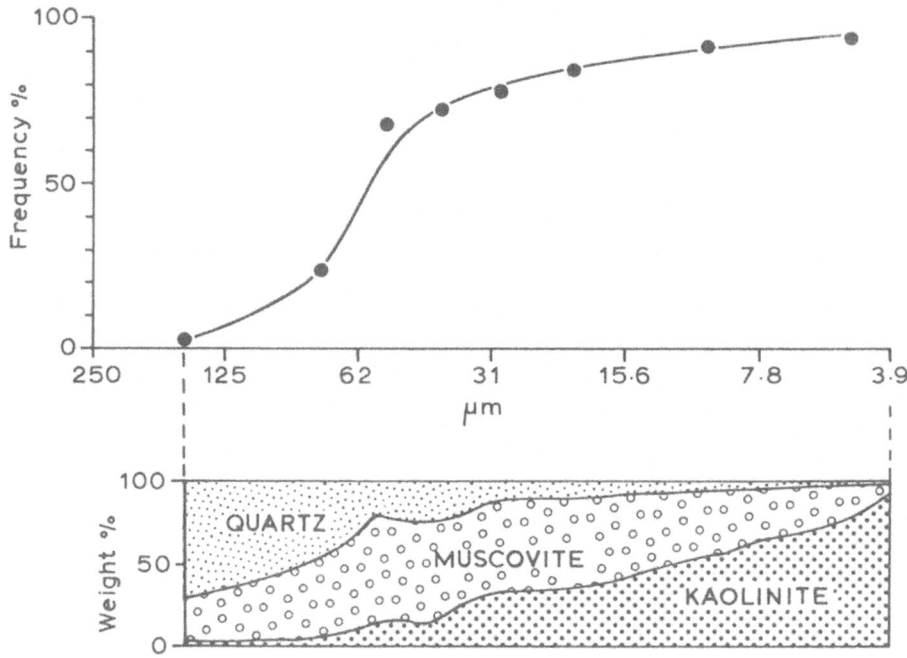

Figure 2.   Grading and mineralogical composition of a typical
micaceous residue.   (Blackpool 4th stage).

In 1968 some 1.4 million tons of fine residue were produced in the St. Austell area. Of this the St. Austell White River received 700 000 tons, while the Par river and mica dams each received 350 000 tons (China Clay Association[1]). Discharge into the St. Austell river was stopped in January 1973, and the waste is now being piped to disused pits in the area. Discharges into the Par river are expected to cease by October 1974. Some of the reasons behind this change in disposal policy emerge from a consideration of the ultimate fate of the micaceous residues formerly discharged into the rivers.

Over the years the quantities of residue produced by the industry have risen steeply as clay production expanded (Fig. 3). Between 1750 and 1852 a total of about 1 million tons of clay was produced. By 1946 some 36 million tons of clay had been produced with a further 26 million tons by 1968. The total estimated output of clay up to this year was 63 million tons, which would have resulted in about 55 million tons of micaceous waste. It has been further estimated that since the commencement of china clay mining some 17 million tons of micaceous residue had been discharged into Mevagissey Bay by the White River (Fig. 4) and that the quantity of sea-bed material in Mevagissey Bay similar to the micaceous residue was of the order of 15 million tons (Portmann[2]).

The St. Austell White River is narrow and fast-flowing, the average flow rate being 30 cusecs (16.2 mgd). No net deposition takes place in the river, all fine residues being carried to the mouth of the river and discharged into the sea. Concentrations of suspended solids in the river were of the order of 40 000 ppm, although some of this may have been carried as bed load. The White River runs out over the beach at Pentewan, and in spite of artificial barriers on either side of the channel it tends to spread out over the lower part of the beach.

The greater part of the suspended load was carried beyond the beach to the sea, where its presence was shown by a dense white plume in the surface water which was carried along the coast or out to sea according to the wind direction. Coarser particles would tend to have been deposited close to the river mouth while finer particles remained suspended in the white plume (Fig. 5). The river water being less dense tended to remain near the surface of the sea, but as fresh and salt water mixed the fine particles were flocculated by the salts in the sea water, to form larger aggregates which settled to the bottom more quickly. Studies of currents in Mevagissey Bay showed that these were weak, less than 0.2 knots and more or less random in direction. Conditions were therefore favourable for deposition of the flocculated material within the bay. A small fraction of the waste remained in suspension long enough to be carried out of the

244

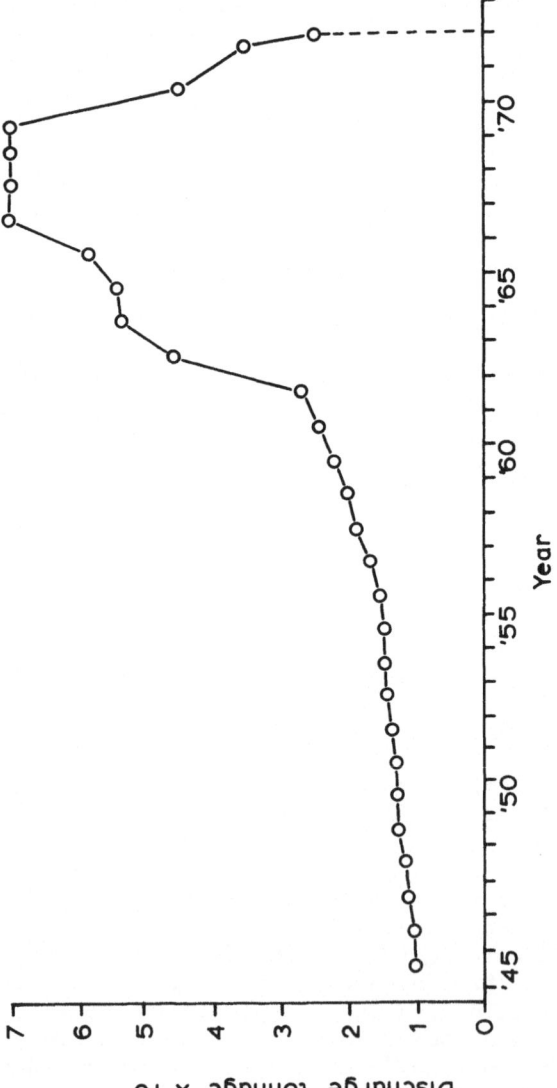

Figure 3. Estimated annual discharge of micaceous residue into the St. Austell White River since 1945. (long tons).

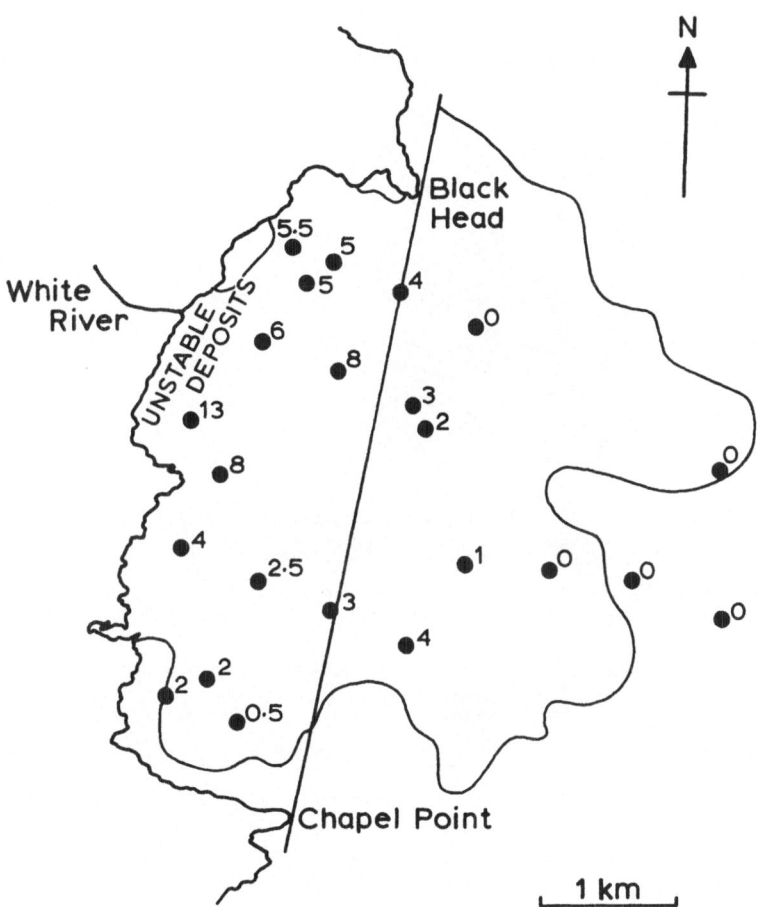

Figure 4.   Extent and depth (in feet) of china clay waste
            deposits in Mevagissey Bay. The unstable inshore area
            which is almost devoid of life is shown.

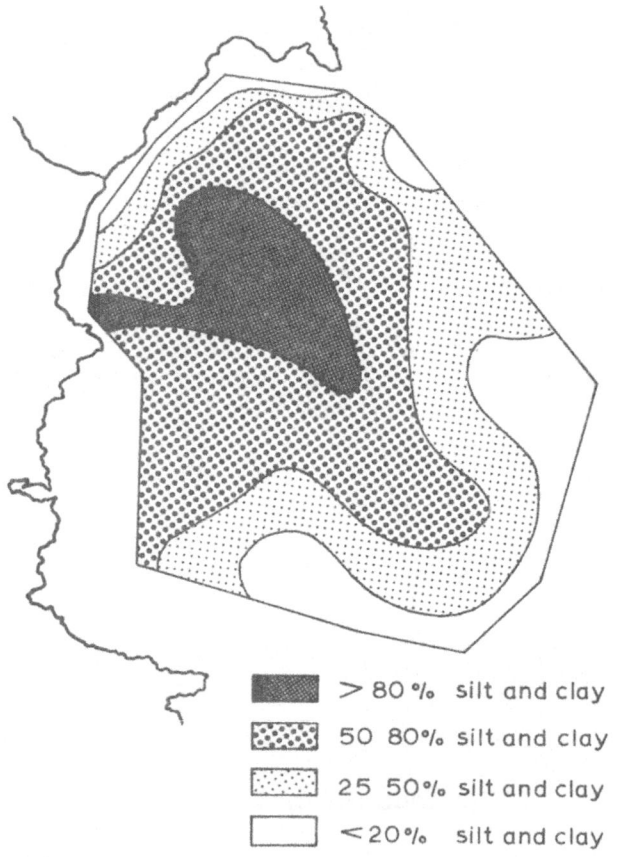

Figure 5. Distribution of the silt-clay fraction on the bed of
          Mevagissey Bay.

bay, and some redistribution of material evidently took place during storms, but the greater part of the solids brought down by the river were deposited within the bay

In the surface waters of Mevagissey Bay considerable variations in suspended load occurred, but away from the immediate neighbourhood of the river outfall these were not exceptionally high. Abrupt boundaries often marked the edges of the white water, and a metre or two below the surface the water was usually much less turbid than at the surface. At a time when the White River was discharging 450 000 tons per annum, suspended solids 400 m off the river mouth approached 40 ppm, while at 1 km distance values between 13 and 30 ppm were recorded. These contrast with 3-4 ppm in outside waters in the Western English Channel, but are comparable with values obtained off the N.E. coast (Moore[3]; Newton & Gray[4]) where the North Sea waters are noticeably more turbid than in the western English Channel.

From old charts it appears that the bed of Mevagissey Bay was formerly covered by deposits in which muddy gravels predominated. Much of the bottom of the bay now resembles in both texture and mineralogical composition the micaceous residues produced by the industry. Despite the somewhat unusual nature of the sediment much of the bay supports a comparatively rich bottom fauna, valuable as food for fish, in which such animals as burrowing worms, bivalve molluscs, brittle-stars, heart-urchin and a species of burrowing sea-cucumber occur (Howell & Shelton[5]; Probert[6]). The fauna of the micaceous wastes is very similar to that which might occur in deposits of similar grading elsewhere in the English Channel.

Close to the mouth of the White River, however, the very high rates of deposition (up to 2000 tons/day were formerly discharged by the White River) resulted in unstable deposits on the sea bed, in which only a few burrowing worms were able to live. Farther out the bottom was more stable probably because deposition was slower, and because the activities of the burrowing animals tended to consolidate it. Some species bind the grains of sand or particles of mud to form tubes in which they live, others pass the sediment through their gut, so that in course of time the sediment particles may acquire coatings of organic material which tend to stabilise the substratum. The sediment also changes through accumulation of dead shells and other animal and plant remains.

An interesting aspect of the properties of fine residues is the effect of mica particles on their porosity and shear strength. It is known from civil engineering studies (Gilboy[7]) that sediment porosity increases with mica content, and that fine particulate mica has a greater effect in increasing porosity than

coarse mica (McCarthy & Leonard[8]). This effect is no doubt due
to the flat platelets of which the mica is composed which dis-
courage close packing of the particles. Quicksand conditions
have been reported locally on beaches near the outflow of the
White River, and this may well be due to the high mica content of
the beach sands.

EFFECTS ON MARINE LIFE

Mevagissey Bay supports quite a variety of marine life. It
is a popular place for small boat fishing, and some trawling and
potting is carried on there. Local fishermen have mixed views on
the effects of the white water, some maintaining that it was good
for fishing, others are not so sure. Fish and other animals
taken in the bay were often very pale in colour, but their flav-
our appeared to have been unaffected.

Experiments were carried out to establish the degree of
toxicity to marine organisms of the fine residues, which may
contain small quantities of substances used to flocculate the
clay in the settling tanks. Results indicated a low level of
toxicity, the majority of the organic additives being biodegrad-
able and therefore unlikely to give rise to long-term toxic
effects. Analyses of marine organisms from Mevagissey Bay showed
no evidence of bioaccumulation when compared with levels found in
the same species from unpolluted areas.

Pollution of the sea by fine micaceous residues is an
example of the disposal of very large quantities of fine particulate inorganic waste having a low toxicity. Similar problems
arise as the result of other quarrying and extractive processes,
so it may be helpful to define the likely results of this method
of disposal. We have seen that china clay wastes have some
peculiar properties due to their high mica content, but most of
the effects noted below could be produced by finely divided
particles of any chemically inert solid substance.

Possible effects on marine life and fisheries are:

1. Effect on primary production. Small drifting plants
   (phytoplankton) which live in the illuminated surface layers
   provide the basic source of food in the sea. If the waters
   are turbid with suspended matter photosynthesis can only
   take place very close to the sea surface so that primary
   production will be drastically reduced. However Mevagissey
   Bay is open to the English Channel, with which interchange
   of water takes place continually, so that any local curtail-
   ment of primary production would be of little importance
   since food materials in the form of phyto- and zoo-plankton

would be regularly renewed from the open sea.

2. An increase in opacity of the water might interfere with the feeding of those fish which feed mainly by sight, such as plaice and dab.

3. The presence of turbid water might affect the distribution or migration of fish. Avoidance of the white water by fish has been reported by fishermen, and an interesting observation was made by the Ministry of Agriculture, Fisheries and Food on the behaviour of a shoal of whitebait (small clupeid fish), which are an important item in the diet of mackerel: a shoal, followed visually and by echo-sounder, clearly avoided the edge of an advancing mass of white water. It has been suggested by local fishermen that whitebait might shelter under the white water, and that this may account for the view that some of the best line fishing is around the edge of the white water, where larger fish feeding on whitebait may congregate.

4. Studies on freshwater fish show that high concentrations of suspended solids may cause damage to the gill tissues. A limited number of experiments, in which plaice and soles were kept in suspensions of micaceous residue for a period of one month, showed no apparent damage to the gills. However the experiments were not extended to include fast-swimming surface living fish, like mackerel, which might be more susceptible to suspended matter than the bottom-living fish tested.

There is no evidence of fish taken from Mevagissey Bay having damaged or abnormal gills.

5. Deposition of large quantities of solid waste on the sea bed may affect the success of different fishing methods. In Mevagissey and St. Austell Bays accumulation of fine residues have made the beaches too soft for shore-seining, and close in off the White River the bottom is soft enough for otter boards to sink into the deposit, with resultant damage to trawling gear. No doubt these effects are partly due to the high mica content of these sediments, but similar conditions might occur whenever deposition rates were high.

Another effect has been the covering of rocky grounds in the bays formerly used for crab or lobster potting. Such grounds may subsequently become smooth enough to be suitable for trawl-fishing.

6. It is sometimes stated that deposition of material on the sea bed will destroy the fishes' spawning grounds. In fact

the majority of commercial fish lay eggs which float freely
in the sea, a notable exception being the herring which lays
its eggs on the bottom on banks of gravel in certain areas.
Fine material would not be deposited in the tide-swept areas
where such spawning takes place.

DISPOSAL CONTROL POSSIBILITIES

Other issues which arise when finely divided solids are
disposed of into the sea are the presence of a plume of discoloured water, and the question of beach pollution. Where an
outfall occurs close to a beach, as at the mouth of the White
River, it is inevitable that the beach will be continually bathed
in turbid water, and that some material will be temporarily
deposited on the beach, at least in calm weather. However
general experience is that on open beaches, material finer than
100 mesh does not remain for any length of time. On the beach at
Blyth in Northumberland enormous quantities of colliery waste
(which includes considerable quantities of fine dust) are tipped
annually on to the beach. The fine fractions are continually
removed by the sea, so that the beach itself is composed of a
mixture of sand and coarser fractions of the coal waste (Harwood
& Wilson[9]).

In 1969 the Consulting Engineers to the China Clay Association (China Clay Association[1]; Holme[10]) proposed a scheme for
piping all the industry's micaceous waste to an outlet on the 18
metre line near Dodman Point, so stopping further pollution of
the white rivers and the inshore bays. It had been found that
the carrying capacity of the White River would no longer permit
it to be used as a vehicle for the ever increasing quantities of
waste produced by the industry, and there had been growing
objections to beach pollution and silting in the bays.

It was planned to pump the residues along a pipeline, of
some 12 in diameter, the material being in a slurry of specific
gravity between 1.23 and 1.40. The outfall point would have been
at the top of a slope descending to the more level bed of the
Channel at 40-50 metres depth, and it was thought that the
residue would spread out over the sea bed, so avoiding the
formation of a white plume at the surface and pollution of nearby
beaches.

In locating a position for the outfall, it was important to
know how exactly the residues would be dispersed from the discharge point. Would they flow down the rocky slope of the reef
as a suspension current, would they slump, or would they be
carried away by currents? A series of observations were made on
currents in the area, which are more vigorous than in the

sheltered bays. The effect of storms however is more difficult
to evaluate - under such conditions material not moved by
currents under normal conditions could be transported considerable
distances. A test was carried out with radioactive material to
see how fine particles would be transported along the sea bed
(Green[11]). A slurry of material similar to the clay waste was
prepared, into which was injected particles of glass treated with
radioactive scandium, which has a half-life of 84 days. A ship
was moored over the selected spot off Dodman Point, and from this
slurry was discharged by a pipe reaching to just above the sea
bed. Pumping of residue continued for several hours, radioactive
material being released on both the flood and ebb tide. Surveys
showed that the greater part of the residue initially settled
close to the outfall, and subsequent surveys showed that it was
only slowly dispersed from this point, mainly in a SW direction.
But no test could be devised which would simulate the very high
rates of discharge from an actual pipeline, nor could the effects
of progressive accumulation of residues on the sea bed be
ascertained. The uncertainty as to the exact behaviour of the
discharged residues on the sea bed, which was not entirely
clarified by a series of tank tests, and doubts as to whether
beach pollution could be entirely avoided resulted in the
abandonment of this scheme.

An example of the apparently successful disposal of fine
solid wastes by submarine pipeline is that from an aluminium
works near Marseilles. The residues, consisting mainly of oxides
of iron, aluminium and silica are discharged at $85m^3/hr$ at 350m
depth at the head of a submarine canyon which descends quite
steeply to beyond 1000m. Deposits of "red mud" from this discharge occur in two areas: a central zone of flow which is devoid
of life, and a surrounding area of deposition in which a bottom
fauna has become established. (Bourcier[12]). In this instance
the steep slope of the bottom ensured dispersal from the outfall,
and the depth was too great for wave disturbance.

Various methods of disposal of solid wastes at sea are
available, although not all may be acceptable because of amenity
considerations. Disposal directly on to the beach, as in the
instance of colliery waste, or into rivers, as formerly with the
china clay wastes, would be unlikely to receive approval if
introduced at the present time. The merits of dumping offshore
whether by barge or submarine pipeline will depend upon local
conditions and the quantities to be disposed of. Careful
consideration should be given to tides, currents and wave action
and trial dumping of material labelled with a radioactive isotope
should be carried out. Siting of an outfall would require
knowledge of local amenities and the location and nature of
fishing grounds. Even so there must always remain an element of
doubt as to the precise paths of dispersal of the wastes and their

consequent effects on amenity and fisheries. In our present
state of knowledge it is not possible to predict with precision
what will happen in the sea, and this element of uncertainty must
be taken into account when weighing the merits of different
disposal methods.

REFERENCES

1. China Clay Association, Report on disposal of micaceous
        residue, Vols. I and II, Binnie and Partners, unpublished,
        1969.

2. Portmann, J.E., The effect of china clay on the sediments of
        St. Austell and Mevagissey Bays, J.Mar.biol.Ass.U.K., 50,
        577, 1970.

3. Moore, P.G., Particulate matter in the sublittoral zone of an
        exposed coast and its ecological significance with special
        reference to the fauna inhabiting help holdfasts, J.exp.
        mar.Biol.Ecol., 10, 1972.

4. Newton, A.J. & Gray, J.S., Seasonal variation of the
        suspended solid matter off the coast of north Yorkshire,
        J.Mar.biol.Ass.U.K., 52, 33, 1972.

5. Howell, B.R. & Shelton, R.G.J., The effect of china clay on
        the bottom fauna of St. Austell and Mevagissey Bays,
        J.Mar.biol.Ass.U.K., 50, 593, 1970.

6. Probert, P.K., The bottom fauna of china clay waste deposits
        in Mevagissey Bay, Ph.D. thesis, University of London,
        1973.

7. Gilboy, C., The compressibility of sand-mica mistures, Proc.
        Am.Soc.civ.Engrs., 54, 555, 1928.

8. McCarthy, D.f. & Leonard, R.J., Compaction and compression
        characteristics of micaceous fine sands and silts, Highw.
        Res.Rec., 22, 23, 1963.

9. Harwood, F.L. & Wilson, K.C., Disposal of power-station ash
        into the sea, Survr.munic.Cty Engr, 116, 1073, 1957.

10. Holme, N.A., Disposal of china clay waste in the English
        Channel, Mem.Bur.Rech.Geol. & Minières (Fr.), 79, 269,
        1972.

11. Green, D.W., An investigation of the disposal of micaceous
     residue through a sea outfall, <u>Proc.Insts.civ.Engrs.</u>, 53,
     127, 1972.

12. Bourcier, M., Écoulement des boues-rouges dans le Canyon de
     la Cassidaigne (Décembre 1968), <u>Tethys</u>, 1, 779, 1969.

MANAGEMENT OF HEAVY METAL WASTES

MANAGEMENT OF HEAVY METAL WASTES – IDENTIFICATION OF PROBLEMS

Mineralized regions have been exploited for their non-ferrous metal ores for centuries, many mines having been worked for zinc, lead, silver, copper and tin since Roman times. Although the scale of the old mining installations was small and did not pose any visual problems in the landscape, modern techniques including strip- and open-pit mining may be much more troublesome. Additionally, the apparatus of modern mining, conveyor belts, cables and engineering structures and buildings can be very unsightly when derelict. Windborne mineral dusts from ventilation shafts and spoil tips may travel 3-4 km downwind affecting vegetation, pastures and grazing animals.

Ore beneficiation is normally carried out by flotation processes involving plentiful supplies of water which becomes heavily polluted. When this enters water courses, potentially toxic elements such as copper, nickel and zinc frequently reduce species diversity. Streams become fishless and devoid of algal or macrophyte growth, with only metal tolerant grass populations surviving along the stream banks.

Even when mines close, the problems do not cease. Local drainage water often finds its way into underground workings and out into local streams. With increasing demand for water supplies for domestic and industrial use, aquatic metal-contamination could present risks to machinery or health. It is not always easy to solve mine drainage problems from old workings because former owners are not always traceable and mine drainage is excluded from water quality legislation in many countries.

Concentrated ores processed at smelters give rise to sulphurous smoke pollution and airborne metal emissions which can spread up to 100 km downwind depending upon the height of the emission stacks. This may give rise to acid rain rich in potentially toxic metals.

Smelter wastes are deficient in nitrogen, phosphorus and potassium and contain concentrations of toxic metals inimical to plant growth. They represent, in an extreme form, problems similar to those posed by metal-mine spoils. Generally speaking, tree planting fails on such substrates which remain permanently bare, or colonize very slowly, usually with metal tolerant genotypes. Although methods have been developed for the artificial colonization of spoil tips and smelter wastes, methods of maintaining the established vegetation are very poorly understood.

THE ORIGIN AND EXPLOITATION OF NON-FERROUS METALS

J.N.M. Firth

Welsh Office, Cardiff, U.K.

ABSTRACT

The utilization by man of the metallic elements in an increasingly industrial society has exposed the ores of these elements to greatly enhanced levels of chemical and microbiological activity. The slow rate of weathering of mineral deposits over geological time has been accelerated by the mining of ore bodies, whilst their processing and waste-disposal has presented the biota with unaccustomed quantities and novel compounds of the metallic elements.

The reactions of living organisms to these stress conditions are complex and can best be understood by an inter-disciplinary approach.

INTRODUCTION

Schroeder[1] has pointed out that in general, there seems to be an inverse relationship between the atomic number of an element in the periodic table and its abundance in the earth's crust i.e. the lighter elements are commoner. Life on earth has evolved over a period of about 3000×10^6 years in contact with these elements, which may partly explain why 99 per cent of the composition of living things is made up of about 14(H, B, C, N, O, F, Na, Mg, Si, P, S, Cl, K, and Ca) out of the first twenty elements in the table. Moreover, with the exception of iodine (Atomic No. 53), the remaining one per cent of living tissue is composed principally of biologically active elements drawn from atomic numbers 22-42 (V, Co, Mn, Fe, Co, Cu, Zn, Se and Mo).

Apart from all these 'nutrient' elements which may often be ultimately toxic in larger amounts, there are a number of non-essential elements, frequently toxic to living things, scattered throughout the periodic table e.g. Be, Al, Ni, Ag, Cd, In, Sn, Sb, Te, Ba, Hg, Pb, U, Pu.

Man now incorporates about 24 elements in his body (Frieden[2]). Some are in relatively large proportions, e.g. calcium, which makes up about 2.5 per cent of our average body weight, whereas others, e.g. cobalt, are equally essential but nevertheless present in microgram quantities. Excesses and deficiencies of any of these essential elements can bring about changes in the living organism. Similarly the introduction of foreign elements capable of competitive substitution with one or more of the essential elements will also modify but in some cases may prove fatal to the organism's metabolism.

One of the most significant factors contributing to the civilization of Homo sapiens has been his ability to recognise, refine and adapt for his own use the metalliferous ores that abound in the crust of the earth. Some of these metals are essential for the metabolism of living organisms, the others having been rejected by them over geological time. Thus, situations developed relatively rapidly where metals and their compounds were produced by man, used to improve man's standard of living and then discarded. All these activities present the human environment with novel stresses. The costs and benefits relationship for man of these economically attractive materials constitutes one of the environmental problems of our society.

THE ORIGIN AND OCCURRENCE OF METALLIFEROUS ORES

Igneous Processes

Both the temperature of the earth and the pressure increase with depth. Increased temperature tends to liquify material whereas increased pressure tends to raise the melting point so that an equilibrium is established. However, various events occur to upset this equilibrium; for example in regions of the crust where earth movements take place, the pressure may be relieved to some extent and the rock in the vicinity becomes molten. This liquid rock, or magma behaves like a viscous liquid and moves through the crust until it is cooled sufficiently to solidfy. The rocks formed after this cooling are known as igneous rocks.

It has been observed that the majority of metalliferous ore deposits occur in regions that have a history of igneous activity and there must obviously be a connection between the two. The sequence of events during the cooling of a magma is often complex but definite stages have been recognised that give rise to characteristic metalliferous ore deposits.

During the initial stages of cooling, the components of highest melting point begin to separate from the liquid magma and can eventually occur as disseminations or consolidate layers in the igneous rock. The main ore of chromium occurs in this context as the mineral chromite ($FeCr_2O_4$) and one of the richest deposits of magnetite (Fe_3O_4) at Kiruna, Sweden is of this type.

As more and more minerals crystallize from the magma on cooling, a liquid phase is left that is very mobile due to the high concentrations of fluxes such as boron, chlorine and fluorine. The fluids can easily migrate into fissures surrounding the magma chamber and eventually solidify into dykes and veins of pegmatite. The crystals in these veins are often of great size and include most of the economic ores of lithium and many of the rarer elements.

During the final stage of consolidation of an igneous magma, residual aqueous solutions and vapours are left which have great mobility over wide variations of temperature and pressure. The hydrothermal veins formed by this process are one of the most important sources of non-ferrous metals which occur usually as the sulphides associated with gangue minerals such as quartz and barytes. Low temperature hydrothermal veins can be found at great distances from their igneous source and some cannot be definitely related to any igneous body.

Metalliferous ores are precipitated from solution under specific conditions of temperature and pressure. Thus different ores are often found in a definite sequence. This simplified idea of a complex process can be observed in the mineral belt associated with the granites of Devon and Cornwall where vertical zoning has given tin ores at the lowest levels, followed by copper, zinc and lead ores. The copper deposits at Butte, Montana show an onion skin zoning with copper containing a little silver in the inner zone, copper with more silver plus lead and zinc in the intermediate zone and gold, silver and zinc with only a trace of copper in the outer zone.

Hydrothermal ores may be precipitated either directly in existing cavities in the country rock or indirectly by replacing some existing minerals, the replaced minerals being carried away in solution. Many of the world's largest metallic ore deposits have been formed by replacement processes in a wide variety of geological settings. Typical examples include the lead ore deposits of Colorado, Utah and Broken Hill, Australia, the zinc lodes of Franklin, New Jersey and the mercury lodes of Almaden, Spain.

This replacement process also occurs at the actual boundary of the magma chamber and relatively small but significant metallic ore bodies have been formed by this process of contact metasomatism. The tin/tungsten ores of Dartmoor and molybdenum ore at Buckingham, Quebec are included in this category.

Surface Processes

The intrusion and emplacement of mineral deposits within the earth's crust is a primary process. In many cases, subsequent earth movements and erosion bring these deposits very close to the surface, and in some cases actually to ˙the surface. It is here that the surface agencies of groundwater, oxidation, bacterial action, mechanical and chemical weathering act upon these primary ores to modify them and generate secondary ore deposits.

The wide variety of primary ores combined with the different environmental conditions found on the surface of the crust gives rise to a number of permutations and combinations that dictate the nature of the secondary ores. The original primary deposit consists of ores mixed with gangue minerals and the two main results of weathering are:-

1. the ores can either be removed and carried to some other area before deposition

2. the gangue may be removed leaving a more concentrated form of the ore.

1. Removal and deposition. High density and chemically resistant minerals are weathered and eroded from their parent rock, transported by water and deposited where hydraulic conditions allow placer deposits to form. These may be exposed at the surface or form lenses in buried sediments. Rivers and beaches are the most likely sites for these deposits and gold and platinum are minerals that have received the most publicity from this source. However, other dense residual material such as cassiterite, limonite and monazite are all concentrated by this hydraulic process which physically changes the shape of the mineral particles without affecting them chemically.

In some cases, existing minerals may be taken into solution by chemical and/or microbiological activity and reprecipitated in other areas where conditions of pH - Eh, temperature and salinity are suitable. This process is responsible for the occurence of many of the non-metallic mineral deposits such as the phosphorites and the evaporites as well as sedimentary iron and manganese ores. Certain copper deposits are almost certainly sedimentary in origin and it is thought that some lead, zinc and uranium ores are formed

by chemical precipitation at the time of accumulation of the
sediment in which they are found (syngenetic), rather than being
introduced afterwards by hydrothermal action (epigenetic).

2. Residual deposits. The chemical decomposition of ultra-
basic igneous rocks such as peridotite results in the concentration
of iron, nickel, cobalt and chromium in the immediate vicinity of
the original rock. Slight layered separation usually occurs with
the iron ore at the top and the other metal ores beneath. The
New Caledonian nickel deposit is of this type.

The emplacement of metalliferous sulphide ore deposits is
often followed by a sequence of uplift and erosion so that the
vein is exposed to the surface. Atmospheric agencies, ground
water and bacteria all act upon the exposed vein to alter and
redistribute the minerals in the uppermost portion of the vein and
a series of zones can be recognised. The top zone, the oxidised
zone, is usually devoid of economically useful ores. It is a
zone of solution where the sulphides have been oxidized and carried
to lower levels, leaving a porous layer of limonitized material,
the "gossan" of old Cornish mining parlance. The lower portion
of the oxidized zone can frequently become a zone of deposition,
especially where carbonate or silica combines with the solubilized
cations to form carbonates and silicates. Lead, zinc and copper
carbonates and silicates are the commoner minerals that are found
in this zone.

Between the base of the oxidized zone and the top of the
sulphide zone is the supragene sulphide zone where the change from
oxidizing to reducing conditions has concentrated greater quantities
of sulphides than were present in the original vein. Replacement
of existing sulphides by those richer in the metallic element,
especially copper, has made the supragene sulphide zone one of
the most attractive sites for mineral exploitation.

THE PRESENCE OR ABSENCE OF NON-FERROUS METALLIC ORES

The presence or absence in a country of ores of the metals
that are vital to 20th century technology has obvious economic
and political overtones. As with oil deposits, few countries
have a sufficiency of all the essential ores to make them independent. An idea of the world distribution of the main ores can
be gained from Table 1, which shows the free world production
percentages by country of the ores of the major non-ferrous metals.

	Antimony	Bismuth	Cadmium	Chromo	Cobalt	Copper	Gold	Lead	Mercury	Molybdenum	Nickel	Tin	Zinc
Asia						7.1							
Australia	24.9	19.7	7.5				2.5	15.1			1.3	6.1	10.9
Bolivia		10.5										16.6	
Canada			28.2		7.9	12.7	5.5	14.8	6.1	17.0	67.1		29.5
Chile						12.9				9.7			
Columbia						4.9	0.5						
Europe													
Finland					4.1			1.0					1.2
France													0.3
Germany F.R.			1.8		2.3			1.6					3.1
Ghana							1.9						
India							0.3						
Indonesia											2.2		
Ireland								2.1					1.9
Italy								1.2	20.3			11.0	2.3
Japan		10.5	4.9				0.6	2.5					6.5
Korea (S)		3.3											
Malaysia												38.8	
Mexico		18.4	15.8				0.4	7.0	12.7				6.2
Morocco	7.7				7.8			3.7			0.1		
Nicaragua	27.0						0.3						
Nigeria												3.4	
Oceania		22.4									22.4		
Peru			9.3			3.9		6.3		1.0			7.6
Phillipines				9.6			1.5						
Rhodesia				9.6			1.3				0.1		

South Africa	34.6		40.1			76.9					1.6		
Spain							2.3	2.7	31.9				2.1
Sweden							2.9						2.5
Thailand			15.8									11.3	
Turkey												1.7	
U.K.		n.a.								3.8			
U.S.A.		16.7			26.7	3.9	22.9	3.2		71.1			11.1
Yugoslavia	5.8	3.3					4.3	8.1					1.9
Zaire		1.3		66.5	7.7	0.4						3.3	2.6
Zambia				10.5	12.8								1.6
Others	11.9	14.6	24.9	0.9	5.8	4.0	9.6	17.7		1.2	1.4	7.8	8.7

Table 1. Percentage production of major non-ferrous metals outside communist countries. 1972

MINING OF ORES

The development of very large scale machinery since the second world war has revolutionized the mining of ores. Prior to this deep mining for relatively high-grade ore bodies was the commoner practice, whilst surface mining was limited to those areas with shallow overburden and medium grade ore beneath. The economic removal of larger quantities of overburden combined with the ability to utilize lower grade ore bodies has now made many un-exploited areas of the world attractive for large scale mining.

Surface mining employs a number of methods to obtain the ore. One of the largest is open cutting in which the overburden is removed for an area a mile or more in diameter and the underlying ore body excavated, often in a concentric terrace pattern, using power shovels and various transport systems before back-filling with the stockpiled overburden. The scale of open cast mining is often vast with hundreds of thousands of tons of ore and waste being transported daily. Even so, it is one of the lowest cost mining operations. The Arizona and Utah copper prophyry deposits are worked by this method, and because of the low cost, very low grade ores can be economically worked to limited depths in the region of 1000 feet.

A variation on the opencast method is used in strip mining where strips of land have the overburden removed to one side and the ore excavated from the bottom of the strip. As the ore is extracted, the strip is backfilled with overburden from the next strip as it is uncovered. This method is mainly used for coal ex-traction but can be used where any gently dipping economic ore occurs near the surface.

Ores occurring in association with friable gangue material can be separated by hydraulicking in which a high pressure water jet is aimed at the deposit and the resulting slurry sluiced through ribbed boxes where the denser metallic ore particles are separated. Tin bearing gravels are worked by this method in Malaysia. Tin ore and other ores are also worked by dredging in which a dredger creates a lake for itself in the placer deposits and sorts the mixed sediment on board before returning the waste material to the already dredged area.

The presence of excessive quantities of overburden, or the existence of a steeply inclined ore body necessitates the under-ground mining of the ore. Vertical or inclined shafts give access to the ore deposit which is then usually worked so that the loosened ore falls under gravity into conveying systems before removal from the mine.

Bedded ores are worked by "room and pillar" and "longwall"

methods in which portions of the original ore body or temporary
supports are left to support the workings.

In most cases, underground mining is carried out beneath the
water table and large volumes of water may have to be continuously
pumped from the workings. This often involves the installation of
very large pumps with high capital and running costs and with a
few mines now getting deeper than 10 000 feet, the pumping costs
can be high. The residence times of this underground water in
contact with freshly disturbed ore bodies is relatively short when
the mine is being pumped, but after closure the residence time
is increased so that bacterial and chemical oxidation take significant quantities of the cations into solution with ensuing
pollution of the neighbouring water courses.

MINERAL DRESSINGS

In the case of high grade ore deposits, it is economic to
pass the bulk of the ore direct to the smelting and refining processes. However, with the exploitation of medium and low grade
ores it is obviously uneconomic to transport and process a vast
bulk containing only a small proportion of the desirable mineral.
Thus the concentration of ores is now a necessary link in the chain
between mining and the end-use of the metal.

In order to obtain a concentrate of the finely disseminated
mineral from its ore, the ore from the workings must be crushed
and milled to a definite particle size depending upon the average
size of the grains of the desired mineral in the unwanted matrix.
The fine particles can then be separated by a number of techniques.

Metallic minerals are nearly always of higher density than the
"gangue" minerals of the matrix in which they are embedded and
some dressing techniques make use of this difference. In some cases
the milled material is placed in vibrating screens through which
water passes upwards, carrying off the lighter gangue material and
leaving the denser minerals on the screens. In other cases con-
centrating tables are used where the finely crushed ore is placed
on a vibrating table fitted with sets of riffles over which water
flows so that again the gangue minerals are combed off and the
denser metallic minerals left against the riffles.

Direct separation by a "sink and float" method can be achieved
by placing the mixed ore in a fluid of high density. Aqueous
suspensions containing galena or barytes are used to float off
the light gangue; this method is especially useful for separating
lead and zinc ores from a limestone matrix.

A widely used method of mineral dressing is that of floatation which utilizes the difference in surface properties between the metallic mineral and the gangue minerals. An aqueous slurry of the finely powdered ore is mixed with an oil or surfactant (typically organic materials such as the xanthates) and passed into a series of flotation cells through which air is pumped. A froth of bubbles forms to which the metallic mineral particles adhere and this surface scum is continuously removed and washed to obtain the ore concentrate. A cascade method using a number of cells removes 90 to 95 per cent of the metallic minerals. By choosing the right oil and surfactant different metallic minerals may be separated from each other in the same mixture by selective flotation.

The magnetic separation of some iron, nickel, cobalt and manganese ores makes use of the unpaired electron in the outer shells of these transition elements, as does the separation of the weakly magnetic iron tungstate, wolfram, from its gangue minerals. Electrical differences in the surface of mineral grains can also be used as in the high tension separation of chromite bearing sands.

Natural leaching of mineral deposits takes place when certain oxidizing bacteria, particularly Ferrobacillus ferro-oxidans and Thiobacillus ferro-oxidans, catalyse the oxidation of pyrite (FeS_2) to ferrous sulphate and sulphuric acid. Further oxidation to the ferric sulphate creates conditions favourable for the dissolution of other metallic cations, particularly copper. The essential ingredients for this leaching process are moisture, oxygen, a source of pyrite, temperatures between 15° and $50^{\circ}C$ and high acidity. The solution of copper, leaching from the low grade ore dumps is passed over scrap iron when the copper is precipitated and recovered for further purification.

Copper ores are the main ores used for bacterial leaching, but millerite (NiS), sphalerite (ZnS), galena (PbS) realgar (AsS) and molybdenite (MoS_2) are also taken into acidic solution by a similar mechanism.

MINERAL SMELTING AND REFINING

The ore concentrates are frequently in their original mineralogical form e.g. as sulphides, tungstates, chromates etc., and require further treatment before the metallic elements can be obtained in a pure form.

Metal smelting is a very old science and the basic methods are still employed, with some modifications. Most sulphide and carbonate ores are first roasted to the oxides which can then be reduced with hydrogen, carbon, carbon monoxide or aluminium to give the metal. However, partial oxidation of sulphides of mercury, antimony, lead and copper yields a mixture in which unreacted sulphide reacts with the oxide to give the metal plus sulphur dioxide e.g.:-

$$2PbS \quad +3O_2 \qquad 2PbO \quad + \quad 2SO_2$$

$$2PbO \quad +PbS \qquad 3Pb \quad + \quad SO_2$$

The smelting stage, whilst producing the desired metal from its oxide, also liberates impurities which are usually removed by the addition of a suitable flux to the smelter charge. The flux generally lowers the melting point of the process and forms a slag with the impurities. The slag floats on the surface of the melt whilst the metal sinks below the melt where it is tapped off. Early smelting processes consisted of a number of separate stages but recent developments have been aimed at obtaining the metal by a more direct route. The Worcra furnace, for example, now produces copper in one furnace from its ore concentrate whilst the Mitsubishi process produces copper continuously from a cascade of three furnaces, each performing a separate function.

Smelting processes produce metals that are relatively impure and the removal of the last small quantities of impurities is achieved either by fire refining of electrolytic refining. In fire refining, the impure metal is melted in a reverberatory furnace in a stream of air. This oxidizes any remaining sulphur to sulphur dioxide, and any remaining metal impurities to a scum of oxides which can be skimmed off. Electrolytic refining has become the more usual refining technique because it is capable of producing a purer metal, whilst at the same time making the recovery of the other metallic impurities a simpler task. In copper refining, anodes of impure copper are suspended in a bath containing aqueous copper sulphate and sulphuric acid. Pure copper cathodes are suspended in the solution to complete the circuit and when the current is passed, further pure copper is deposited at the cathode whilst the impurities such as gold, silver, selenium and bismuth, either pass into solution or are deposited as a mud near the anode.

WASTE MATERIALS

The mining, concentration and smelting of metallic ores are all processes that produce a number of waste products. Mining operations involve the disposal of often vast quantities of overburden that has to be removed before the ore body can be worked. Mineral dressing again separates the gangue minerals from the economic ore and in addition produces tailings that are allowed to settle out in ponds. The smelting processes inevitably give rise

to a series of slags that contain the residual impurities from
the ore concentrate and some of the economic ore as well.

At every stage in the processing of an ore body, the metallic
minerals have a much greater surface area exposed than in the
original ore body. Chemical oxidation and bacterial activity that
were normally confined to the upper levels of the ore body can
now take place throughout the entire waste deposit and metallic
elements pass from the insoluble form into aqueous solution.
Flowing water, and wind in drier climates, can also distribute the
finer solid particles over a wide area where subsequent chemical
and bacterial activity can take the metallic ions into solution.

The problems associated with metallic mineral wastes thus
occur in the mineral waste accumulations themselves, throughout
the drainage system and in the flood plains that border them.

Certain bacteria function as catalysts in oxidative reactions
involving sulphide minerals. The commonest of these are the
Thiobacilli and Ferrobacilli, and their use in the Bingham Canyon
copper deposits increased the rate of leaching of copper by a
factor of a thousand. Their activity is based upon their ability
to oxidise iron sulphide to ferrous sulphate and sulphuric acid.
Further oxidation yields ferric sulphate and eventually ferric
hydroxide plus more sulphuric acid. The bacteria require oxygen
and become inactive at depths where the oxygen supply is limited.
The optimum temperature is about $35^{\circ}C$ at pH 2.0 to 3.5 and they
become inactive above $50^{\circ}C$ and pH 9.0. There is a marked inverse
relationship between the rate of oxidation and the particle size
of the waste, so that tailing dumps are very susceptible to this
form of leaching.

Many ores can be leached by these oxidizing bacteria. The
copper ores chalcocite, bornite and tetrahedrite all yield copper
sulphate solutions (Bryner et al.[3]) as does chalcopyrite (Razzell
& Trussel[4]). In the zinc ores, smithsonite and sphalerite can be
oxidized (Malouf & Prater[5]), whilst the nickel sulphide, millerite
(Razzell & Trussell[4]) and the molybdenum sulphide, molybdenite
(Bryner & Jameson[6]) have also been reported to oxidise in the
presence of Thiobacillus ferro-oxidans. Covellite, arsenopyrite,
phrrhotite and galena have all been oxidized in a similar manner
(Ivanov, Lyalikova & Kuznetsov[7]).

The analysis of waste material resulting from metalliferous
activity shows that the individual metals occur in two distinct
forms. Firstly a form that is insoluble in acidic ground water
and not readily complexed by conventional agents such as ethylene
diamine tetracetic acid, and secondly a form that is soluble and
can be complexed. This latter form is available to be taken up by
the root systems of vegetation.

Plants growing on ground underlain by metallic ore bodies have developed a tolerance to the elevated available levels in the soil and often reflect this tolerance by higher metal levels in leaves and roots. However these localities, because of their geological age, have also acquired normal populations of soil bacteria that provide nitrogen and phosphorus essential for necessary plant growth. In waste material from metalliferous activity these two essential nutrients may be deficient, and even though tolerant strains of grasses may colonise the fringes of the waste deposits the vast bulk may be still free of vegetation after many years.

The Lower Swansea Valley Project revegetation survey (Street & Goodman[8]) indicated that soil micro-organisms can tolerate high levels of toxic metals and, given supplements of nitrogen and organic matter, will build up populations large enough to support tolerant strains of grasses on smelting waste tips. There is a danger, however, that this bacterial activity will also solubilize more metals over a prolonged period of time and raise the levels to a point where the plants will be killed.

The leaching of metallic cations from waste material contaminates the water courses with four types of hazard, namely metals in solution, particulate material in suspension, metals bound to inorganic particles, and metals complexed with organic compounds. The effects of these different forms on the biota are complex, and in addition their toxicity depends upon such parameters as pH, hardness, temperature and particle size. Different organisms have different degrees of tolerance to these varied forms and the overall effect on the biota is often a reduction in the species diversity. Fertility and growth rates can also be affected and in some cases the effect on predatory species is greater than that on organisms lower down the food chain, with the resulting explosions in population of the more tolerant lower organisms. In addition, mixtures of different metals result in reduced, additive or synergistic toxic effects on different organisms.

Some organisms have adapted to tolerate environments that contain high metal loads and numerous examples can be found in the literature of plants, algae and molluscs that have accumulated higher levels of metals than exist in their surrounding aquatic environment. If these organisms are used as a food source for higher organisms that cannot tolerate enhanced metal levels then a reduction in species diversity will again result.

The assimilation of metals by organisms in the aquatic environment does not rely solely on their being in solution. A recent study in the New Lead Belt of Missouri (Gale et al.[9]) has shown that no significant problem arises in that area from lead and zinc in solution. However, large quantities of particulate lead and zinc compounds escape trapping in the flotation and tailing reservoirs and are assimilated by the local aquatic biological community, with a resultant inverse relationship between the lead content of lower organisms and the distance downstream from the source of pollution.

REFERENCES

1. Schroeder, H.A., The biological trace elements, J. chronic. Dis., 18, 217, 1965.

2. Frieden, E., The chemical elements of life, Sci.Amer., 227, 52, 1972.

3. Bryner, L.C., Beck, J.V., Davis, D.B. & Wilson, D.G., Micro-organisms in leaching sulphide minerals, Ind.Eng.Chem., 46, 2587, 1954.

4. Razzell, W.E. & Trussell, P.C., Isolation and properties of an iron oxidizing Thiobacillus, J.Bacteriol., 85, 595, 1963.

5. Malouf, E.E. & Prater, J.D., Role of bacteria in the alteration of sulphide minerals, J.Metals, 13, 353, 1961.

6. Bryner, L.C. & Jameson, A.K., Micro-organisms in leaching sulphide minerals, Appl.Microbiol., 6, 281, 1958.

7. Ivanov, M.V., Lyalikova, N.N. & Kuznetsov, S.I., The role of sulphur bacteria in the decay of mountain rocks and sulphide ores, Izv. Akad. Nauk. CSSR., Ser. Biol. 2, 183, 1958.

8. Street, H.E. & Goodman, G.T., Revegetation techniques in the Lower Swansea Valley, In The Lower Swansea Valley Project, (ed. K.J. Hilton), Longmans, London, 1967.

9. Gale, N.L., Hardie, M.G., Jennett, J.C. & Aleti, A., Transport of trace pollutants in lead mining waste waters, In Trace Substances in Environmental Health VI, (ed. D.D. Hemphill), Univ. Missouri, Columbia, 95, 1973.

A SIMPLE TECHNIQUE FOR THE MONITORING OF AIRBORNE HEAVY METALS PRIOR TO REVEGETATION

G.D.R. Parry†, G.T. Goodman* and S. Smith‡

Department of Applied Biology, Chelsea College,
London, U.K.

ABSTRACT

Counteracting the toxicity of old metal-mine and smelter wastes during revegetation procedures is usually a costly process. It is therefore wasteful of effort to embark upon this only to find that plant growth eventually fails, not because of the failure of the revegetation treatments themselves, but because of hitherto unrecognized air pollution by metals. Potentially toxic, airborne metals, emitted from nearby urban-industrial sources, can be intercepted by the newly establishing vegetation and eventually kill it.

The paper proposes a simple inexpensive method of obtaining relative rates of airborne metal deposition onto the ground and vegetation. Use is made of the bog-moss Sphagnum acutifolium agg., specially cleaned and washed free of any naturally

Present addresses: † Mersey County Council Planning Department,
Liverpool, U.K.

* Beijer Institute, Royal Swedish Academy of
Stockholm, Sweden.

‡ Monitoring and Assessment Research Centre,
Chelsea College, London, U.K.

occurring metals and sewn into flat, nylon-mesh envelopes. When
an array of these are exposed vertically, or preferably horizontally, for 1 - 6 weeks and subsequently analysed chemically, they
give a picture of metal interception in an area. Preliminary
work suggests that, per unit area, metal interception of a horizontal bag is similar to that of a short-grass-sward.

INTRODUCTION

Dumps of metal-mining waste and metal-refining waste pose
serious problems to those who attempt to reshape, stabilize and
revegetate them. They are usually nutrient deficient, often
toxic, and generally hostile to plant growth and soil forming
organisms. The introduction of vegetation onto such material is
one of the preferred methods of cosmetic treatment although it is
often found that vegetation growing on such metal-waste substrates
is toxic to animals and great care must be taken in deciding the
uses to which such reclaimed land is finally put.

Added to the obvious problems of unsightliness and general
dereliction of these large masses of waste, is a far more cryptic
problem, viz the continual deposition of potentially toxic heavy
metals from the air and their interception by soils and plants.
Metals may enter the air from a number of sources such as fume
and dust from metalliferous ore crushing, concentration and
smelting processes; incineration of domestic refuse; the burning
of coal, oil, and leaded petroleum and numerous other chemical
industrial activities.

A great deal of evidence has accumulated in recent years to
show that the trace metal content of plants and soils may be
highly elevated above normal levels by the fallout of such
elements from the air. Purves[1-4] has described the enhancement
of the copper, boron, and lead content of soils which he attributes to general urban air contamination. Others, (Burkitt et al.[5];
Djuric et al.[6]; Costescu & Hutchinson[7] and Goodman & Roberts[8])
have described how soil and plants are enriched in heavy metal
content as a result of aerial fallout from metallurgical industrial activity. Automobile exhausts are an important source of
lead, both in the inorganic and organic form, to the atmosphere,
(Motto et al.[9]; Dedolph[10]; Everett et al.[11] and Lagerwerff &
Specht[12]).

Metal fallout onto soil and vegetation has contributed to
livestock illness and death in the Lower Swansea Valley, South
Wales, U.K., where horses pastured on land free from metal waste
dumping died from clinically diagnosed lead poisoning after eating
grass and hay contaminated by heavy metal fallout from nearby
industry, (Goodman & Roberts[8]).

RATIONALE

However, it is _not_ the intention of this paper to estimate
the extent of the risks to the health of wildlife, livestock,
crops or man from the various levels of potentially-toxic airborne metals prevailing around urban-industrial sources. But, it
is clearly wasteful of effort and resources to revegetate spoils
and other substrates by adding organic matter, lime, fertilizer
and other amendments, only to find that plant growth eventually
fails because of the toxic effects of airborne metal pollution
emitted from ongoing industrial activities and received by the
newly established plant cover. This means that some attempt must
be made to find an inexpensive way to quantify the deposition
rates of such metals on a reclamation site _before_ embarking on
costly revegetation treatment.

THE MOSS-BAG METHOD

Following on the work of Rühling & Tyler[13],[14] in Sweden,
naturally occurring _Hypnum cupressiforme_ var. _filiforme_, surface
soils and _Festuca rubra_ were collected by Goodman & Roberts[8] in
the Swansea district (S. Wales) and analysed for their metal
content. The results showed that all three agents could be used
as indicators and integrators of airborne metallic burdens, but
the method suffered from one serious drawback, a _Hypnum_ desert
zone existed around the city of Swansea and its industrial complex presumably caused by decades of severe aerial pollution. This
led the authors in 1969 to 'transplant' _Hypnum_, in nylon mesh
bags, from 'clean' areas to the denuded area. It was found that
the moss, whether alive or dead, still accumulated heavy metals
from the air and from this procedure was developed the 'moss bag
method' of determining metals in the air.

In continuation of the early moss-bag studies of Goodman &
Roberts[8] and the developments of Roberts[15] using _Hypnum_ moss-bags,
a further investigation was made in the Swansea district.   The
bog-moss _Sphagnum_ spp. was collected from a rural site at Cefn
Bryn, Gower, S. Wales and washed three times in 0.5N HNO_3
followed by three successive double-distilled water washes to
remove any metals initially present in it. _Sphagnum_ was used
instead of _Hypnum_ because it is much more abundant and has been
shown to have a high cation exchange capacity[16]. Approximately
15g (equivalent to 2.5g dry weight) of the washed moss was sewn
into flat nylon mesh bags, 10cm square.

As an illustration of the kind of information which could be
obtained around any centre of urban industrial activity (see also
Roberts[15]), about 100 of these bags were suspended around the
Swansea Valley region for 4 weeks in April 1971.   The area

surveyed covered the district surrounding the two main metal refining industries in the Swansea Valley at that time viz. a zinc-lead smelter and a nickel refinery. Additionally, about 7 000 000 cubic yards of metal-smelter waste had been deposited in the Valley over the last two centuries. This was still bare and liable to be blown by strong winds during dry periods. After exposure for 4 weeks the moss contained in the bags was analysed for its metal content. After correction for the original very small metal content of the moss, the amount of metal deposited and retained on the bags was calculated in terms of ng metal cm^{-2} bag day^{-1}. By mapping the sampling sites and joining up sites of equal deposition it was possible to construct 'isopleth maps', giving 'contour lines', of relative metal fallout burdens surrounding the two emission sources in the area studied.

RESULTS OF MOSS BAG SURVEYS IN SWANSEA

The 'contour' map of lead deposition onto the moss bags (Fig. 1) shows a high centre of 1000ng cm^{-2} day^{-1} of about 2km in diameter lying directly over the site of the zinc works. The

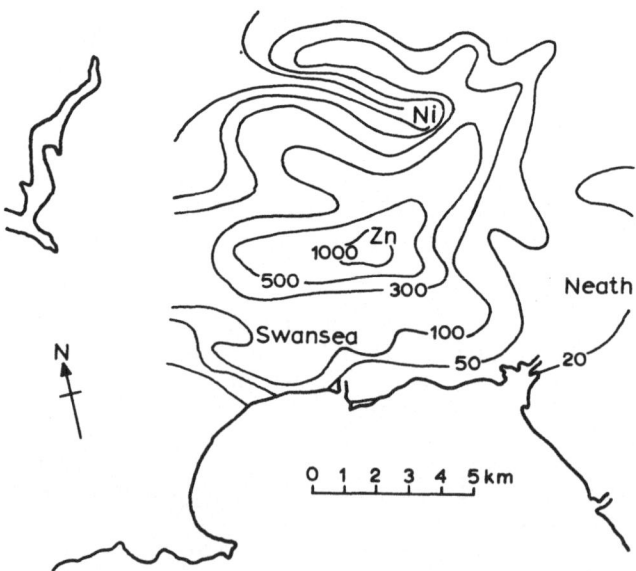

Ni   Nickel works
Zn   Zinc smelter

Figure 1.   Lead deposition on to Sphagnum moss-bags exposed during April 1971 (expressed as ng.$cm^{-2}day^{-1}$).

isopleth lines show a gradual decrease in metal deposition with distance from the source until 'background' levels are reached 10 - 15km away. This somewhat unexpected result suggested that lead in fumes and as dust was being transported appreciable distances from its site of origin.

A similar situation exists for the nickel analyses (Fig. 2). However, in this case the centre is not in the same place as before, but lying over the site of a nickel works 3km further up the Valley. In this case, the highest deposition level was 300ng cm^{-2} day^{-1} and the contours show a similar decrease in metal deposition with distance from the source until a 'background' contour of 10ng cm^{-2} day^{-1} is achieved again 10 - 15km away.

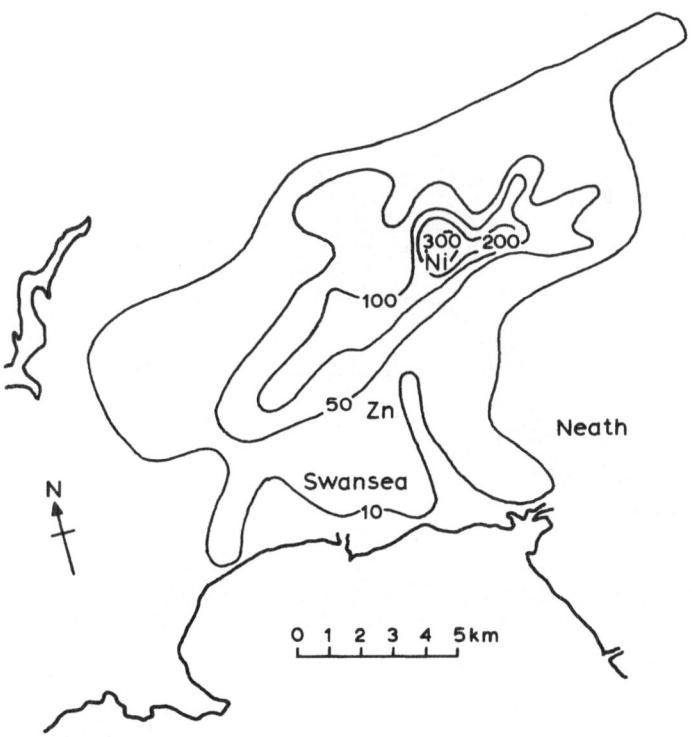

Figure 2. Nickel deposition on to <u>Sphagnum</u> moss-bags exposed during April 1971 (expressed as ng.cm^{-2}day^{-1}).

Shortly after the above study was carried out, the Swansea zinc smelter closed down and so in December 1971 the experiment was repeated. This showed (Fig. 3) a great reduction, both in the amount of lead deposited on the bags, and in the movement of lead away from the original source-area. A high centre of only 50ng cm^{-2} day^{-1} was found to coincide with the site of the closed zinc works as opposed to that of 1000ng cm^{-2} day^{-1} found in the previous study. The radius of this centre was similar to that described in Fig. 2, but background levels of lead deposition were found 6-8km away. Under these conditions a new high centre of 100ng cm^{-2} day^{-1} was noted 3km SW of the works. This coincided with the position of the old smelter-waste tips whose contribution to the atmospheric Pb burden had hitherto been swamped by emissions from the works.

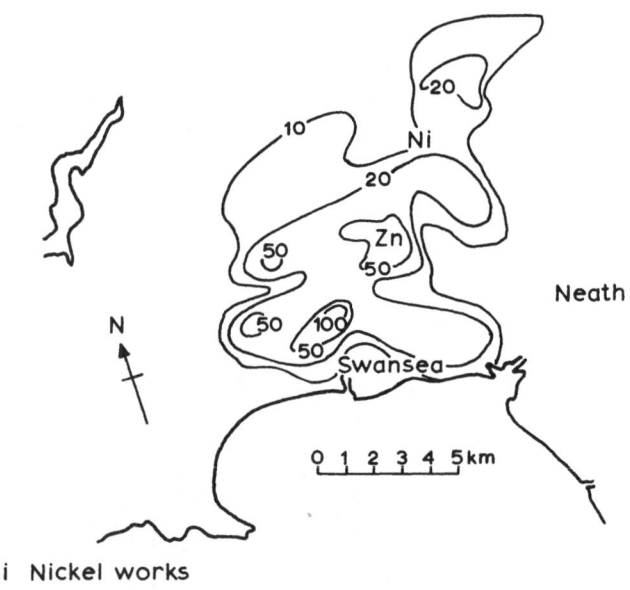

Ni Nickel works
Zn Zinc smelter

Figure 3.   Lead deposition on to Sphagnum moss-bags exposed during December 1971 (expressed as ng.cm^{-2} day^{-1}).

In the case of both surveys, the contour lines appear to be extended in a SW – NE direction along the line of the prevailing winds in the area.

SOIL ANALYSIS AND MOSS-BAG SURVEYS IN U.K.

The above study showed that by using moss-bags, source identification and relative deposition rates of metals could be determined simply and inexpensively. A further study was carried out in 1972 in an attempt to see how a method could be worked out to identify significant metal pollution areas in the U.K. Work has already been quoted describing how surface soils and epiphytic mosses can indicate metal fallout.

Samples of soil were collected from six districts in the U.K: Cornwall, Bristol, Mid-Wales, Midlands, Sheffield and The North. These all lay along a transect belt extending from Whitby in North-East England and then running through Sheffield, Birmingham and Bristol, and finishing up in Cornwall. Two branch lines ran westwards, one from Whitby to Windermere and one from Birmingham into Mid-Wales. The soils were air dried and passed through a 1mm mesh sieve. They were then extracted with atomic absorption grade nitric acid at 100^0C for 1 hour and then subjected to peroxide oxidation. Analysis of the extracts were made by atomic absorption spectrophotometry. The ranges of metals in the samples are shown in Tables 1 - 6.

The soils sampled in the north of England (Table 1) gave a good indication of low background levels of the six metals studied and may be regarded as reference controls. The metal content of the soils sampled in Mid-Wales and Cornwall (Tables 2, 3) were in general agreement but somewhat higher than those of the North. This situation is a reflection of the greater miner-alisation of these regions and was particularly illustrated in the soils taken at the two mine sites, a Cornish tin mine and a Welsh lead mine.

In relation to the soils of the above regions, those in and around the Sheffield area (Table 4) were slightly higher in their cadmium, zinc and cobalt content and distinctly higher in their lead, nickel and copper content. The soils of the Midlands (Table 5) showed the same trends, whilst those of the Bristol area (Table 6) were very much higher in their lead, cadmium and zinc content than any other area sampled and their copper content was distinctly higher than those of the Northern soils. Cobalt and nickel were not elevated but the cadmium figures in Bristol were particularly striking.

In all areas where soils were taken from woodland sites they had an elevated metal content. This may arise from the fact that trees act as a 'dust trap', the atmospheric dust being washed from the leaves or falling with them onto the soil.

Analysis of variance on the data showed there was a differ-

a) The North

O.S. map reference	Depth of soil sample (cm)	Locality	ppm metals in dry soil					
			Pb	Cd	Zn	Cu	Co	Ni
45/920110	2	Sea cliffs	114.0	1.7	76.8	24.2	6.1	13.1
	20		114.0	1.0	77.2	19.4	6.1	26.1
45/850080	2	Moorland	114.0	0.9	52.3	9.7	12.2	6.5
	20		57.0	0.7	37.6	9.7	0.0	6.5
44/550820	2	Field	0.0	0.9	52.3	14.5	15.3	0.0
	20		28.5	0.7	52.3	14.5	23.0	6.5
34/870960	2	Field 200 yd from road	57.0	0.7	52.3	29.0	7.7	6.5
	20		85.5	1.0	37.6	24.2	23.0	6.5
34/410990	2	Copse	57.0	1.7	139.0	33.9	23.0	39.2
	20		85.5	1.6	139.0	38.8	23.0	45.7
34/480720	2	Hillside above village	28.5	1.6	77.2	29.1	23.0	19.6
	20		114.0	1.2	77.2	33.9	23.0	19.0

Table 1. 'Total' metal content of soil samples.

b) Mid Wales

O.S. Map reference	Depth of soil sample (cm)	Locality	ppm metals in dry soil					
			Pb	Cd	Zn	Cu	Co	Ni
32/360740	2	Field	57.0	1.4	83.8	33.9	15.3	39.2
	20		57.0	1.7	99.7	43.6	15.3	39.2
22/820740	2	Moorland	114.0	1.7	37.6	19.4	15.3	19.6
	20		57.0	1.6	51.5	19.4	15.3	19.6
22/720720	2	Cwm Ystwyth Lead Mine	339.0	4.9	1390.0	174.5	23.0	26.1
	20		2478.0	2.3	1128.0	92.1	18.4	39.2
22/650560	2	Field	57.0	1.7	58.9	14.5	7.7	19.2
	20		57.0	1.3	58.9	19.4	7.7	13.1

Table 2. 'Total' metal content of soil samples.

282

c) Cornwall

O.S. Map reference	Depth of soil sample (cm)	Locality	ppm metals in dry soil					
			Pb	Cd	Zn	Cu	Co	Ni
10/360250	2	Field	57.0	2.5	125.9	29.1	12.2	16.3
10/38040	2	Disused tin mine	114.0	1.9	220.1	4119.0	42.8	27.2
10/560360	2	Woodland	199.4	2.5	417.0	37.8	24.5	38.0
10/705265	2	Copse	57.0	1.0	106.0	48.5	6.1	27.2
10/805204	2	Moorland	57.0	0.9	65.4	33.9	12.2	27.2
10/125195	2	Moorland	57.0	0.8	81.8	77.5	6.1	21.7
10/710120	2	Sea cliff	114.0	2.2	78.5	19.4	6.1	21.7

Table 3. 'Total' metal content of soil samples.

d) Sheffield

O.S. map reference	Depth of soil sample (cm)	Locality	ppm metals in dry soil					
			Pb	Cd	Zn	Cu	Co	Ni
43/350000	2	Woodland	142.5	1.4	52.3	53.3	12.2	19.6
	20		342.0	2.6	63.8	96.9	6.1	26.1
43/372452	2	Field	57.0	1.4	104.6	48.5	18.4	32.7
	20		57.0	1.6	88.2	43.6	18.4	45.7
43/386469	2	Field downwind	85.5	1.7	104.6	43.6	18.4	32.7
	20	Minestack	85.5	1.8	137.3	48.5	18.4	26.1
43/395902	2	Nr. shotblasting	256.0	0.6	456.0	130.8	18.4	71.9
	20	works	142.2	1.4	86.7	53.3	12.2	19.6
43/403894	2	Industrial	513.0	6.6	1046.0	378.0	30.7	163.0
	20	chimneys	285.0	3.2	261.0	252.0	67.3	111.0
43/450990	Dust	Roadside nr. foundry	1386.0	45.0	6246.0	407.0	53.7	340.0
43/428908	2	Nr. foundry	199.4	2.3	150.4	96.9	18.4	26.1
	20		85.5	1.4	99.7	38.8	18.4	13.1
43/436928	2	Power station	199.4	5.2	420.0	92.1	38.3	26.1
	20		199.4	2.6	163.5	63.0	12.2	19.6
43/458948	2	Nr. foundry	114.0	2.3	157.0	48.5	18.4	19.6
	20		57.0	1.4	63.8	33.9	12.2	26.1
43/486963	2	Field above	228.0	1.4	63.8	87.2	6.1	32.7
	20	industry	28.5	1.3	31.1	14.5	6.1	6.5
43/527987	2	Field above town	85.5	2.2	88.2	145.0	18.4	19.6
	20		57.0	1.4	88.2	63.0	12.2	13.1
43/398922	2	Nr. works smokey	313.0	0.7	150.4	145.4	18.4	32.7
	20		180.0	1.3	78.5	232.6	12.2	6.5

Table 4. 'Total' metal content of soil samples.

e) Midlands

O.S. map reference	Depth of soil sample (cm)	Locality	ppm metals in dry soil					
			Pb	Cd	Zn	Cu	Co	Ni
32/450850	2	Field	57.0	1.3	112.8	38.8	23.0	32.7
	20		57.0	1.7	112.8	33.9	15.3	39.2
32/712835	2	Disused rly. track	71.2	1.7	170.0	67.9	18.4	21.7
	20		57.0	1.7	163.5	96.9	18.4	16.3
32/742788	2	Field	85.5	1.7	125.9	436.0	12.2	10.9
	20		57.0	1.2	65.4	339.0	12.2	5.4
43/726965	2	Woodland	114.0	0.3	604.0	9.7	15.3	56.5
	20		57.0	0.9	508.0	4.8	7.7	65.3
43/830759	2	Canal bank behind	313.0	2.6	431.0	140.6	18.4	21.7
	20	Carpet factory	171.6	2.9	271.0	106.6	18.4	27.2
43/858997	2	Disused airfield	114.0	1.7	714.0	33.9	23.0	19.6
	20		256.0	2.2	2420.0	43.6	23.0	26.1
43/978903	2	Football field	142.5	1.5	455.0	475.0	18.4	32.7
	20	behind works	142.5	2.6	613.0	436.0	12.2	32.7
42/042874	2	Somerfield park	228.0	3.5	620.0	252.0	12.2	45.7
	20	ctr. Birmingham	171.0	2.3	280.0	485.0	18.4	32.7
32/008004	2	Power station	313.0	6.4	1848.0	155.0	30.6	48.9
	20		227.9	4.3	1638.0	121.2	30.6	52.3
32/048030	2	Nr. sandstone	85.5	2.9	24.5	82.4	12.2	13.1
	20	quarry	57.0	0.6	44.1	53.3	6.1	6.5
32/098083	2	Field	57.0	1.4	78.5	43.6	6.1	13.1
	20		57.0	1.4	24.5	33.9	6.1	13.1
32/101103	2	Field	57.0	1.6	88.1	43.6	12.2	10.9
	20		85.5	1.7	31.1	29.1	6.1	10.9

Table 5. 'Total' metal content of soil samples.

Figure 4. Map of the Bristol area

f) <u>Bristol</u> Map ref. of R.T.Z. Smelter 31/523794.

O.S. map reference	Distance from smelter (km)	Depth of soil sample (cm)	Locality	ppm metals in dry soil					
				Pb	Cd	Zn	Cu	Co	Ni
31/518795	0.3	Dust	Roadside dust outside smelter	2251.0	827.0	309015.0	2084.0	42.8	43.5
31/558800	4.0	2	Field N.W. smelter	227.9	10.8	1112.0	92.1	24.5	43.5
		20		85.5	2.9	286.0	33.9	24.5	43.5
31/535804	2.0	2	Rough ground S.W. smelter	2350.0	89.9	11216.0	213.0	24.5	38.8
		20		512.0	143.9	5640.0	82.4	24.5	38.0
31/537784	1.2	2	Field near smelter	855.0	64.7	3597.0	111.5	24.5	43.5
		20		313.4	34.2	1700.0	92.1	18.4	38.0
31/465772	6.4	2	Parkland	170.9	19.8	494.0	33.9	6.1	16.3
		20		199.4	8.1	706.0	58.2	12.2	21.8
31/536722	7.2	2	Golf course	328.0	7.3	327.0	19.4	18.4	21.8
		20		228.0	12.6	319.0	24.2	12.2	21.8
31/474666	13.6	2	Woodland	541.0	16.2	706.0	33.9	18.4	21.8
		20		512.0	12.6	533.0	33.9	6.1	27.8
31/564746	6.4	2	Woodland	598.0	12.6	541.0	53.3	6.1	21.8
		20		1253.0	12.6	613.0	43.3	18.4	21.8
31/524778	1.4	2	Park near smelter and road	99.7	57.5	5477.0	58.2	18.4	27.2
31/517781	1.6	2	Traffic island near smelter	1111.0	52.2	8666.0	232.0	18.4	32.6
		20		313.0	58.4	1431.0	29.1	18.4	16.3
31/532793	0.8	2	Right behind smelter waste land	1390.0	37.8	5477.0	966.0	36.7	32.6
		20		712.0	8.2	2453.0	286.0	29.4	45.7
31/590885	11.2	2	Field downwind smelter	68.4	2.3	152.0	38.8	12.2	27.2
		20		106.8	4.5	320.o	48.5	18.4	27.2
31/613885	12.8	2	Parkland	114.0	2.6	216.0	33.9	12.2	21.7
		20		85.5	2.3	125.9	29.1	18.4	21.7

Grid ref		Depth	Location						
31/632869	13.2	2	Hedge in field	142.5	2.3	304.0	101.8	18.1	32.6
		20		170.9	3.5	481.0	101.8	18.1	32.6
31/582811	6.2	2	700 yds from motorway plantation N.W. smelter	883.0	54.0	1373.0	67.9	12.2	27.2
		20		114.0	23.0	830.0	33.9	12.2	21.7
31/545836	4.8	2	Field near smelter	142.5	2.6	468.0	48.5	12.2	21.7
		20		85.5	3.2	176.6	19.4	18.4	21.7

Table 6. 'Total' metal content of soil samples.

ence between the lead, zinc and cadmium contents of the soils of
the different areas, statistically significant at 0.1 per cent
probability for zinc and cadmium, and 1 per cent for lead. Application of Duncan's new multiple range test, Duncan[17] showed that
the soils of the Bristol area were significantly different from
all other areas in their zinc, lead and cadmium content.

Another feature of interest was that the soil samples taken
at 2cm below the surface from the Sheffield and Bristol districts,
showed a marked tendency to be higher in Pb, Cd, Zn and Cu than
corresponding samples taken at 20cm below the surface, whereas
only a slight tendency was observable in the Midlands samples, and
none at all in the soils from the North of England and Mid-Wales.
This suggested surface enrichment from atmospheric fallout of Pb,
Zn, Cd and Cu from urban industrial emissions in the Sheffield and
Bristol districts. However it was not possible to decide whether
this was from an earlier period of emission which had now ceased
or whether metal fallout was still going on, at the same or a
reduced rate, as formerly.

In order to clarify this point, the sampling was followed by
a moss-bag survey in the Bristol area (Fig. 4) on two occasions,
firstly when a zinc smelter in the area had closed for internal
modificiation, and secondly a month later when normal operation had
been resumed. The results of the first survey showed that elevated levels of zinc, lead and cadmium deposition were found to
extend eastward from the works marked on the contour maps (Figs 5a,
6a, 7a). Centres of 1000, 500 and 30ng cm^{-2} day^{-1} for zinc, lead
and cadmium respectively were found to occur at the site of the
smelter. Analysis also showed that cobalt and copper deposition
onto moss-bags (Figs 8a, 9a) was slightly elevated in a localized
area around the smelter.

Similar trends for zinc, lead and cadmium deposition were
found in the second survey after the smelter re-opened, (Figs 5b,
6b, 7b) except that there was a 3000ng cm^{-2} day^{-1} centre for zinc,
100ng cm^{-2} day^{-1} centre for cadmium, but there was no elevation in
lead deposition during this survey. Elevated levels of cobalt and
copper (Figs 8b, 9b) again occurred in bags placed close to the
smelter. There was a general trend eastwards in the contour and
proportional circle plots of all metals in this survey as there
was in the first. The exact shape of the isopleths is probably
affected by wind direction and strength. The prevailing winds in
this area are from the SW.

These results indicated ongoing metal deposition, although
it is impossible to say from the present study whether the rate
is reduced compared with earlier deposition rates.

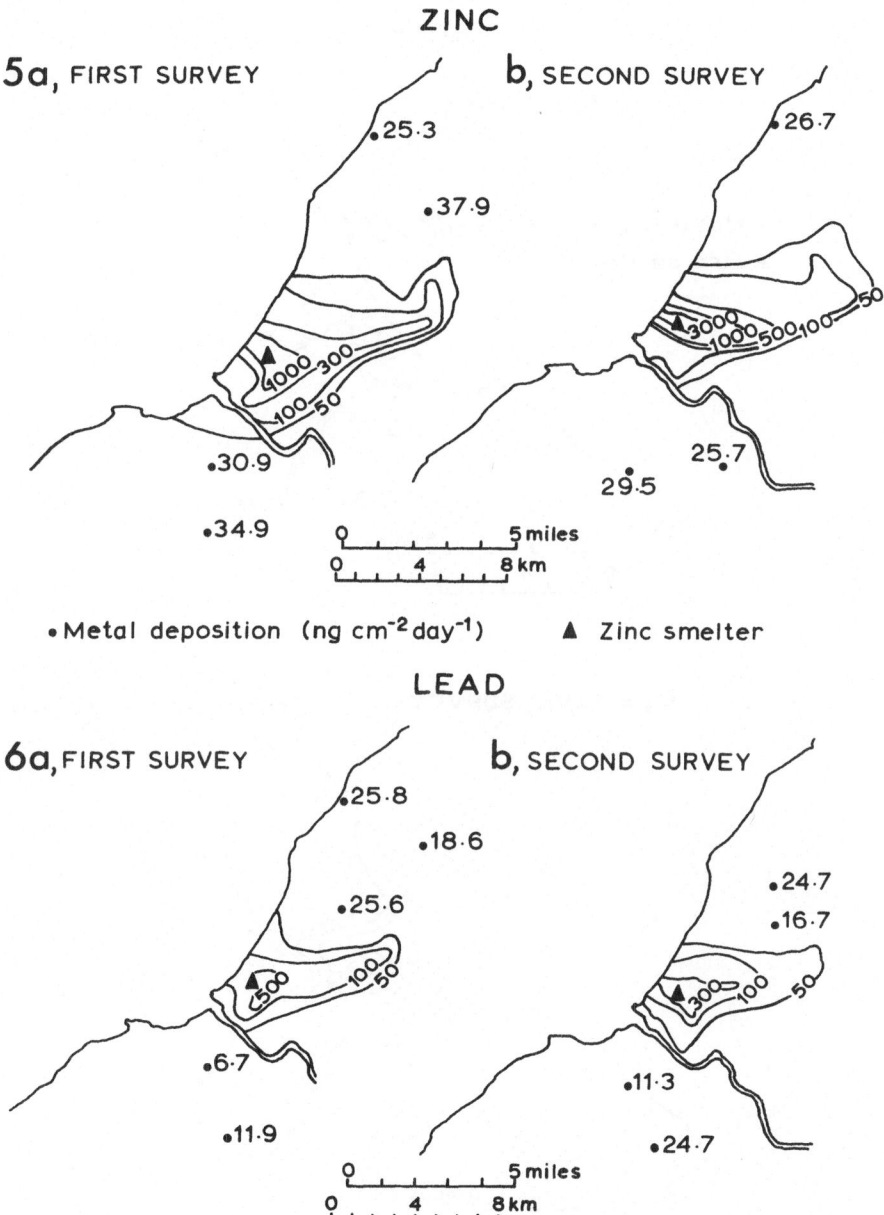

Figures 5 & 6. Zinc and lead deposition in the Bristol area.

CADMIUM

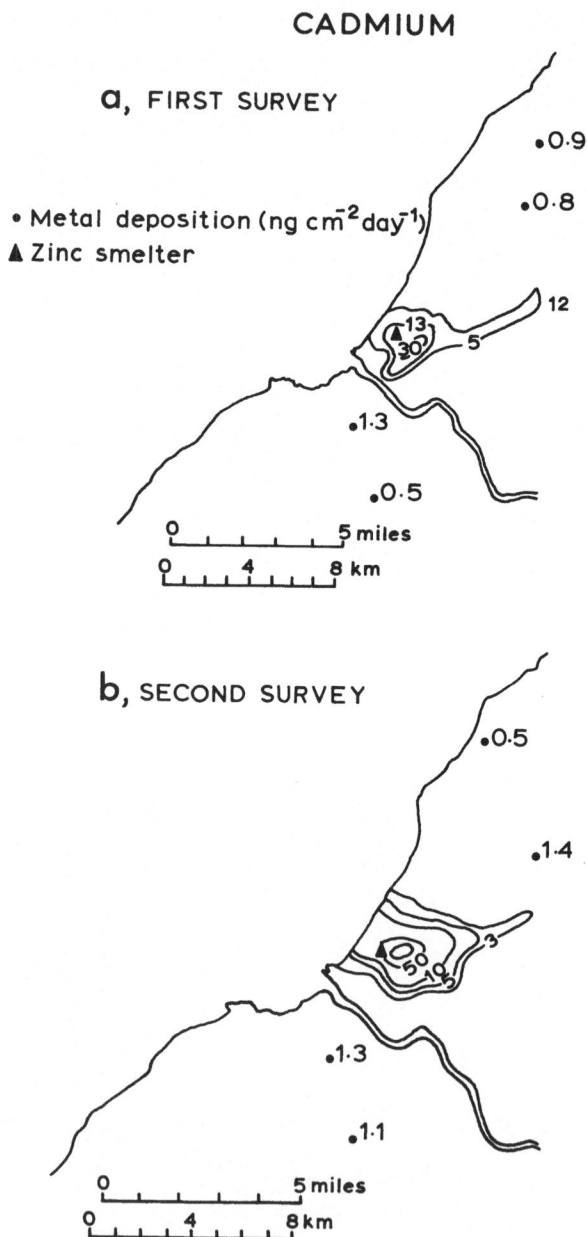

Figure 7.   Cadmium deposition in the Bristol area.

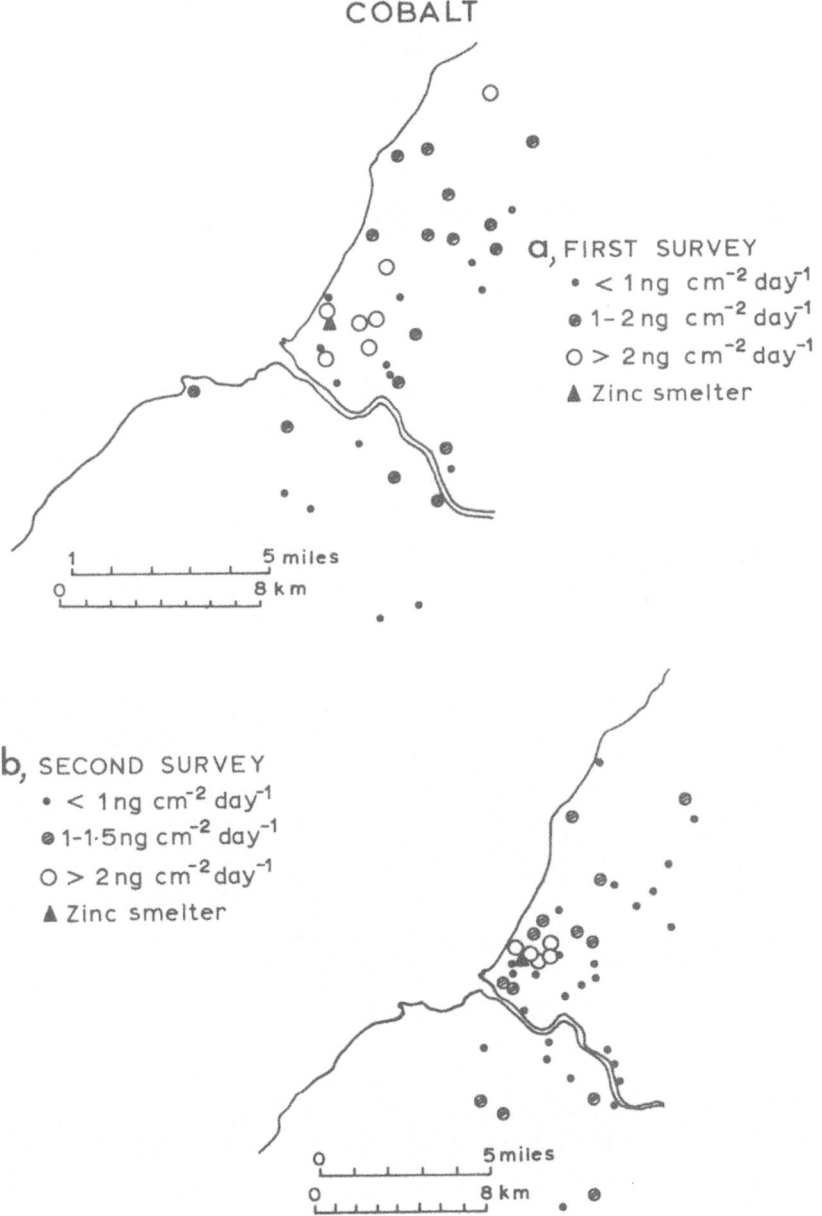

COBALT

a, FIRST SURVEY
• < 1 ng cm^{-2} day^{-1}
⊙ 1- 2 ng cm^{-2} day^{-1}
○ > 2 ng cm^{-2} day^{-1}
▲ Zinc smelter

1   5 miles
0   8 km

b, SECOND SURVEY
• < 1 ng cm^{-2} day^{-1}
⊙ 1-1·5 ng cm^{-2} day^{-1}
○ > 2 ng cm^{-2} day^{-1}
▲ Zinc smelter

0   5 miles
0   8 km

Figure 8. Cobalt deposition in the Bristol area.

292

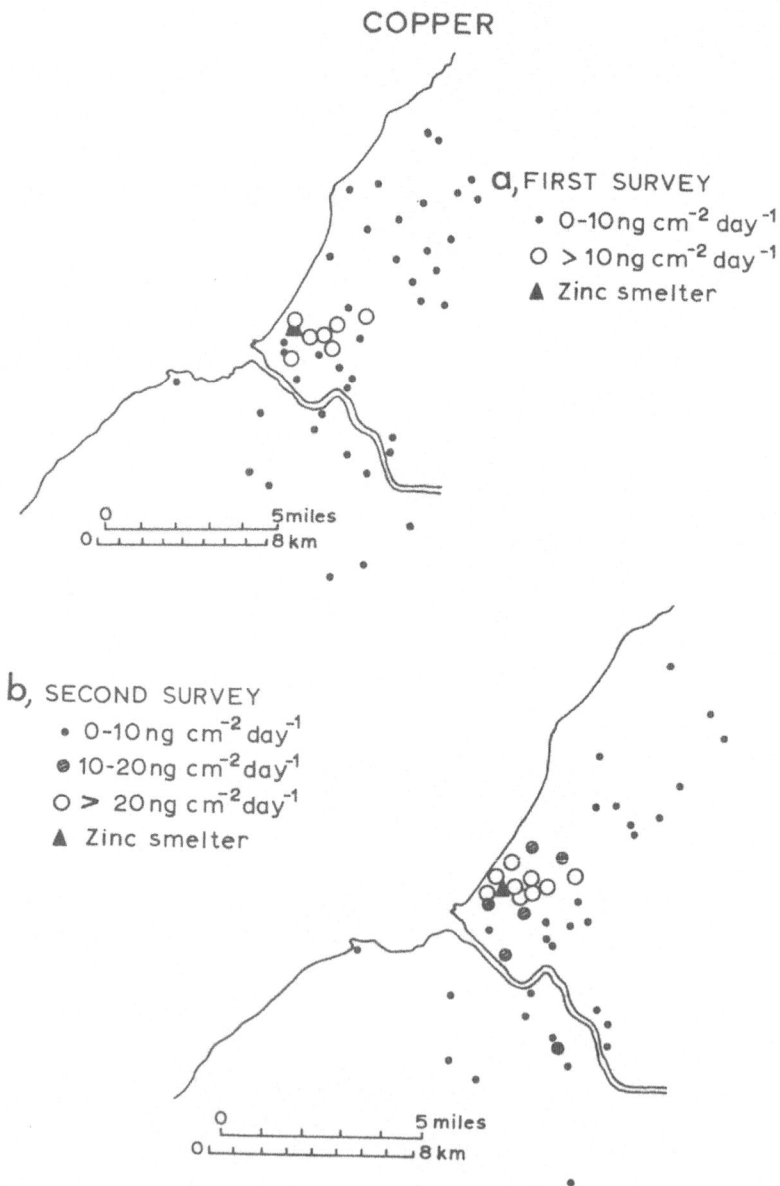

COPPER

a, FIRST SURVEY
• 0-10 ng cm^{-2} day^{-1}
O > 10 ng cm^{-2} day^{-1}
▲ Zinc smelter

5 miles
8 km

b, SECOND SURVEY
• 0-10 ng cm^{-2} day^{-1}
● 10-20 ng cm^{-2} day^{-1}
O > 20 ng cm^{-2} day^{-1}
▲ Zinc smelter

5 miles
8 km

Figure 9. Copper deposition in the Bristol area.

MOSS-BAGS AS SEMI-QUANTITATIVE OR RELATIVE DEPOSITION GAUGES

Because of the small amounts of metal usually present in the air, conventional aerial metal analysis is expensive in terms of apparatus and complex in subsequent chemical analysis. Electrical power supply is required to drive filter pumps. Its on-site availability as well as the cost of the air-filtration equipment severely limits the number of sites which may be sampled at one time. However, to assess the geographical distribution of metal deposited from the air it is desirable to have as many sampling sites as possible, often in remote places. Using the methods described in this paper a simple and relatively inexpensive means of defining geographically the area of metal contamination is available. This could well be coupled with conventional air monitoring techniques which would provide additional quantitative information.

It is becoming increasingly evident from work in progress that a real relationship exists between data collected from vertically, or particularly from horizontally exposed moss-bags and that achieved by measurements of total deposition, dry deposition, and air concentration, as measured by AERE, Harwell at a sampling station at Trebanos in S. Wales sponsored by the Natural Environment Research Council. Data collected over a period of six months in 1973 has shown that there is a very significant correlation between results of all four methods of monitoring (Table 7). In fact, work to date suggests that a moss-bag placed horizontally at the same height as a short-grass-sward intercepts the same amount of metal per unit area and time. Detailed work to verify this preliminary conclusion is continuing.

Table 7. Correlation coefficients (r) comparing four measures of airborne metals.

Moss bag/ total deposition	Moss bag/ dry deposition	Moss bag/ air concentration
0.841 ++++	0.891 ++++	0.739 ++++

Total deposition/ air concentration	Dry deposition/ air concentration	Total deposition/ dry deposition
0.768 ++++	0.736 ++++	0.820 ++++

Significant at 0.1%   ++++

CONCLUSION

Our environment is being continually subjected to fallout from metals in the air but in some cases, mainly in those areas close to intensive urbanization, ongoing industry or old industrial dumps, the fallout is increased. Any further planning for the revegetation of newly created or old derelict sites must involve attempts to neutralise the toxic effects of the land itself. Additionally, an estimate must be made of airborne contamination, prior to embarking on reclamation schemes. Monitoring by the methods described will greatly facilitate such studies and give some clues to their control.

ACKNOWLEDGMENTS

We wish to thank the Natural Environment Research Council for financial support for this work and the Royal Society for the provision of items of analytical equipment.

REFERENCES

1. Purves, D., Contamination of urban garden soils with copper and boron, Nature, Lond., 210, 1077, 1966.

2. Purves, D., Contamination of urban garden soils with copper, boron and lead, Pl. Soil, 26, 380, 1967.

3. Purves, D., Trace-element contamination of soils in urban areas, Trans. int. Congr. Soil Sci., Adelaide, No. 9, 351, 1968.

4. Purves, D., Enhancement of trace-element content of cabbages grown in urban areas, Pl. Soil, 33, 483, 1970.

5. Burkitt, A., Lester, P & Nickless, G., Distribution of heavy metals in the vicinity of an industrial complex, Nature, Lond., 238, 327, 1972.

6. Djuric, D., Kerin, Z., Graovac-Leposavic, L., Novak, L. & Kop, M., Environmental contamination by lead from a mine and smelter, Archs. env. Hlth., 23, 275, 1971.

7. Costescu, L.M. & Hutchinson, T.C., The ecological consequences of soil pollution by metallic dust from the Sudbury smelters, Proc. 18th ann. Mtg. Inst. env. Sci., New York, 18, 540, 1972.

8. Goodman, G.T. & Roberts, T.M., Plants and soils as indicators of metals in the air, Nature, Lond., 231, 287, 1971.

9. Motto, H.L., Daines, R.H., Chilks, D.M. & Motto, C.K., Lead in soils and plants: its relationship to traffic volume and proximity to highways, Environ. Sci. Technol., 4, 231, 1970.

10. Dedolph, R., Ter Haar, G., Holtzman, R. & Lucas, H. Jr., Sources of lead in perennial ryegrass and radishes, Environ. Sci. Technol., 4, 217, 1970.

11. Everett, J.L., Day, C.L. & Reynolds, D., Comparative survey of lead at selected sites in the British Isles in relation to air pollution, Fd. Cosmet. Toxicol., 5, 29, 1967.

12. Lagerwerff, J.V. & Specht, A.W., Contamination of roadside soil and vegetation with cadmium, nickel, lead and zinc, Environ. Sci. Technol., 4, 584, 1970.

13. Rühling, Å. & Tyler, G., An ecological approach to the lead problem, Bot. Not., 121, 321, 1968.

14. Rühling, Å. & Tyler, G., Ecology of heavy metals - a regional and historical study, Bot. Not., 122, 248, 1969.

15. Roberts, T.M., The spread and accumulation in the environment of toxic non-ferrous metals from urban and industrial sources, Ph.D. thesis, Univ. Wales, 1972.

16. Clyrno, R.S., Ion exchange in Sphagnum and its relation to bog ecology, Ann. Bot. Lond. (N.S.), 27, 309, 1963.

17. Duncan, D.B., Multiple range and multiple F tests, Biometrics, 11, 1, 1955.

HEAVY METALS IN RELATION TO PLANT GROWTH ON MINE AND MILL WASTES

H.B. Peterson and Rex F. Nielson

Department of Agricultural and Irrigation Engineering
and Department of Plant Science, Utah State University,
Logan, Utah, U.S.A.

ABSTRACT

In the Western United States studies have been conducted in
the laboratory, greenhouse and the field to identify and solve the
problems relating to stabilizing metal mine tailings with vegetation
in an arid climate. The major objective has been to establish a
protective cover.

The wastes have been found to be highly variable in physical
and chemical characteristics. Variability is related to type of
mining operation, ore composition, amount of pyrite and other sulphides present, type of milling process, quality and quantity of
water used to transport the waste and age and degree of exposure
of the tailings.

Before establishing vegetation on the sites, it is necessary
to make an evaluation of the environmental conditions. This is difficult because of the analytical problems and the changes that take
place in the waste piles. Sulphides oxidize, increasing the acidity,
salinity, and solubility of heavy metals. In addition, there is
frequently heavy-metal enrichment from outside the pile (from smelter stacks or water irrigation).

Vigorous, non-succulent, unpalatable species that are not sensitive to nutrient deficiencies or nutrient imbalances and are adapted to the prevailing climate are most successful. In arid climates
where some irrigation water is available, Agropyron elongatum and
Festuca arundinacea are used effectively. Among the legumes tried,
Melilotus officinalis has been most successful.

Many of the wastes are low in ion exchange capacity and buffer capacity. All waste pile materials studied are deficient in plant nutrients and are acidic or become so. Few species will grow in tailings of pH 5.5 or less when heavy metals are present in appreciable amounts. Lime incorporated into the tailings and fertilizers are needed in order to facilitate plant growth. Lime raises the pH and reduces the solubility of the metals but in so doing, increases the availability of Molybdenum if present. Uranium tailing ponds can usually be covered with top soil in which plants may easily be established.

Drainage water from the waste pile containing soluble heavy metals, constitutes a pollution hazard. Plants grown on the waste piles may contain amounts of heavy metals and/or Molybdenum that are toxic to grazing animals.

Some means for control of wind erosion is usually necessary. Cost is a major factor determining the method used.

In arid climates some irrigation is usually necessary in order to establish an adequate vegetative cover. It is likely some supply of irrigation would be profitable in more humid climates during periods of drought.

INTRODUCTION

Solid mill wastes from heavy-metal mining operations may constitute an important environmental problem in the Western United States. The large dumps of material detract from the aesthetic appearance of the landscape and can contribute greatly to air and water pollution. Vegetation establishment is one of the preferred methods of stabilization, but conditions favorable for plant growth seldom exist in arid climates. Unfortunately, it is not possible to use vegetation for stabilization of active waste piles.

The composition and quantity of waste material varies considerably, great differences existing at each of the several mine and mill waste sites. For this reason, it is necessary to study each site separately to identify the adverse characteristics in order to determine the treatment required to enhance plant growth. Simple solutions to the problems are rarely found. Working with mill tailings from copper, uranium, lead and zinc mining operations most sites were found to have conditions unfavorable for plant growth caused by one or more of the following: (i) salinity; (ii) acidity; (iii) erosion from sand blasts; (iv) improper water relations; (v) poor aeration; (vi) nutrient deficiencies; (vii) toxic substances; (viii) selective grazing by animals; (ix) interactions of any of these that militate against plant growth.

Although the literature on the levels and adverse effects was reviewed for the following elements: Al, As, Ba, B, Cd, Co, Cr, Cu, Fe, Hg, Mn, Mo, Ni, Li, Pb, Rb, Sr, V, and Zn, attention is confined here to B, Cu, Fe, Mo, Mn, Pb and Zn.

SITE EVALUATION

In appraising which adverse site conditions are present, the most difficult task is the measurement of nutrient deficiencies and potential toxicities.

There are problems of chemical analysis (e.g. it is difficult to measure concentrations of an element present at deficiency thresholds in the presence of great excesses of other elements). The greatest difficulties however are encountered as a result of changes in the concentration of toxic materials following the development of acidity. One estimate of potential toxicity in relation to acidity can be made by treating a sample with acid or buffer extractants of different pH (Table 1).

Most of the wastes coming from the mill, are near neutral or alkaline. In the fresh condition, the concentration of soluble heavy metals is low and is unlikely to be harmful to plants. If Molybdenum is present, as it is soluble at near neutral conditions, it is readily taken up with little or no harm to the plants. In many of the wastes, there are potential acid forming materials, principally sulphides of iron or copper which may also release heavy metals in soluble form.

Table 1. Comparison of the amounts (ppm) of Cu, Fe, Mn and Zn extracted by various extractants*.

Elements	0.1 N HCl c.pH 1.0	NH_4OAc pH 4.8	NH_4OAc pH 7.0	Water
Cu	640	500	122	<0.5
Fe	790	275	2.1	<1.0
Mn	28	19	6.0	<1.0
Zn	24	18	0.6	<0.1

*20 g copper tailings shaken for 2 hours with 50 ml of extractant, filtered and analysed by absorption spectrophotometry.

The reactions may be conveniently represented as:

$$FeS_2 + H_2O + 3\ 1/2\ O_2 \rightarrow FeSO_4 + \underline{\underline{H_2SO_4}}$$

$$2\ FeSO_4 + H_2SO_4 + 1/2\ O_2 \rightarrow Fe_2\ (SO_4)_3 + H_2O$$

$$Fe_2\ (SO_4)_3 + 6\ H_2O \rightarrow 2\ Fe\ (OH)_3 \downarrow + \underline{\underline{3\ H_2SO_4}}$$

Chalcocite reacts chemically with ferric sulphate:

$$Cu_2S + 2\ Fe_2\ (SO_4)_3 \rightarrow \underline{\underline{2\ CuSO_4}} + 4\ FeSO_4 + S$$

The elemental sulphur can be oxidized by <u>Thiobacillus</u> making only a small contribution to total acidity:

$$2\ S + 3\ O_2 + 2\ H_2O \rightarrow \underline{\underline{2\ H_2SO_4}}$$

Copper oxide reacts with ferric sulphate:

$$3\ CuO + Fe_2\ (SO_4)_3 \rightarrow \underline{\underline{3\ CuSO_4}} + Fe_2O_3$$

Chemical changes are also possible as a result of enrichment from irrigation water, or emissions from stacks where coal is burned or from smelter stacks. Soils around coal-burning installations have been found to be enriched in Ag, Cd, Cr, Fe, Hg, Ni, Ti and Zn[1]. Around smelters enrichment by such toxicants as arsenic, fluorides, and the oxides of heavy metals may occur. In addition, large amounts of oxides of sulphur are released.

With the many possible means of site deterioration outlined above, it is not uncommon for early plant establishment successes to be followed by the cessation of plant growth and destruction of the vegetation. The changes can be more or less qualitatively anticipated, but quantitative the enrichment potential is difficult to estimate.

TOXICITY TO PLANTS

In spite of the large literature on the effect of various levels of metals on plant growth[2] there is a paucity of information indicating exact metal levels toxic to plants and animals, particularly in relation to valency states and acute and chronic damage. Usually more than one element is involved and synergistic or antagonistic effects have been demonstrated[3, 4, 5]. In the presence of organic matter, such as dead plant roots, metals may become complexed. The changed availability produces an effect on growing plants different from that of soluble inorganic salts of the heavy metal. The anions accompanying the heavy metals also influence the absorption of the metal. For example, more lead is absorbed by plants in the presence of high nitrate levels than in the presence of high levels of chloride[5].

Plants are usually relatively easy to establish on fresh tailings if soluble salts are leached and the nutrient status, water level and protection from the wind is adequate. Soon after establishment, however, acidity may increase and the plants die. On copper tailings all plants died when the pH fell to 5.5. At this pH copper compounds in the tailings become highly soluble (Fig. 1). These pH changes may occur most rapidly at or near the root surface.

The relationship found between pH and the solubility of two heavy metals from tests made on several copper and uranium wastes is given in Table 2. In the absence of lime and in material where the buffer capacity is very low, acidity can increase rapidly and toxic amounts of soluble metal salts are formed.

Table 2. pH and metal content of water saturation extracts of mine waste materials.

Source of Waste	pH	Saturation Extract (ppm)				
		B	Cu	Fe	Zn	Pb
Copper	2.2	10	600	3000	21	5.00
Copper	5.2	0.35	195	0.59	2.40	5.00
Copper	7.6	0.18	1.00	0.50	0.15	5.00
Uranium	1.9	24	600	5000	77	30
Uranium	2.9	1.40	475	45	103	16
Uranium	7.8	0.18	1.00	1.00	0.15	5.00

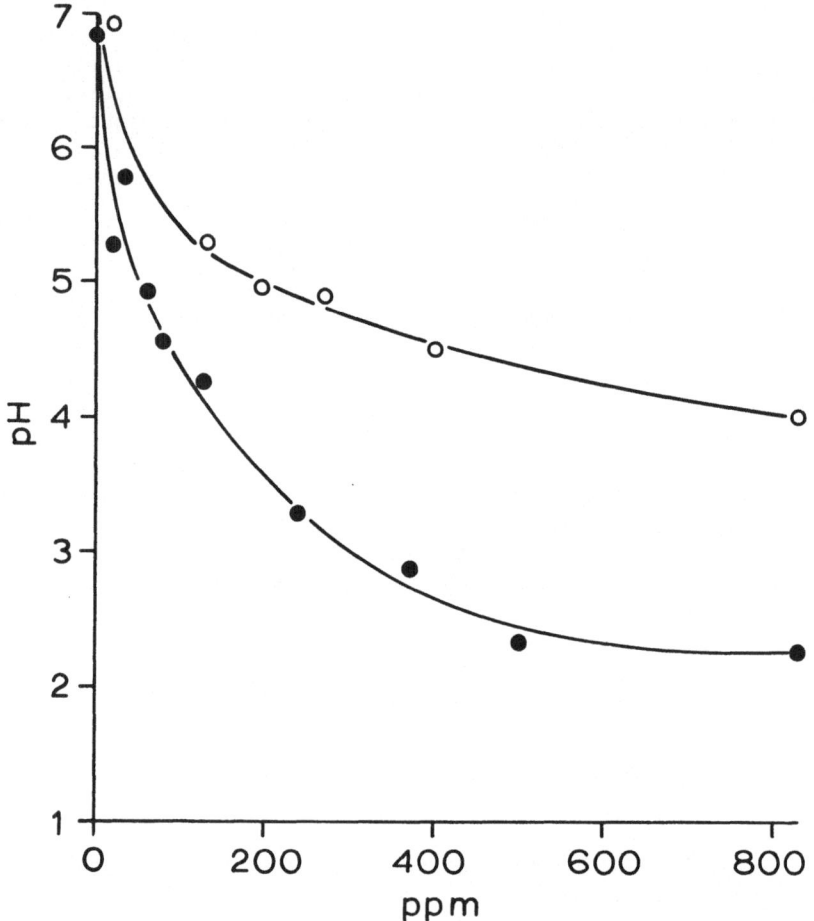

Figure 1. The solubility of total copper and total iron in tailings as affected by induced pH change.

In humid areas, the rainfall may be sufficient to leach waste piles and remove the acid and soluble heavy metals salts as they are formed. In arid areas, it is necessary to apply irrigation water for the crop. This may leach out the accumulating salts and soluble compounds of the heavy metals. If the water applied contains soluble calcium salts, it may regulate the pH and the formation of toxic amounts of the soluble heavy metals.

The use of plant tissue analysis to determine the relationship between substrate metal concentrations and poor growth presents certain difficulties. Many heavy metals are accumulated in the roots and translocation rates to the shoots are low. High concentrations cause progressive root death and with an inadequate system, the plants are not able to absorb enough nutrients or water.

FERTILITY AND LIME TREATMENTS

With few exceptions, wastes are low in nitrogen, phosphorus
and potash and inorganic fertilizers may effectively be used. The
waste materials of low exchange and buffer capacity must be fertilized frequently. Manure, sewage treatment sludge and cannery
wastes can serve as sources of nutrients and organic matter. These
organic materials provide for slow release of plant nutrients. In
many areas, it is not possible to use sewage effluent or sludges
because of the health hazards from possible virus infection or the
size and remoteness of the tailings ponds.

Minor element deficiencies are likely to be induced when treatments are made to reduce toxicities (e.g. reduction of copper toxicity may induce zinc or iron deficiencies). It has been found in
greenhouse studies that when manure is added to reduce phosphorus
deficiency, chlorosis is induced which appears to be due to zinc
deficiency.

It is helpful to allow plant residues to accumulate on the
surface. The dead material serves as a mulch that improves the
moisture relations, reduces damage from wind, serves as a source of
plant nutrients for new plant growth and possibly as a source of
chelating agents.

The lime requirement of the material is frequently difficult to
determine accurately as the waste material is not uniform in composition when it comes from the mill and is separated or stratified
when deposited. In copper tailings the slimes may contain less
sulphides and more calcium and magnesium carbonates than is found
in the coarse sandy fractions. Acid deposition from stacks can
further add to the problem of measuring the amount of lime necessary
to prevent acidification. Smelter plants near a tailings pond where
we conduct research may emit large quantities of sulphur dioxide per
day, as much as 500-2,500 tons per day (Thomas[6]). If scrubbers are
installed, these amounts may be reduced. With these uncertainties,
as a general guide about 4 tons of finely ground limestone should
be well mixed with the tailings before crops are planted. If the
tailings materials are acidic at the time of treatment, there is
some evidence that the lime should be added before the phosphate is
applied otherwise the phosphate is less effective. Additional lime
should be added periodically, as indicated by tests. When a perennial species is to be maintained, all the lime required should be
applied initially. Large amounts of lime may otherwise be necessary
to prevent the acidification of the materials at the root surfaces.
However the problem of when to apply the lime to counteract the
developing acidity and the exact amounts required has not been
solved.

CROP SELECTION

Experience has shown that the most successful plants are likely
to be vigorous, not sensitive to nutrient deficiencies or nutrient
imbalances and not especially attractive to grazing animals. In
arid climates, salt and drought tolerant plants of a non-succulent
nature seem best adapted. Tests with native plants have not been
encouraging but Festuca arundinacea or Agropyron elongatum meet
these requirements. Sweet clover (Melilotus officinalis) has been
the most satisfactory legume tested. Where salty and wet slimes
are prevalent, tamarisk (Tamarix pentandra) is well-adapted. This
species exhibits a degree of salt tolerance and adapts well to the
prevailing spoil conditions. It appears to translocate the metals
through the plant and excrete a certain amount through salt glands
in the leaves. Table 3 shows that the greatest concentrations are
found in the leaves.

As no sites have been found in the Western United States where
plants have been exposed to toxic waste conditions for several generations, little or no opportunity exists to find species populations
with tolerance to heavy metals. Additionally there is seldom only
one toxic element present in excess. Tolerant plants such as the
Agrostis spp. (Bradshaw[7]), are not adapted to arid climatic conditions. If they were adapted, they are too palatable to animals.

Plants tolerant to soluble aluminum have been tested, but show
no specific tolerance to copper and iron.

WIND EROSION CONTROL

It is essential that wind erosion be controlled while plant
cover is being established in arid conditions. This is usually difficult and expensive. In South Africa, stems of reeds have been
used as windbreaks (James[8]). The costs for such barriers may however be prohibitive. Good control has been accomplished by the use

Table 3. Metal content (ppm) of Tamarisk seedlings grown in tailings

	Cu	Fe	Zn	Mn
Whole plant	35	155	24	1.5
Leaves	65	240	31	4.5
Stems	20	90	20	0
Roots	35	105	18	1.5

of petroleum binding agents such as "Coherex" which has been used
without deleterious effects on seedlings or plants at an estimated
cost of approximately $60 per acre for the material.

In Montana, straw has been partially incorporated into the surface by a special disc plough. Wood chips and bark have also been
found to be effective in control of erosion on some berm sites. For
straw or bark mulches to be effective, there must be no unprotected
area from which material can be blown to cover the mulching material.

In most waste piles in arid climates some irrigation water is
necessary in order to reduce wind erosion and ensure rapid establishment and maintenance of a vegetative cover. It is likely that some
irrigation would be desirable in humid climates to ensure against
damage to plants during periodic droughts. The trickle system of
irrigation is one attractive method of applying water to trees on
steep slopes.

SECONDARY PROBLEMS

Potential water pollution is another important problem. When
soluble salts of the heavy metals are leached from the root zone,
they move out of the tailings and into streams, drains or ground-
water where ultimately they may have an adverse effect on man and
other organisms. Only minute amounts of trace elements are permissible in drinking water (Table 4). Even the amounts tolerable in
irrigation water (Table 5) are readily exceeded. Too often these
facts have been and still are neglected when leaching wastes.

Animals may be damaged as a result of eating vegetation polluted with heavy metals. Some very high concentrations of Molybdenum
have been found in alfalfa and grass grown on tailings. The lead
and Molybdenum in the tailings may not adversely effect plant growth,
but when ingested by grazing animals via plants may be harmful to
them. Horses seem particularly sensitive to lead poisoning. This
may be due to a greater stomach acidity or the ingestion of small
amounts of spoil during grazing.

In an area down stream from a smelter, horses were being severely damaged by suspected metal toxicity. A study revealed that
material from the mill waste had been carried by water to the area
where the water was used to irrigate the pasture. Apparently, the
lead and other metals became soluble as a result of the oxidation
of sulphides. Table 6 shows a high content of metals in the soils
and the grass eaten by the horses. No soluble metals were ever
found in the animal drinking water. It was established that there
was lead poisoning in the horses, but it was also possible there
was damage from zinc and copper.

There seems to be no problem of mercury toxicity to plants or to animals from the tailings waste piles. Some of the potential problems of cadmium contamination have not yet been explored.

Table 4. Criteria for trace elements in public water supplies.

Metal	Permissible Criteria	Desirable Criteria
	mg/1	
Arsenic	0.05	Absent
Barium	1.0	Absent
Boron	1.0	Absent
Cadmium	0.01	Absent
Chromium	0.05	Absent
Copper	1.0	Virtually absent
Iron (filterable)	0.3	Virtually absent
Lead	0.05	Absent
Manganese	0.05	Absent
Selenium	0.01	Virtually absent
Silver	0.05	Virtually absent
Zinc	5	Virtually absent

Note: Virtually absent implies that the substance is present in very low concentrations and is used where the substance is not objectionable in these barely detectable concentrations.

Table 5. Trace element tolerances for irrigation waters[7].

Element	For Water Used Continuously on Any Soil	For Short-Term Use on Only Fine-Textured Soils
	mg/1	mg/1
Aluminum	1.0	20.0
Arsenic	1.0	10.0
Beryllium	0.5	1.0
Boron	0.75	2.0
Cadmium	0.005	0.05
Chromium	5.0	20.0
Cobalt	0.2	10.0
Copper	0.2	5.0
Lead	5.0	20.0
Lithium	5.0	5.0
Manganese	2.0	20.0
Molybdenum	0.005	0.05
Nickel	0.5	2.0
Selenium	0.05	0.05
Vanadium	10.0	10.0
Zinc	5.0	10.0

Table 6. The content in soils and grass of four metals for
'contaminated' and reference sites.

Location of 'Contaminated' Soils	Parts per million			
	Zn	Cu	Pb	Fe
{Bridge & plant	640	15	1650	300
{Bridge	250	3.0	790	450
Bottom pasture	335	<1.0	580	600
Railroad pasture	340	<1.0	790	475
Reference Soils				
Farmington	6.5		<1.0	6
Smithfield	1.3		<1.0	11
'Contaminated' Grass Clippings a)	190	24	310	3.99
b)	354	25	135	6.27

Ammonium acetate extract in pH 4.8 for soils.
Acid soluble metals in grass clippings.
*Data from personal communication from Dr James L. Shupe.

REFERENCES

1. Klein, D.H. & Russell, P., Heavy metals: fallout around a power
    plant, Envir. Sci. Technol., 7, 357, 1973.

2. Chapman, H.D. (Ed.), Diagnostic Criteria for Plants and Soils,
    Univ. California, Div. agric. Sciences, 1966.

3. Hemphill, R.D. (Ed.), Trace Substances in Environmental Health
    III, Univ. Missouri, 1968.

4. Hemphill, R.D. (Ed.), Trace Substances in Environmental Health
    IV, Univ. Missouri, 1970.

5. National Academy of Sciences, Biological Effects of Atmospheric
    Pollutants: Lead, Washington, 1972.

6. Thomas, M.D., The effects of air pollution on plants and animals,
    in Ecology and the Industrial Society, Wiley, New York, 1965.

7. Bradshaw, A.D., McNeilly, T.S. & Gregory, R.P.G., Industrial-
    ization, evolution and the development of heavy metal tolerance
    in plants, in Ecology and the Industrial Society, Wiley,
    New York, 1965.

8. James, A.L., Stabilizing mine dumps with vegetation, Endeavour,
    25, 154, 1966.

9. Federal Water Control Administration, Water Quality Criteria,
    Washington, 1968.

THE VALUE OF HEAVY METAL TOLERANCE IN THE REVEGETATION OF
METALLIFEROUS MINE WASTES

A.D. Bradshaw, M.O. Humphreys and M.S. Johnson

Department of Botany, University of Liverpool, United
Kingdom

ABSTRACT

The adverse visual and environmental affects from heavy metal
mining and processing may be counteracted by vegetational
establishment on waste material. Toxicity, nutrient deficiency
and physical checks to growth may be overcome by a combination of
spoil amendments (inorganic fertilizers or organic wastes) and the
use of metal tolerant plant material. Some commercial varieties
of metal-tolerant grass species are now available in the United
Kingdom.

INTRODUCTION

Mining for heavy metals and processing of crude mineral ores
produces areas of derelict land which are both visually unattractive and possible sources of environmental contamination. In many
parts of the world these problems are accentuated because the
derelict areas are often adjacent to agricultural land or in areas
of great aesthetic and amenity value. There is a great need to
stabilise these areas. There are many different methods of which
the establishment of vegetation is one. It has considerable
advantages in terms of cost and permanence. However metalliferous
wastes are very hostile environments for plants. Several major
problems have to be overcome. Recently it has been pointed out
that metal tolerant plants can provide a simple and economical
solution to many of the problems encountered. In this account the
origins of tolerant plants and the genetical and physiological
characteristics of metal tolerance are therefore considered, and
the potential value of metal tolerant material in restoration work

and the limitations are discussed.

THE NEED FOR RESTORATION

The abandoned spoil tips which form an integral part of
derelict heavy metal mines are sources of environmental contamination, because of their metal content (Table 1). On older mine
sites the mineral wastes produced by the ore concentration
processes occur either as surface spoil tips or as deposits in
crude lagoons. Present day operations produce large tailings
dams and, particularly in open pit operations, large dumps of
waste rock.

All of these materials contain heavy metals and are sources
of heavy metal pollution from physical erosion, oxidation and
hydrolysis of the unextracted minerals. This is encouraged by
the absence of effective retaining barriers around the perimeter
of derelict sites, and in many older mines by the proximity of
natural and artificial watercourses which were essential to the
original processing operations.

Whilst major pollution incidents are rare, massive erosion of
tailings dams by water in times of storm is always possible,
particularly where dam walls have been made of hydrocycloned
tailings. A severe storm was responsible for the destruction of
30 ha. of high class permanent agricultural pasture in a valley in
North Wales below an abandoned tailings dam[1]. Continuous low
level pollution by water erosion is a persistent problem of many
sites.

Wind transmission of airborne material appears at first sight
to be of lesser significance. But occasional substantial dispersal
can occur which builds up quite high levels in the surrounding
land[2]. Wind erosion can be particularly serious from disused
modern tailings dams because of their considerable size and flat
surface topography.

Percolating drainage water is a metal transport system
common to old and newer mines. The drainage from some disused
mines working is a very serious source of pollution[3]. The
precipitation falling on waste heaps and tailings dams must
normally find its way eventually into watercourses and constitute
a further source of pollution.

Movement of the waste is made particularly easy by the
absence of a vegetation cover; even many years after the cessation
of mining activities many derelict sites are either devoid of
vegetation or at best only sparsely colonised.

Total content (µg/g.)

Site reference	pH	Pb	Zn	N	P	K	Ca
Glebe (SK 219764)	7.7	30300	17700	23	12	1360	219800
Laporte (1 & 2) (SK 208753)	7.6	8400	1970	31	74	1080	119800
Laporte (3) (SK 204752)	7.4	7300	1130	47	62	1890	167200
Parc (SH 603787)	7.3	1300	2030	58	128	688	68800
Trelogan (1) (SJ 123864)	6.8	8625	35000	78	186	677	77080
Trelogan (2) (SJ 127806)	7.0	39800	95000	126	160	1070	138500
Darley (SK 263623)	7.6	6250	46400	32	93	1550	257300
Halkyn (SJ 205706)	7.6	5440	11300	10	207	818	288800
Frongoch (SN 722743)	4.4	2800	5770	64	115	1260	475
Y-Fan (SN 875943)	4.5	40900	6190	97	188	3065	600
Goginan (SN 688816)	5.0	16800	2700	120	103	458	708

Table 1. Chemical composition of metalliferous mine wastes and fluorspar tailings in Britain.

The establishment of a vegetation cover would go a long way towards providing the necessary surface stability. It would also intercept and return to the atmosphere by evapotranspiration a considerable proportion of the precipitation thereby reducing pollution from drainage water. However the slow and often incomplete natural colonisation indicates that the wastes possess severe limitations to plant growth and development.

PHYSICAL PROPERTIES

Both old and modern mines produce wastes of contrasting particle size. Usually two or three types of waste can be distinguished: the very coarse waste rock (2 - 20 cm. dia.) consisting mainly of unmineralized country rock and low grade, sub-marginal ores; the intermediate wastes (2.0 - 0.2 cm.), mainly found in older mines; the fine-grained tailings (<2mm) from the final stage separators. Each of these can present important restrictions to plant growth and development.

The coarse spoils retain very little water and are subject to surface drought. This is enhanced by the movement of fine-grained particles through the surface layers and their settlement at depth. Very fine-grained spoils (<0.2 mm) retain water better, but they are devoid of the clay minerals and organic matter which contribute to the water-retaining properties and cation exchange capacity of normal soils. They are liable to surface compaction, and even cementation processes which can combine to form an impenetrable superficial barrier leading to surface drainage and gulley erosion.

CHEMICAL PROPERTIES

The principal chemical limitations to plant growth are produced by the metals left behind in the wastes many of which are very toxic to plants, and by the extremely low nutrient status. The relative importance of these varies considerably between sites. Metal wastes are invariably deficient in essential macronutrients, but their heavy metal concentrations range from as low as 0.1 - 0.5% w/w in modern, relatively innocuous tailings materials to >5% w/w in the potentially more toxic spoils from long disused mines (Table 1). In general, total concentrations of the order of 1 per cent zinc, 1 - 2 per cent lead and 0.5 per cent copper are limiting to normal plant populations. However phytotoxicity is also influenced by other factors including particle size, phosphorus and calcium status: threshold concentrations, above which only metal tolerant plant populations will grow satisfactorily, are therefore not easily defined. Assessment of toxicity by the determination of plant-available metals (e.g. extractable by 0.5M CH_3COOH) is usually more accurate than total analysis and predictive chemical techniques are in general of limited value compared to bioassay investigations[4].

One of the most important factors influencing the availability of metal residues in mine spoil is the associated gangue mineral (e.g. quartz, calcite) present in the matrix. Plant uptake of metals from calcareous materials is restricted by the formation of hydroxides and carbonates or calcium - heavy metal complexes. In acidic materials the metallic components from weathering remain in solution and are more readily accumulated by vegetation. Thus metal toxicity thresholds are usually lower for acidic than for calcareous materials. In mine spoils containing pyrite (FeS_2) toxicity may be accentuated by the progressive release of sulphuric acid and ferric sulphate from the weathering of pyrite. These enhance the rate of conversion of heavy metal sulphides to the relatively soluble and phytotoxic sulphates.

Excessive salinity may also be important because of inter-actions between the sulphuric acid and native carbonates to produce soluble salts, mainly calcium and magnesium sulphates. This will be important only in arid conditions where natural leaching of salts does not occur and they accumulate at the surface by evaporation.

The low nutrient status of metalliferous mine wastes, particularly for nitrogen and phosphorus, is of major importance to vegetation establishment[5,6]. Other major nutrients which tend to be deficient on a more site-specific basis are potassium and calcium. Micronutrient deficiencies are rarely encountered. In many spoil tips the concentrations of micronutrients and other non-essential associated elements (e.g. cadmium, arsenic) are markedly elevated in comparison with normal, uncontaminated soils, but the levels rarely impose additional restrictions upon plant growth except where they constituted major components of the original crude ore.

THE NATURAL VEGETATION OF MINE WASTES

Despite the severe limitations to plant growth produced by both physical and chemical factors, some natural vegetation does occur on many metalliferous mine wastes. A very characteristic flora is found and certain plants have even been used as indicators when prospecting for heavy metals[7]. However, very few species have a distribution which is restricted entirely to metal contaminated soils (obligate metallophytes). Most species found on metalliferous wastes also occur on normal soils (pseudo- or facultative metallophytes). The number of species is fairly limited with members of the family Gramineae often a dominant component of the vegetation (Table 2). Plants collected from

Calcareous sites (pH 6.5 - 8.0)	Acidic sites (pH 6.5)
Agrostis stolonifera (creeping bent)*	*Agrostis tenuis* (common bent)*
Agrostis tenuis (common bent)	*Anthoxanthum odoratum* (sweet vernal grass)
Anthoxanthum odoratum (sweet vernal grass)	*Deschampsia flexuosa* (wavy hair grass)*
Armeria maritima (sea pink)	*Festuca ovina* (sheep's fescus)*
Deschampsia caespitosa (tufted hair grass)*	*Holcus lanatus* (Yorkshire fog)
Festuca rubra (red fescus)*	*Rumex acetosella* (sheep's sorrel)
Holcus lanatus (Yorkshire fog)	*Silene vulgaris* (bladder campion)
Campanula rotundifolia (harebell)	
Minuartia verna (vernal sandwort)	
Plantago lanceolata (ribwort plantain)	
Rumex acetosa (common sorrel)	
Thlaspi alpestre (alpine pennycress)	
Thymus praecox (wild thyme)	

* major components of the natural flora

Table 2. The main higher plant species found on metalliferous mine wastes in Britain

mine sites have been shown to possess tolerance to the particular metals present in their native substrate[8]. Levels of tolerance vary considerably and correlate, to some extent, with metal levels in the substrate. Plants from uncontaminated soils are usually non-tolerant. No species has yet been found to be tolerant throughout its distribution.

Differences in heavy metal tolerance between plants of the same species have been shown to be highly heritable[9,10,11,12]. Seed multiplication of metal tolerant selections for commercial use is therefore possible without loss of tolerance. Evidence concerning the precise genetic control of heavy metal tolerance is still rather inconclusive despite some recent detailed investigations [10,11,12,13]. For all practical purposes, metal tolerance, as measured by various rooting tests (based on Wilkins[14]), may be regarded as a quantitative character. Genetic analysis requires the use of biometrical techniques. However there is some evidence[13] that in some species at least, the fundamental tolerance mechanism may be controlled by two or three major genes together with a larger number of modifier genes of small effect.

Metal tolerant populations are the result of true evolution by natural selection. It appears that, in some species, genes producing metal tolerance occur at a low frequency in populations from uncontaminated sites. Plants possessing these genes are easily identified using a simple seed screening technique [15,16]. Thus species seem to have an inherent genetic potential for evolving tolerant populations. This, together with the very high selection pressures which operate on metal contaminated wastes[17] can result in the very rapid evolution of metal tolerant populations[18]. The rapid production of metal-tolerant selections using seed screening techniques has also been demonstrated[15].

Heavy metal tolerance is largely metal specific[8]. This specificity indicates that tolerance to different metals is under independent genetic control and involves different physiological mechanisms and can be combined by natural or artificial selection. However, there is some slight evidence[19,20] that tolerant plants possess a common low level of tolerance to a range of heavy metals which is greater than that of non-tolerant plants. This may result from a secondary physiological adaptation to the low nutrient status of many mine wastes.

PHYSIOLOGICAL BASIS

The physiological and biochemical basis of heavy metal tolerance is not fully understood. Evidence supporting two main hypotheses has been presented. The metal-binding hypothesis states that tolerant plants avoid the toxic effects of free metal

ions by removing them from the metabolic system as stable
complexes.

Turner & Marshall[21,22] demonstrated that the roots of zinc-
tolerant and copper-tolerant plants take up more zinc and copper
respectively than do the roots of non-tolerant plants. They also
showed that zinc uptake into the cell wall fraction of Agrostis
tenuis roots was highly correlated with the level of zinc tolerance in the plants. Zinc uptake in copper -tolerant plants was
similar to that found for non-tolerant plants. The nature of the
metal binding is still not clear. Wynn-Jones, Sutcliffe &
Marshall[23] found that 66 per cent of bound zinc was released by
cellulose digestion of cell walls. Peterson[24] found more zinc in
the pectate extract from zinc tolerant Agrostis tenuis and A.
stolonifera grown in ^{65}Zn solutions, than in similar extracts
from non-tolerant plants. This evidence indicates that metal-
tolerant plants are able to detoxify heavy metal ions by forming
stable complexes in root cell walls.

Although the capacity of this binding mechanism may have
been underestimated, it appears that appreciable quantities of
metal are not complexed but pass up into the shoots of tolerant
plants and can have a direct effect upon the metabolic system.
This has led to the hypothesis that metal-tolerant plants possess
enzymes which are adapted to function efficiently in the presence
of heavy metal ions. The differential inhibition of cell wall
acid phosphatases by copper in copper-tolerant and non-tolerant
Agrostis tenuis [25] supports this hypothesis. Cox[26] has recently
found a similar situation in zinc tolerant Anthoxanthum odoratum.
Soluble acid phosphatases from root extracts of zinc tolerant
plants were less inhibited by zinc than those from non-tolerant
plants. This difference was only apparent after exposure of the
roots to a culture medium containing sub-lethal concentrations of
zinc.

Further work is required in order to understand fully the
physiological nature of heavy metal tolerance. It is clear that
a variety of mechanisms is involved and their relative importance
is likely to vary between plant species and for different metals.

ASSOCIATED CHARACTERISTICS IN METAL TOLERANT PLANTS

As outlined earlier, there are a number of environmental
factors which limit plant growth on metaliferous wastes. Metal-
tolerant plants from these sites possess various physiological
and morphological adaptations which affect these factors.

Jowett[27] demonstrated that tolerant Agrostis tenuis plants
from an acidic lead mine were adapted to low levels of both

calcium and phosphate. Similarly Khan[28] showed that plants from a copper mine were adapted to low levels of phosphate. Mine plants generally appear to have a slower growth rate than non-mine plants[29]. There is evidence to suggest that low growth rate is an adaptation to nutrient limitations and other stress factors[30,31].

Recent observations indicate that metal tolerant plants are more drought resistant than non-tolerant plants on non-toxic soils. Further investigations are necessary to establish the cause of this but there are various morphological characteristics which may be of importance (e.g. cuticle thickness, leaf size and hairiness).

It is apparent that heavy metal tolerant plants possess an integrated complex of adaptations to cope with a range of environmental stresses. This is maintained by a balanced and stable genetic system built up and adjusted by natural selection. Such material is very suitable to be used in restoration work designed to produce a continuous cover of vegetation to improve the appearance of derelict land and to reduce erosion and dispersal of toxic material. With this in mind, three metal tolerant varieties have been produced for commercial use (Table 3).

Species	Variety	Application
Festuca rubra	'Merlin'	Calcareous, lead-zinc contaminated wastes
Agrostis tenuis	'Goginan'	Acidic, lead contaminated wastes
Agrostis tenuis	'Parys'	Acidic, copper contaminated wastes

Table 3. Metal-tolerant varieties available on a commercial scale

PRACTICAL VALUE OF METAL TOLERANCE

Controlled bioassay experiments designed to compare the performance of metal tolerant and non-tolerant commercial varieties of grasses sown on phytotoxic substrates have been consistent in their outcome. The yield data depicted in Fig. 1

320

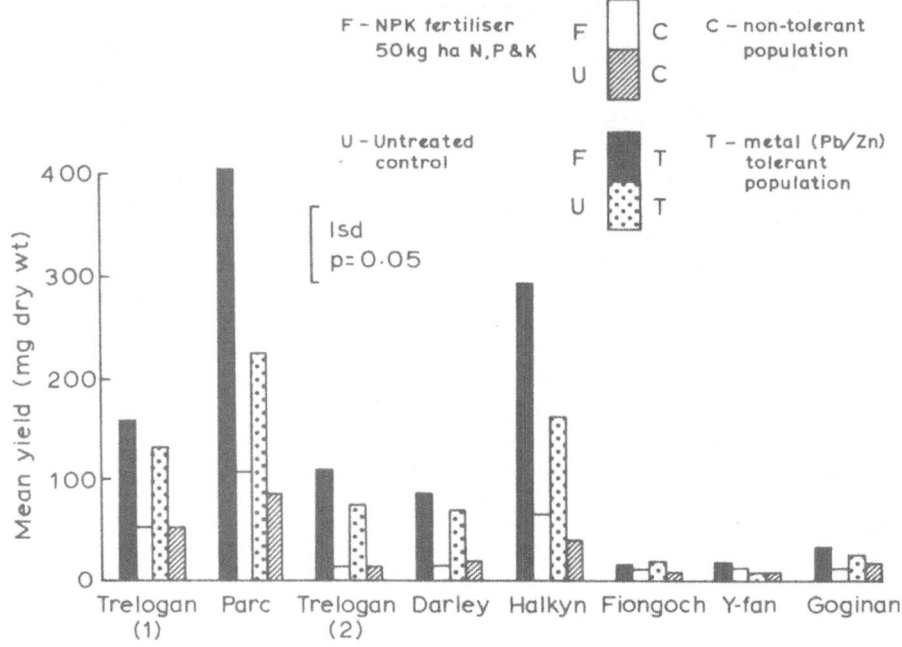

Figure 1. Bioassay of a range of metalliferous wastes and
fluorspar tailings using lead/zinc tolerant and non-
tolerant Festuca rubra.

summarises the results of a short-term (10 week) pot experiment
carried out under glasshouse conditions, in which growth of the
lead/zinc tolerant variety of Festuca rubra (Merlin) was compared
with that of a commercial variety (S.59) on a range of metalliferous mine wastes. On the calcareous substrates (pH 6.8 - 7.6)
growth of the tolerant population was generally superior to that
of the non-tolerant material, the distinction being emphasised
where 50 kg/ha N, P and K were applied as a balanced, inorganic
fertiliser to ameliorate the low fertility of the growth media.
However, both the tolerant and non-tolerant varieties failed on
the three acidic lead/zinc spoils (pH 4.4 - 5.0) irrespective of
the fertiliser treatment. This differentiation reflects the
importance of the associated gangue minerals in the matrix, and
the need to select the species/populations appropriate to individual sites, prior to implementing large scale revegetation
schemes.

That growth of tolerant material is ultimately superior to
that of unadapted non-tolerant plant material on metal wastes is
well established[5] but the distinction between them is not always
evident in the short term. Thus, the apparently satisfactory

growth, in some cases, of the non-tolerant material supplied with essential nutrients (Fig. 1) is misleading. A similar growth pattern has been observed in large scale experiments conducted under field conditions, using vegetative material[32] and direct seeding methods[1], but only the metal tolerant material is capable of forming a permanent, stable vegetation cover[5,6]. The rapidity with which differences between varieties develop depends largely on substrate toxicity, and can be delayed by regular application of fertiliser; phosphorus and organic matter are particularly significant in this context as the formation of organometallic complexes and precipitation of heavy metal phosphates alleviates toxicity on a temporary basis[32,33].

Whilst the eventual superiority of the tolerant varieties is invariably apparent on older, toxic mine wastes (Fig. 2), this is not necessarily true of modern mine tailings. Improvements in ore processing technology have decreased the levels of the valuable metallic minerals in the tailings, in some cases to the

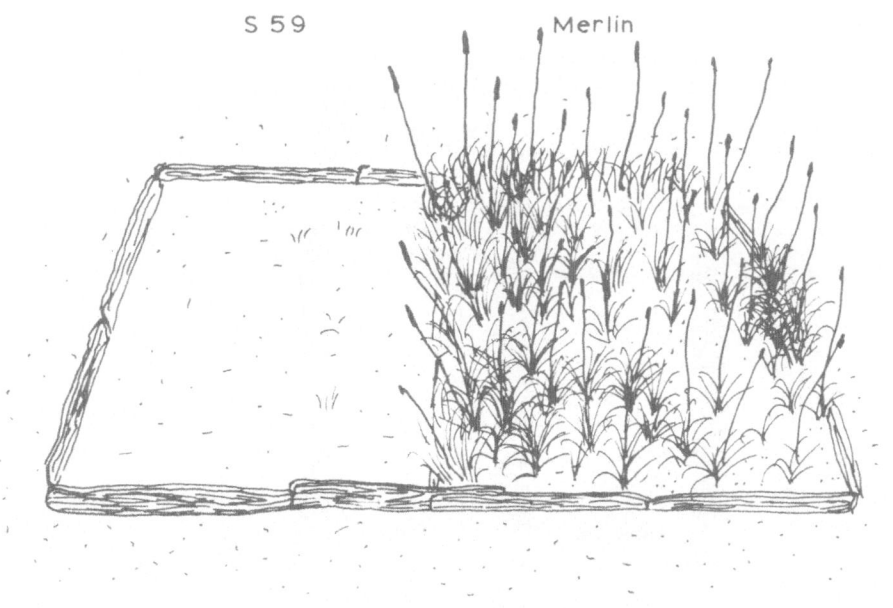

Figure 2. The growth of lead/zinc tolerant (cultivar Merlin) and non-tolerant (cultivar S.59) Festuca rubra on lead/ zinc waste on a mine site at Trelogan, Clwyd, three years after sowing.

322

extent that metal tolerant varieties have no growth advantage
compared to non-tolerant varieties, even in the long term. This
is evident from the results of pot experiments harvested after
eighteen months growth on metalliferous tailings of different
ages (Fig. 3). Clear distinctions between populations at the
same fertiliser level are evident only on the oldest and
potentially most toxic of the substrates. The response to inorganic fertiliser (NPK) confirms a consistent deficiency of
essential macronutrients, irrespective of the age of the material.

However metal tolerant populations tend to translocate less
heavy metals into their aerial parts than non-tolerant populations
under these conditions[5] (Table 4). There can be a reduction of
50 per cent. This can be important if there is any possibility
that a reclaimed area might be used for grazing, since ingestion
of metals which could have serious effects on the grazing animal
would be reduced.

In order to ensure the persistence of a developing vegetation
cover, a continuous supply of nutrients to the established sward

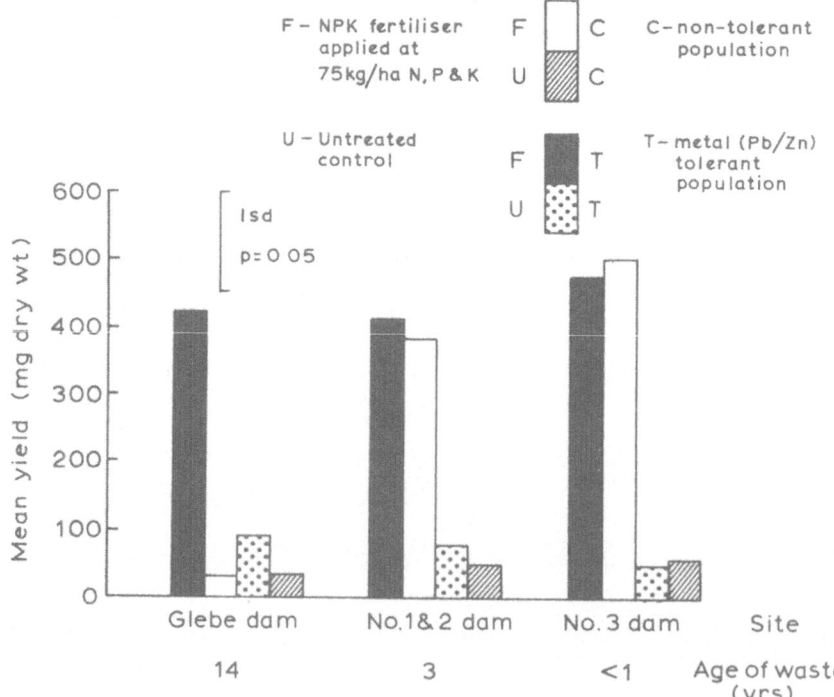

Figure 3. Growth of lead/zinc tolerant and non-tolerant populations of <u>Festuca</u> <u>rubra</u> on metalliferous fluorspar
tailings produced at different times (representing
different levels of efficiency of extraction).

Species	Population	Total content (µg/g)			
		Lead		Zinc	
		+	−	+	−
Agrostis	Commercial *	243	395	779	3160
tenuis	Metal-tolerant	185	264	744	2995
Festuca	Commercial *	500	1040	2260	5140
rubra	Metal-tolerant	315	628	1600	4150

* non-tolerant
+ treated with standard NPK fertiliser
− untreated control

Table 4. Heavy metals in shoots of different species/populations established on lead/zinc waste from Trelogan mine.

is essential. This is often difficult to achieve because applied nutrients are readily leached from mine waste in percolating drainage water and the fraction recovered by the sward may be re-cycled only very slowly if organic residues accumulate within the developing soil ecosystem[34,35]. The former problem is apparent from the variation in yield of lead/zinc tolerant Festuca rubra after six and eighteen months growth on a recently disused tailings dam, when supplied with various essential macronutrients as a single pre-seeding treatment using standard compound fert-ilisers, slow-release granular fertiliser or an organic manure (Fig. 4). When supplied with nitrogen, phosphorus and potassium, the clear differences between plant performance according to the source of nutrients are largely due to rapid leaching of nitrogen provided as inorganic fertiliser, a common phenomenom on derelict land materials [36]. On metalliferous waste this problem is emph-asised by the absence of metal tolerant populations of legumes, the establishment of which would maintain a supply of nitrogen through symbiotic fixation. However, this limitation must be viewed in context of the ultimate land use objectives.

Inevitably, the uses to which revegetated spoil tips can be put will be severely restricted by the physical and chemical nature of the rooting environment. Even though tolerant plants have some capacity for immobilising heavy metals in root cell walls, elevated concentrations are invariably found in the shoots (Table 4). Management of revegetated areas as permanent grazing pasture may therefore be impracticable, so maximum productivity and the restrictions on legume establishment and organic matter breakdown may not be of primary importance. On undisturbed mine

324

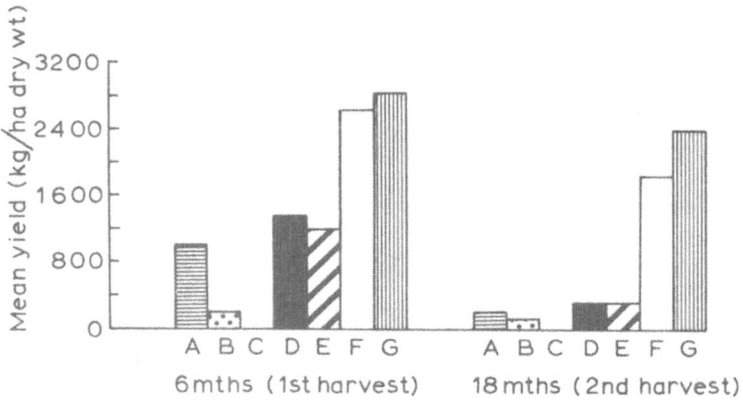

A - Inorganic fertilisers (N)   D - Standard compound fertilisers (NPK)
B - Inorganic fertilisers (P)   E - Standard compound fertilisers (NP)
C - Inorganic fertilisers (K)   F - Granular, slow release fertilisers (NP)
        G - Organic manure (sewage sludge) - NPK

Figure 4.   Yield of lead/zinc tolerant Festuca rubra on metalli-
            ferous fluorspar tailings supplied with different
            sources of N, P and K (75 kg.ha^{-1}) alone and in
            combination.

sites the natural flora persists despite the low fertility of the
substrate so where the major objectives are vegetative stabilisation and amenity improvement, tolerant plant populations often
present the most suitable and certainly the most economical
restoration technique.   In consequence, a simple reinstatement
scheme has been advocated based upon direct seeding of tolerant
plants and fertiliser amelioration[5].   In large scale work this
can be implemented using either traditional agricultural techniques or hydroseeding, depending on site-specific parameters
including topography, accessibility and location[37].

    Seeding rates will vary from 40-100 kg.ha^{-1} of which 25 - 50
per cent w/w will be metal tolerant populations, the remainder
being non-tolerant material included to assist sward establishment
and growth in the short term.   The optimum fertiliser treatment
will also vary according to the physical and chemical characteristics of the substrate.   In most cases an initial application of
300-500 kg.ha^{-1} of standard 17:17:17 NPK fertiliser will be

appropriate, together with annual maintenance treatments (200 – 400 kg.ha^{-1}) for two or three years after establishment. Intensive management programmes should be unnecessary subsequently for although the level of productivity may gradually decline, an effective stabilising sward will persist on a permanent basis.

DEVELOPMENT OF FURTHER TOLERANT MATERIAL

The varieties at present commercially available are adapted to a range of environmental conditions and to different metals. But they are species of temperate climates and will not be suited to tropical or arid regions. They also do not cover the full range of metals that can occur in metal wastes in toxic quantities nor the full range of combination of metals. There is therefore considerable need to develop further tolerant material. This can be shrub or herbaceous material to be propagated either by seed or vegetative means. There are four possible approaches, which are summarised in Fig. 5.

The Discovery and Utilisation of Naturally Occurring Tolerant Material

New cases of metal tolerance are constantly being discovered. At the same time a wide range of species is being found associated with metalliferous soils and must be deduced to be tolerant, (e.g. Nicolls et el.)[38]. As a result it is clear that metal tolerance is a characteristic which is widely distributed among the Angiosperms and indeed in other members of the plant kingdom[7]. It is not restricted to a few selected families (Table 5). It has been specifically tested for and found to occur in material from tropical and sub-tropical regions[39,40].

Tolerant populations may be found on natural metalliferous soils or on recently produced metalliferous wastes. Species growing on natural metalliferous soils will have had maximum opportunity to evolve tolerance. But metal tolerance can evolve rapidly and species found on recently produced wastes may also possess tolerant populations. These could be very valuable for reclamation and stabilisation purposes. Besides possessing metal tolerance the latter will obviously be adapted to the special edaphic conditions pertaining to the waste either because they are populations of species which have colonised the wastes because of pre-existing appropriate characteristics or because they are populations which have become adapted by selection on the wastes themselves. They will also be likely to possess superior abilities for colonisation and vegetative establishment. Examples of such species are Cynodon dactylon and Pennisetum clandestinum which are found widely on mine wastes in a manner similar to Agrostis tenuis, but have not yet been investigated

Families known to have species showing tolerance:

Campanulaceae	Linaceae
Caryophyllaceae	Plantaginaceae
Compositae	Plumbaginaceae
Cruciferae	Polygonaceae
Ericaceae	Scrophulariaceae
Gramineae	Violaceae
Labiatae	

Families having species found in high concentrations of metals:

Amaranthaceae	Olacaceae
Anacardiaceae	Papaveraceae
Caprifoliaceae	Philadelphaceae
Combretaceae	Primulaceae
Cyperaceae	Rubiaceae
Euphorbiaceae	Rutaceae
Lauraceae	Salicaceae
Leguminosae	Velloziaceae
Meliaceae	

Table 5. Families known to have species showing heavy metal tolerance, or found in areas of high concentration of metals and likely to be tolerant (from Antonovics, Bradshaw & Turner[7]).

for metal tolerance.

Nevertheless examination of material from undisturbed metalliferous areas should be very rewarding especially since several legumes such as <u>Tephrosia</u> spp. can be found and may have associated metal tolerant Rhizobia.

If the material is strongly vegetative it can be propagated and used as vegetative material without resort to seed. Once tolerant material has been identified it can be used without fear of loss of adaptive characteristics due to segregation. However in the cases investigated so far metal tolerance has been found to be highly heritable. This suggests that seed material can readily be used, without the problems that can occur when dealing with characters with low narrow-sense heritability.

Artificial Selection in Non-tolerant Material

Since metal tolerance can evolve rapidly[18], there is no reason why attempts to select for it should not be made in material not previously exposed to metal toxicity. Wild or cultivated material can be used, chosen for its adaptation to the general characteristics of the site being reclaimed. The evidence so far does not provide any means of predicting the material in which selection will be successful : the occurrence of selectable variation in metal tolerance in different species appears to be random[16]. But screening trials can easily be set up on metal wastes and the occurrence of survivors observed. Seed can be saved from these plants to form the basis of the next generation. Such simple mass selection is easily carried out. If genetic variability for tolerance exists, selection will occur and changes in the constitution of the populations will be rapid, as has been recorded in experiments on adaptation to climate and soil type[41]. In such mass selection experiments characteristics of adaptation to edaphic factors other than metal toxicity will also be selected.

Other methods of selection based on pair-cross or polycross techniques aimed at producing a synthetic variety[42] can also be used. However the extra complication is unlikely to be with the gain in selection efficiency, especially since mass selection techniques work particularly well where extreme selection pressures are operating.

Hybridisation of Existing Tolerant Material.

It is possible that tolerance already occurs in material available, but is not a low level or is not of appropriate combinations of metal specificities. In such cases it may be worth while hybridising appropriate material and carrying out selection on F_2 and subsequent generations to select for new or improved genetic combinations. This could be used for instance to produce a variety of Agrostis tenuis tolerant to lead, zinc and copper together. Superior tolerance to a single metal may be obtained by hybridisation of material of different geographical origins. In these cases it is not expected that the hybrid itself will be used for reclamation but selected segregates from it.

Development of Tolerance to Characteristics of Metal Wastes Other than Metal Toxicity.

Metal wastes can be difficult for plant growth for many different reasons, for example low nutrient status or salinity. It has already been suggested that adaptation to these will automatically be selected during mass selction for metal tolerance. However there is no reason why selection programmes should not be designed specifically for adaptation to these characteristics following the methods given above. Salt tolerance for instance can be found in specific natural populations[43] and selectable variation exists in many different species both wild and cultivated.

Populations which are salt tolerant as well as appearing metal tolerant have been identified in Paspalum vaginatum, Cynodon dactylon and Dactyloctenium geminatum and other species in Rhodesia[44] and are already being used for the stabilisation of tailings in a large scale on the Trojan nickel mine in Rhodesia, planted vegetatively. Such an approach could be very important for the stabilisation of tailings in arid areas where salinity can be a more important factor than metal toxicity in preventing plant growth. It could be equally important where excess pyrite leads to extreme acidity.

CONCLUSIONS

The metal toxicity of wastes produced by mining for heavy metals require that the wastes are properly stabilised. Vegetative stabilisation provides a method which is cheap, permanent and visually attractive. Where plants are sown directly into the waste the costs are not high; indeed they are lower than other means of stabilisation. The permanence of a vegetative cover if given appropriate aftercare is considerably greater than any chemical method of stabilisation and is only exceeded in permanence by inert coverings.

On some tailings it may be possible to establish non-tolerant normal commercial material with the aid of fertilisers and/or sewage sludge [45,46,47]. But there are a wide range of modern tailings and old mine wastes on which plants cannot be established unless they are tolerant or unless deep inert coverings are used to isolate the metal waste[37]. For these, tolerant material offers exciting possibilities for reclamation. Even if tolerant seed or plant material is more expensive than normal commercial material the increased cost is a fraction of the cost of inert coverings.

Tolerant material has value also on wastes which are sufficiently non-toxic to allow non-tolerant commercial material to be used. Such sites can degrade due to the release of metals by breakdown and oxidation of ore particles. Against this tolerant plants represent an insurance. But it is likely that most

Examination of waste
materials to identify
major edaphic char-
acteristics and
toxicities

Search for tolerant
plant material already
occurring naturally on
waste material

Search for toler-
ant plant material
occurring on simil-
ar toxic sites
elsewhere

Acquisition of app-
ropriate non-toler-
ant commercial var-
ieties or wild
material

Collection of seed or
vegetative material

(Assessment by
tolerance test)

Hybridisation of
potentially useful
material

Establishment of
mass selection trials

Establishment of
selection trials

Collection of toler-
ant survivors

Collection of tol-
erant segregants

Assessment by small scale
pot or field trials

Multiplication of successful
tolerant material

Large scale field trials
examining seeding methods,
fertiliser treatments etc.

Establishment of methods
for using the material

Figure 5. The steps necessary in a programme for the development
of tolerant plant material to be used for the stabilisation and cover of metal wastes : the programme is
suitable for the development of material tolerant to
metal toxicity or to other edaphic factors.

tolerant material will have other attributes which will make it
more successful than normal material on metal wastes, attributes
such as drought and nutrient deficiency tolerance.

The identification and development of metal tolerant material
is not expensive and can be carried out by semi-skilled labour,
since the collection of material, arrangement of assessment trials,
and organisation of mass selection experiments are not difficult.
An effective programme using natural material can be completed in
three years.   It is therefore a technique for the stabilisation
and reclamation of mine wastes which can be applied widely.

REFERENCES

1. Smith, R.A.H., The reclamation of old metalliferous mine
    workings using tolerant plant populations, Ph.D. thesis,
    Univ. Liverpool, 1973.

2. Johnson, M.S., Roberts, D. & Firth, N., Heavy metals in the
    terrestrial environment around derelict lead/zinc mines in
    Wales, Sci. tot. Env., (in press), 1977.

3. Blessing, N.V., Lackey, J.A. & Spry, A.H., Rehabilitation of an
    abandoned mine site, in Minerals and the Environment, (ed.
    M.J. Jones), 16, Instn Min. Metall. London, 1974.

4. Johnson, M.S., Establishment of vegetation on metalliferous
    fluorspar mine tailings, Ph.D. thesis, Univ. Liverpool, 1977.

5. Smith, R.A.H. & Bradshaw, A.D., Stabilisation of toxic mine wastes
    by the use of tolerant plant populations, Trans. Instn Min.
    Metall., 81(A), 230, 1972.

6. Goodman, G.T., Pitcairn, C.E.R. & Gemmell, R.P., Ecological
    factors affecting growth on sites contaminated by heavy
    metals, in Ecology and Reclamation of Devastated Land, (ed.
    R.J. Hutnik & G. Davis), 149, Gordon & Breach, New York,
    1973.

7. Antonovics, J., Bradshaw, A.D. & Turner, R.G., Heavy metal
    tolerance in plants, Adv. ecol. Res., 7, 1, 1971.

8. Gregory, R.P.G. & Bradshaw, A.D., Heavy metal tolerance in
    populations of Agrostis tenuis Sibth. and other grasses, New
    Phytol., 64, 131, 1965.

9. McNeilly, T. & Bradshaw, A.D., Evolutionary processes in pop-
    ulations of copper tolerant Agrostis tenuis Sibth., Evolution,
    22, 108, 1968.

10. Urquhart, C., Genetics of lead tolerance in Festuca ovina, Heredity, 26, 19, 1971.

11. Gartside, D.W. & McNeilly, T., Genetic studies in heavy metal tolerant plants I. Genetics of zinc tolerance in Anthoxanthum odoratum, Heredity, 32, 287, 1974.

12. Gartside, D.W. & McNeilly, T., Genetic studies in heavy metal tolerant plants II. Zinc tolerance in Agrostis tenuis, Heredity, 33, 303, 1974.

13. McNair, M.R., The genetics of copper tolerance in Mimulus guttatus (Scrophulariaceae), Ph.D. thesis, Univ. Liverpool, 1977.

14. Wilkins, D.A., A technique for the measurement of lead tolerance in plants, Nature Lond., 180, 37, 1957.

15. Walley, K., Khan, M.S. & Bradshaw, A.D., The potential for evolution of heavy metal tolerance in plants I. Copper and zinc tolerance in Agrostis tenuis, Heredity, 32, 309, 1974.

16. Gartside, D.W. & McNeilly, T., The potential for evolution of heavy metal tolerance in plants II. Copper tolerance in normal populations of different plant species, Heredity, 32, 335, 1974.

17. Antonovics, J., The genetics and evolution of differences between closely adjacent plant populations with special reference to metal tolerance, Ph.D. thesis, Univ. Wales, 1966.

18. Wu, L., Bradshaw, A.D. & Thurman, D.A., The potential for evolution of heavy metal tolerance in plants III. The rapid evolution of copper tolerance in Agrostis stolonifera, Heredity, 34, 165, 1975.

19. Bradshaw, A.D., The evolution of heavy metal tolerance and its significance in land reclamation, in Heavy Metals and the Environment, (ed. Hutchinson, T.C.), Toronto Univ. Press, Toronto, 1977.

20. Jowett, D., Populations of Agrostis spp. tolerant to heavy metals, Nature, Lond., 182, 816, 1958

21. Turner, R.G & Marshall, C., The accumulation of ^{65}Zn by root homogenates of Zn-tolerant and non-tolerant clones of Agrostis tenuis, New Phytol., 70, 539, 1971

22. Turner, R.G. & Marshall, C., The accumulation of zinc by subcellular fractions of roots of Agrostis tenuis Sibth. in relation to zinc tolerance, New Phytol., 71, 671, 1972.

23. Wyn-Jones, R.G., Sutcliffe, M. & Marshall, C., Physiological and biochemical basis for heavy metal tolerance in clones of Agrostis tenuis. In Recent Advances in Plant Nutrition, (ed. R.M. Smith), Gordon & Breach, New York, 1971.

24. Peterson, P.J., The distribution of zinc-65 in Agrostis tenuis Sibth. and A. stolonifera L. Tissues, J.exp.Bot., 20, 863, 1969.

25. Wainwright, S.J. & Woolhouse, H.W., Physiological mechanisms of heavy metal tolerance in plants, In The Ecology of Resource Degradation and Renewal, (ed. M.J. Chadwick & G.T. Goodman), 231, Blackwell, Oxford, 1975.

26. Cox, R., Properties of some enzymes of zinc tolerant and non-tolerant clones of Anthoxanthum odoratum, Ph.D. thesis, Univ. Liverpool, 1976.

27. Jowett, D., Adaptation of a lead tolerant population of Agrostis tenuis to low soil fertility, Nature, Lond., 184, 43, 1959.

28. Khan, M.S.I., The process of evolution of heavy metal tolerance in Agrostis tenuis and other grasses, M.Sc. thesis, Univ. Wales, 1969.

29. Humphreys, M.O. & Bradshaw, A.D., Genetic potential for solving problems of soil mineral stress -heavy metal toxicity, In Proc. Workshop on Mineral Stress in Problem Soils, Cornell Univ., Ithaca, 1977.

30. Parsons, R.F., The significance of growth rate comparisons for plant ecology, Amer. Nat., 102, 595, 1968.

31. Grime, J.P. & Hunt, R., Relative growth rate : its range and adaptive significance in a local flora, J. Ecol., 63, 393, 1975.

32. Gadgil, R.L., Tolerance of heavy metals and the reclamation of industrial waste, J.appl. Ecol., 6, 247, 1969.

33. Newton, L., Pollution of the rivers of west Wales by lead and zinc mine effluent, Ann.appl.Biol., 31, 1, 1944.

34. Griffiths, A.J., Hughes, D.E & Thomas, D., Some aspects of microbial resistance to metal pollution, In Minerals and

the Environment, (ed. M.J. Jones), 387, Inst. Min. and
Metall., London, 1975.

35. Williams, S.T., McNeilly, T. & Wellington, E.M.H., The
    decomposition of vegetation growing on metal mine waste,
    Soil Biol. Biochem., 9, 271, 1977.

36. Bradshaw, A.D., Dancer, W.D., Handley, J.F. & Sheldon, J.C.,
    The biology of land revegetation and the reclamation of
    the china clay wastes of Cornwall, In The Ecology of
    Resource Degradation and Renewal, (ed. M.J. Chadwick &
    G.T. Goodman), 363, Blackwell, Oxford, 1975.

37. Johnson, M.S. & Bradshaw, A.D., Prevention of heavy metal
    pollution from mine waste by vegetative stabilisation,
    Trans. Inst. Min. Metall., 86(A), 47, 1977.

38. Nicolls, O.W., Provan, D.M.J., Cole, M.M. & Tooms, J.S.,
    Geobotany and geochemistry in mineral exploration in the
    Dugald River Area, Cloncurry District, Australia, Trans.
    Inst. Min. Metall., 74, 695, 1965.

39. Ernst, W., Ecophysiological studies on heavy metal plants in
    South Central Africa, Kirkia, 8, 125, 1972.

40. Erdoma, E.L., Copper pollution in Rwenzori National Park,
    Uganda, J. Ecol., 11, 1043, 1974.

41. Sylven, N., The influence of climatic conditions on type
    composition, Imperial Bureau Plant Genetics, Herb. Bull,
    21, 8, 1937.

42. Allard, R.W., Principles of Plant Breeding, Wiley, New York,
    1960.

43. Tiku, B.L. & Snaydon, R.W., Salinity tolerance within the
    grass species Agrostis stolonifera L., Pl. Soil, 35, 421
    1971.

44. Wild, H. & Wiltshire, G.H., The problem of vegetating Rhode-
    sian mine dumps examined, Chamber of Mines J. Rhodesia, 13,
    no. 11, 26, no. 12, 35, 1971.

45. Peters, T.H., Using vegetation to stabilise mine tailings,
    J. Soil Water Cons., 25, 65, 1970.

46. Ludeke, K.L., Vegetative stabilisation of copper mine tailing
    disposal berms of Pima Mining Company, In Tailings Disposal
    Today, (ed. C.L. Alpin & G.O. Argall), 377, San Francisco,
    1973.

334

47.   Johnson, M.S., Bradshaw, A.D. & Handley, J.F., Revegetation
      of metalliferous fluorspar mine tailings, Trans. Inst.
      Min. Metall, 85(A), 32, 1976.

PROBLEMS OF GRASSLAND MAINTENANCE ON METALLIFEROUS SMELTER WASTES

R.P. Gemmell*   and   G.T. Goodman+

Department of Botany, University College of Swansea,
Wales.

ABSTRACT

Revegetation trials in the Lower Swansea Valley Project were
designed to obtain a rapid cover of grass to stimulate redevelop-
ment of derelict sites. Under these circumstances maintenance
considerations were of less importance than when more or less
permanent grass cover is required. Later, a series of experiments
on zinc, copper and steel wastes were undertaken to investigate a
range of species, fertilizer and cutting treatments for long-term
maintenance.

Using commercial grass varieties a high level of organic
matter must be maintained to ensure acceptable swards. If metal-
tolerant grass seed is available for use, only standard fertilization and cutting treatments are required.

INTRODUCTION

An important aspect of derelict-land reclamation which has
been largely neglected until recently is the problem of vegetat-

Present addresses: *   Joint Reclamation Team of Greater Manchester
                       and Lanchashire County Council, Wigan, U.K.

                   +   Beijer Institute, Royal Swedish Academy of
                       Sciences, Stockholm, Sweden.

ion after-management[1]. Local authorities engaged in derelict-land reclamation in the U.K. are showing increasing concern about the persistence of planted vegetation and now recognise the need for more research on maintenance techniques.

One of the commonest causes of plant-growth deterioration is the loss of fertility resulting from the leaching of the fertilizers originally added during vegetation establishment. Apart from this, an important factor is the presence of iron pyrites which generates acidity, or of metalliferous sulphides or other ores and wastes which release toxic metal ions into the soil solution; in both cases oxidative reactions are involved. On pyritic colliery spoils and metalliferous smelter wastes regression may occur soon after planting, resulting in serious and rapid dieback within a few growing seasons only.

A major feature of experimental grass trials conducted on zinc, copper and iron smelter wastes in the Lower Swansea Valley was the appearance of chlorosis and yield regression in the second and third growing seasons after planting[2]. Although research conducted by Pitcairn[3] showed that an important contributory cause of this deterioration was macro-nutrient deficiency, principally nitrogen, it was found that the reappearance of metal toxicity was of major importance on the treated potentially toxic non-ferrous smelter wastes. The behaviour with time of the various treatments investigated led to the conclusion that problems of revegetation could not be solved by short-term evaluations only and a long-term monitoring of the effects of different initial and maintenance treatments was considered to be essential.

PREVIOUS RESEARCH

During the 19th and early 20th centuries Swansea was a world centre for the smelting of non-ferrous metal ores. By 1930, the smelting of pig-iron and the manufacture of steel had almost entirely replaced non-ferrous ore smelting but, in time, the ferrous metal industry itself disappeared. This left the Lower Swansea Valley in a state of severe industrial dereliction. Apart from roofless factory ruins, there was widespread soil erosion following the destruction of the vegetation cover by sulphurous smoke pollution. Additionally, the Valley contained vast deposits of industrial wastes, chiefly residues from the smelters. Most of the tips consist of zinc and copper smelter wastes and blast furnace slag; it has been estimated that approximately 7 - 10 000 000 tonnes of waste have been deposited on the valley floor.

In 1962 the University College of Swansea embarked upon a multi-disciplinary investigation of the factors inhibiting re-use

of the Lower Valley. This study, the Lower Swansea Valley
Project, was intended to produce a working plan for reclaiming
the valley landscape. Because the tips were obviously inimical
to vegetation establishment, it was necessary to produce practical
recipes for use in large-scale reclamation, not so much for
permanent grassland and trees as for temporary revegetation to
green-up the landscape and stimulate re-development in the short
term.

The approach used was largely empirical during the course
of the Project. For example, it was discovered that the simple
act of producing a cosmetic cover of grass or tree-planting would
stimulate an interest in derelict sites by potential developers.
For this reason, it was essential to develop recipes for quick
inexpensive grass-cover in the knowledge that the new vegetation,
if successful in this role, would quickly be replaced by permanent site developments. However, when it became clear that some
areas should remain under more or less permanent grass or trees,
a longer-term, more experimental approach was required to ensure
that the methods devised were practically feasible on a large-
scale, capable of producing visually acceptable vegetation over
large areas, and were able to achieve persistent growth which
could be sustained by conventional types of maintenance. The
botanical investigations were therefore continued after completion
and publication of the Project Report with a new emphasis on
after-management techniques and treatments of long-term viability.

The nature of the waste substrates

The Valley floor was covered with three principal types.

Zinc Waste. This waste was found to be extremely phytotoxic,
the principal toxin being zinc. Other metal levels were greatly
elevated relative to normal soils. The pH, at about 6.5, was
satisfactory for growth and physical factors were relatively
unimportant except for the poor retentive capacity of the material for moisture and nutrients. Major nutrient concentrations
were all very low, NPK deficiency being regarded as an important
but secondary growth limiting factor in the waste[6].

Copper waste. The toxicity of this waste was caused by high
copper in solution, the situation being aggravated by slight
acidity. Other metal toxicities were absent. However, macro-
nutrient deficiency was the initial factor limiting growth,
copper toxicity being operative only in the presence of adequate
soil fertility[6].

Blast furnace slag. Although this waste contained fairly
high levels of certain non-ferrous metals, these did not exert
toxicity. The material was strongly calcareous with pH values
in the alkaline range. The primary growth limiting factor was
nitrogen deficiency. Phosphorus and potassium were 'very low'
and 'low' respectively, being of secondary importance to nitrogen.

Details of the chemical factors responsible for the adverse
growth properties of the three wastes are summarised in Table 1.

Waste	Element	Concentration (plant available)	Remarks
Zinc smelter	Zn	26 000 ppm (0.5N acetic acid sol.)	In toxic range
	Zn	72 - 130 ppm (water sol.)	In toxic range
	pH	6.3	Satisfactory
	N	< 1 ppm	Very low
	P	2.6 ppm	Very low
	K	25 ppm	Low
Copper smelter	Cu	2 000 ppm (E.D.T.A. sol.)	In toxic range
	Cu	0.8 - 3.0 ppm (water sol.)	In toxic range
	pH	5.5	Rather low
	N	4 ppm	Very low
	P	15 ppm	Low
	K	42 ppm	Low
Blast furnace	pH	8.7	High
	N	< 1 ppm	Very low
	P	0.9 ppm	Very low
	K	110 ppm	Low

Data from Street & Goodman[2].

Table 1. Chemical analysis of substrate toxicity factors in
metalliferous smelter wastes

Techniques of revegetation

The metal-complexing properties of organic matter[4],[5] were utilised to alleviate the toxicities of the zinc and copper smelter wastes. The cheapest and most readily available materials of high organic matter content were sewage sludge and coarsely sieved domestic refuse. These were applied to the wastes as top-coverings of various thicknesses, zinc waste requiring a greater depth than copper waste to overcome the toxicity. The toxicities of both wastes were also lowered by lime addition, the most effective treatments involving organic materials and lime in combination.

Nutrients were applied as chemical inorganic fertilisers after the addition of organic matter. Although both sewage sludge and domestic refuse provided some nutrients, further macro-nutrient treatments were required for good growth.

These methods enabled grass swards to be established from standard grass seeds mixtures. Later experiments showed that tillers of metal-tolerant ecotypes made good growth when trans-planted into waste containing little or no organic matter, prov-ided that nutrients were applied[7]. This suggested that sowings of metal-tolerant varieties could be an alternative and, if enough seed could be obtained, possibly a cheaper method of overcoming toxicity for long-term swards of grass.

On blast furnace slag, fertilizers alone were sufficient to establish plant growth. However, organic matter did improve the growth potential of the waste, mainly because of its nutrient content. For high grass yields on the waste, organic matter plus fertilizers were necessary.

A summary of the amendment treatment required to establish acceptable growth on the wastes is given in Table 2.

MAINTENANCE PROBLEMS

Although the treatments described in Table 2 were successful in the short-term, some regression appeared in the second and third growing seasons of the trials. The level of deterioration was so rapid that most of the zinc waste treatments which had produced acceptable growth in the first year were clearly unacc-eptable in the second[2]. Many of the copper waste treatments and some on blast furnace slag showed a similar decline. The avail-able evidence suggested that this was caused by the gradual loss of the added organic-matter by microbial oxidation[6]. The addit-ion of NPK fertilizers rejuvenated the grass swards on blast furnace slag but effected only partial or temporary recovery of

Waste	Amendment	Amount	Effect
Zinc smelter	Sewage sludge or domestic refuse	5 - 10 cm	Complexes zinc and provides nutrient
	Lime	1,000 kg.ha^{-1}	Antagonizes zinc toxicity
	NPK fertilizer (10:10:18)	1,000 kg.ha^{-1}	Corrects NPK deficiency
Copper smelter	Sewage sludge or domestic refuse	2.5 - 5 cm	Complexes copper and provides nutrients
	Lime	1,000 kg.ha^{-1}	Corrects low pH and precipitates copper from solution
	NPK fertilizer (10:10:18)	1,000 kg.ha^{-1}	Corrects NPK deficiency
Blast furnace	Sewage sludge or domestic refuse	2.5 cm	Provides nutrients
	NPK fertilizer	1,000 kg.ha^{-1}	Corrects NPK deficiency

Data from Street & Goodman[2]

Table 2. Amendment treatments required for establishment of growth on metalliferous smelter wastes.

the copper- and zinc-waste swards[3,6]. On the latter waste, the planting of metal-tolerant grasses appeared to be more promising and these did respond to nutrients.

Further to the above work, the following lines of investigation were pursued in the present study:

1. Assessment of the effects of maintaining soil fertility by
    nutrient applications to sustain growth after different
    initial amendment treatments.

2. The importance of clover in maintaining soil nitrogen fertil-
    ity on blast furnace slag.

3. The influence of cutting treatments on yield regression.

4. The effects of soil sprinkling to introduce micro-organisms
    into the wastes.

5. The effects of layering treatments with pulverised fuel ash
    to provide waste-free rooting material for plant growth.

6. Evaluation of metal-tolerant grass swards established by
    seeding direct into untreated or minimally amended zinc and
    copper wastes. Investigations of mixed non-tolerant and
    tolerant sowings to avoid regression on the re-appearance of
    metal toxicity.

MAINTENANCE INVESTIGATIONS

Grass swards established on organically amended wastes

Experimental design. A split-plot experiment was set up on
each of the zinc, copper and blast furnace wastes in May 1965.
The waste treatments before planting were: 5 cm added sewage
sludge, 5 cm added domestic refuse, (1.3 cm sewage sludge and
domestic refuse added on blast furnace slag) and a control with
no organic material. All the main plots were treated with lime
at 1000 kg.ha^{-1} and NPK fertilizer (10:10:18) at 1000 kg.ha^{-1}
but lime was omitted on the calcareous blast furnace slag. The
main plots, each of 0.1 ha, were split to receive three seeding
types via: a grass ley (Table 3), Festuca rubra and Agrostis
tenuis all sown at 67 kg.ha^{-1} in May 1965. Full details have
been described by Gadgil[8] and summarised by Street & Goodman[2].

During the maintenance period, the following treatments were
commenced in further sub-divisions of the original split-plots:

1. Initial versus annual major nutrient additions with NPK
    fertilizer (20:10:10) at 600 kg.ha^{-1}.

2. Cutting and clearing once annually in autumn versus cutting
    and clearing twice annually in June and autumn.

3. Soiling at 4 m^3.ha^{-1} in spring 1968 to introduce micro-
    organisms.

Species	%
Lolium perenne S.23	14
Lolium multiflorum S.22	12
Dactylis glomerata S.143	50
Phleum pratense S.48	8
Festuca rubra S.59	10
Trifolium pratense (English)	6

Table 3. Composition of grass ley sown on organically amended wastes.

Results.   Serious yield regression was exhibited by all the zinc waste swards during the period 1966 to 1969 (Table 4); this occurred despite annual additions of nutrients.   In fact, further laboratory, glasshouse and field results strongly indicated that zinc toxicity was aggravated by the added fertilizer effecting a partial exchange release of unavailable zinc in the substrate. The only swards which remained visually acceptable in 1969 were those established on sewage sludge, sown to F. rubra, and treated annually with fertilizer.

In absolute terms, yield regression was increased by the addition of nutrients.   In relative terms, however, the rate of regression or decline in yield was independent of fertilization, the overall yield reduction from 1966 to 1969 being in the order of 75 - 85 per cent which was similar for all the grass types. Thus, maintaining soil fertility neither prevents nor reduces regression; it merely increases by a year or two the longevity of the swards in visual and yield acceptability terms.   Examination of the swards in 1970 showed that even the annually fertilized swards of F. rubra on sewage sludge amended waste had deteriorated below the level of visual acceptability.

On copper waste the situation was very similar (Table 5). Swards of F. rubra and grass ley both exhibited a yield decline of about 80 per cent during the period 1966 to 1969; this was again independent of soil fertilization.   There was evidence, however, that regression of A. tenuis swards was reduced in both absolute and relative terms by annual fertilization in the case of sewage sludge amended waste.   On domestic refuse, the absolute and relative yield declines of A. tenuis were less than for the other grass types but growth was much poorer initially.

| Treatments | Sward | Fertilizer | Dry Weight $(g.m^{-2})$ | | | % reduction (1966 - 1969) |
			1966	1968	1969	
Sewage sludge	F. rubra	Initial NPK	398	83	97	76
		Annual NPK	605	319	134	78
	A. tenuis	Initial NPK	358	164	84	77
		Annual NPK	482	327	110	77
	Ley	Initial NPK	575	122	91	84
		Annual NPK	563	300	127	77
Domestic refuse	F. rubra	Initial NPK	36	1	3	92
		Annual NPK	166	33	47	72
	A. tenuis	Initial NPK	5	9	7	—
		Annual NPK	135	49	46	66
	Ley	Initial NPK	83	42	15	82
		Annual NPK	287	197	77	73

Values are means of autumn yields for total annual growth

Table 4. Effects of annual nutrient additions on the growth of zinc-waste swards from 1966 to 1969.

By 1969 the copper waste swards of F. rubra and A. tenuis on domestic refuse without additional nutrients and A. tenuis on domestic refuse fertilized annually were no longer acceptable visually. The remaining swards were still acceptable but declining rapidly. The rather longer lifetime of the copper waste swards compared with those on zinc waste is probably a reflection of the relative toxicities of the two tip types, zinc waste being the more toxic. In neither case, however, does maintenance of soil fertility affect the decline of the swards to any significant extent.

On blast furnace slag the overall situation was completely different. As indicated by the data in Table 6, maintaining soil nutrient status by annual fertilization did significantly reduce

Treatments	Sward	Fertilizer	Dry weight$(g.m^{-2})$			% reduction (1966-1969)
			1966	1968	1969	
Sewage sludge	F. rubra	Initial NPK	504	231	85	83
		Annual NPK	679	322	136	80
	A. tenuis	Initial NPK	432	250	124	71
		Annual NPK	447	326	233	48
	Ley	Initial NPK	517	226	107	80
		Annual NPK	512	332	108	79
Domestic refuse	F. rubra	Initial NPK	147	115	29	80
		Annual NPK	434	310	151	65
	A. tenuis	Initial NPK	67	116	40	30
		Annual NPK	200	218	105	48
	Ley	Initial NPK	266	128	48	82
		Annual NPK	427	370	126	71

Values are means of autumn yields for total annual growth

Table 5. Effects of annual nutrient addition on the growth of copper waste swards from 1966 to 1969.

yield regression and promoted real increases in some cases.

In the absence of metal toxicity on blast furnace slag, the observed declines in yield when soil fertility was maintained (Table 6) must be due to other factors. The results are explicable in terms of nutrient release from the organic amendments during the early stages of sward development; Pitcairn[3] has pointed out that the growth responses to these amendments can be attributed to their NPK content. Clover also has an effect on soil fertility because of nitrogen fixation, accounting for the observed increases in yield of the grass ley established on waste without organic matter treatment and nutrient additions.

Examination of the effects of cutting regimes in relation to soil fertility on all types of waste (Table 7) revealed that F. rubra and A. tenuis produced higher June yields in 1969 in response to twice annual cutting during the preceding years. This effect occurred only when soil fertility was maintained by

| Treatments | Sward | Fertilizer | Dry weight (g.m^{-2}) | | | % reduction (1966-1969) |
			1966	1968	1969	
Sewage sludge	F. rubra	Initial NPK	260	72	72	73
		Annual NPK	430	354	205	52
	Ley	Initial NPK	390	185	133	66
		Annual NPK	469	364	331	29
Domestic refuse	F. rubra	Initial NPK	62	116	72	+16
		Annual NPK	130	327	234	+80
	Ley	Initial NPK	370	262	110	70
		Annual NPK	463	382	270	42
No organic matter	F. rubra	Initial NPK	1	23	7	-
		Annual NPK	77	316	200	+160
	Ley	Initial NPK	26	240	81	+212
		Annual NPK	288	333	204	29

Values are means of autumn yields for total annual growth.

A. tenuis omitted due to weed infestation and encroachment by F. rubra and ley swards.

Table 6.   Effects of annual nutrient additions on the growth of blast furnace slag swards from 1966 to 1969.

annual fertilization, presumably due to increased tillering supported by high nutrient status. In contrast, yields of the grass ley on zinc waste were depressed by increased cutting, even though nutrients were applied. This suggests that cutting may hasten the re-appearance of toxicity.

Soil sprinkling to introduce micro-organisms had no effect on growth, irrespective of soil fertility or other factors. The presence of earthworms in many of the unsoiled plots indicated that soil organisms were probably introduced naturally or via the amendment materials.

In conclusion, whilst nutrient deficiency was a major growth-limiting factor on all three organically amended wastes throughout the after-management period, responses to fertilizers

Waste	Sward	Annual NPK		Initial NPK	
		Cut once	Cut twice	Cut once	Cut twice
Zinc	F. rubra	92	140	50	38
	A. tenuis	56	84	36	26
	Ley	116	30	26	28
Copper	R. rubra	90	178	30	36
	A. tenuis	104	143	34	48
	Ley	88	112	36	46
Blast furnace slag	F. rubra	144	167	21	36
	Ley	283	170	69	88

The means were compared by Duncan's New Multiple Range Test.
Any two means underscored by the same broken line are not
significantly different at the 5% level; any two underscored by
the same continuous line are not significantly different at the
1% level.

Table 7. Effect of cutting frequency on the June 1969 yields of
swards established on organically amended wastes.

on the zinc and copper wastes became more limited with time
because of the re-appearance of metal toxicity. Thus, in 1969,
the fifth year of the trials, toxicity rather than nutrient
deficiency had become of principal importance on the non-ferrous
wastes. In contrast, nitrogen fertility alone appeared to be
growth-limiting on blast furnace slag.

Grass swards established on wastes layered with pulverised fuel
ash

Experimental design. This experiment was run concurrently

with the preceding one. Split-plots were established on each of
the three tip types. The main plots, again of 0.1 ha, were sub-
divided to give three levels of ash addition (7.5, 15.0 and 22.5
cm depths). All the plots were sown at 67 kg.ha^{-1} to a grass ley
containing 40% Lolium perenne (S.23), 30% Festuca rubra (S.59)
and 30% Trifolium repens (S.100). The split-plots were further
sub-divided as in the previous experiment to accommodate the
fertilizer, cutting and soiling treatments.

Results. As found in the organic matter experiments, severe
yield regression occurred on the zinc and copper wastes throughout the period of study (Table 8). Regression was less marked on
blast furnace slag but still of substantial magnitude. In spite
of the declining productivity, however, the swards remained
visually acceptable on all the tip types although limited areas
of chlorosis and poor growth were beginning to appear on zinc
waste amended with 7.5 cm of ash. In general, the appearance
of the swards was similar to those of F. rubra established on
organically amended wastes.

The post-June yields presented in Table 8 give a poor indication of the effect of soil fertility as influenced by nutrients
applied in the spring. Looking at the total autumn yields of the
zinc and copper waste swards with no June cutting (Table 9) it is
evident that regression was reduced or eliminated by maintaining
soil fertility on the two tip types. On copper waste 22.5 cm of
ash was needed for yields to be maintained by fertilization but
only 15 cm of ash was required on zinc waste although the latter
was the more toxic material.

Investigations on establishment and management of metal-tolerant
grass swards

Experimental design. Preliminary small-scale randomised
trials were set out on zinc and copper wastes amended with a thin
layer of pulverised fuel ash (1.75 - 2.0 cm depth) sufficient to
form a seed-bed. This was treated with a 20:10:10 NPK fertilizer
at 500 kg.ha^{-1} and three sowing treatments were examined as
follows:

1. Non-tolerant Agrostis tenuis sown densely at 72 g.m^{-2}.

2. Metal-tolerant Agrostis tenuis from mine spoil sites, sown
    sparsely at 6 g.m^{-2}.

3. Mixed sowings of non-tolerant and tolerant A. tenuis at the
    above rates respectively.

The m^2 plots were seeded in September 1967 and fertilized

Treatments	Ash depth (cm)	Fertilizer	Dry weight (g.m^{-2}) 1966	1968	1969	% reduction (1966 - 1969)
Zinc waste	7.5	Initial NPK	259	50	28	89
		Annual NPK	328	96	40	88
	15.0	Initial NPK	253	98	31	88
		Annual NPK	365	166	75	80
	22.5	Initial NPK	389	191	49	88
		Annual NPK	337	131	141	58
Copper waste	7.5	Initial NPK	356	185	29	92
		Annual NPK	397	261	55	86
	15.0	Initial NPK	400	228	26	94
		Annual NPK	501	231	56	89
	22.5	Initial NPK	563	269	73	87
		Annual NPK	493	258	125	75
Blast furnace slag	7.5	Initial NPK	155	202	86	45
		Annual NPK	263	146	84	68
	15.0	Initial NPK	250	238	87	65
		Annual NPK	344	173	124	64
	22.5	Initial NPK	232	214	62	73
		Annual NPK	127	176	78	39

Values are given only for those swards cut twice annually.

Table 8. Effects of pulverised fuel ash layering treatments and nutrients on post-June growth of grass-ley swards from 1966 to 1969.

annually in April with a 20:10:10: NKP fertilizer at 675 kg.ha^{-1}. Dry matter production was measured in autumn 1968 and 1969.

Further trials were established at a later date using larger plots, thinner layers of pulverised fuel ash, and different proportions and rates of tolerant and non-tolerant seed.

Treatments	Ash depth (cm)	Fertilizer	Dry weight(g.m^{-2})			% reduction (1966 - 1969)
			1966	1968	1969	
Zinc waste	7.5	Initial NPK	269	140	82	70
		Annual NPK	415	601	278	33
	15.0	Initial NPK	314	282	61	81
		Annual NPK	433	659	430	1
	22.5	Initial NPK	301	359	130	57
		Annual NPK	335	744	384	+16
Copper waste	7.5	Initial NPK	298	311	37	88
		Annual NPK	626	539	250	60
	15.0	Initial NPK	460	369	43	91
		Annual	577	574	231	60
	22.5	Initial NPK	535	495	112	79
		Annual NPK	661	623	662	0

Blast furnace slag yields omitted due to weed infestation.

Values are means of autumn yields with no June cutting.

Table 9. Effects of pulverised fuel ash layering treatments and nutrients on total yields of grass-ley swards from 1966 to 1969.

Results. Assessment of the preliminary trials (Table 10) showed that on both types of waste the growth of the non-tolerant swards, although satisfactory in 1968, deteriorated markedly in 1969. Conversely, the metal-tolerant swards which produced less dry matter than the non-tolerant ones in the first year improved in 1969 and greatly out-yielded them by this time.

The aim of the mixed sowings was to test the feasibility of large-scale revegetation using the minimum amount of tolerant seed. Because the latter is not commercially available in sufficient quantities, Street & Goodman[2] suggested sowing metal-tolerant seed in combination with a high proportion of non-tolerant seed of the same species. This could reduce the time

Waste	Seed	Dry wt. (g.m^{-2}) 1968	1969	Remarks
Zinc	Non-tolerant	134	22	Rapid initial development but chlorosis and dieback in 1969
	Zn-tolerant	62	349	Slow initial development but excellent growth in 1969
	Combined	96	191	Medium initial development and improving in 1969. Inferior to tolerant sowing due to swamping of seedlings in 1968
Copper	Non-tolerant	225	123	Rapid initial development but chlorosis and dieback in 1969
	Cu-tolerant	103	196	Slow initial development, improving during 1969
	Combined	359	330	Rapid initial development and good growth in 1969. Possesses attributes of both types of sowing

Table 10. Growth and assessment of small-scale trials on wastes amended with a thin layer of pulverised fuel ash and sown in September 1967 to tolerant and non-tolerant strains of _Agrostis tenuis_.

and cost of a seeds multiplication programme because, from the evidence of Bradshaw et al.[9], the non-tolerant plants would act as a "nurse crop" and provide a pollen and ovule source for hybridisation with the tolerant strain to produce metal-tolerant seedlings. The non-tolerant plants would provide most of the initial grass cover, thereby combining the rapid development characteristics of non-tolerant strains with the long-term metal tolerance of the appropriate ecotypes.

From Table 10 it can be seen that the combined sowings on zinc waste resulted in intermediate growth compared with the pure non-tolerant and tolerant sowings. On copper waste the mixed swards were superior to the non-tolerant swards in the first year and to the tolerant in the second. On both wastes, however, although flowering and seeding occurred in the first year, no new seedlings appeared in the second. Development of the swards from both the tolerant and mixed sowings was entirely vegetative. This was confirmed by further observations in 1970, the tolerant and mixed swards on both tip types being very similar with no new seedlings appearing.

The major problem involved in establishing tolerant swards by combined sowings is that the sowing rates and soil conditions must be very carefully controlled to prevent the competitive elimination of the tolerant seedlings in the early establishment phase. Such competitive effects were observed to be occurring in the zinc waste swards. To investigate this problem, further small-scale trials were conducted using reduced levels of pulverised fuel ash (1.25 cm depth) on zinc waste, ground limestone on copper waste, and different proportions of tolerant to non-tolerant seed at sparse and dense sowing rates of 6 and 30 $g.m^{-2}$ respectively.

FACTORS INFLUENCING REGRESSION

Changes in species composition

The massive declines in dry matter production of the large-scale trials, even when soil fertility was maintained, point to the re-appearance of metal toxicity as the causal factor of regression. It is possible, however, that some of the swards were adjusting to the adverse tip soil conditions by changes in species composition with concomitant reductions in yield.

To examine this possibility the grass ley or mixed swards were sampled for species composition in 1967 and 1969 using the vegetation core technique described by De Vries and De Boer[10]. On the organically amended zinc and copper wastes, Dactylis glomerata was dominant throughout the trials but declined from 77 per cent of the total dry matter in 1967 to 69 per cent in 1969. Phleum pratense increased from 7.7 per cent to 14.6 per cent. F. rubra remained unchanged with L. perenne and clover either disappearing or contributing a few per cent only. Although the measured changes did vary somewhat in relation to tip type, amendment treatments and cutting, they were always too small to account for the observed declines in yield.

On blast furnace slag, however, quite marked changes in
species composition occurred. D. glomerata decreased from 59 per
cent to 45 per cent while P. pratense increased from 10.4 to 20.2
per cent and F. rubra from 11.9 to 33.6 per cent. Clover either
declined or disappeared, persisting only in those plots receiving
no annual nutrient additions. L. perenne had vanished by 1969.
The most marked change was the increase in F. rubra, probably on
account of its tolerance of high pH[11], the increasing dominance
of this species in the earlier randomised block trials confirming
that this was a significant long-term trend. The fact that the
large-scale F. rubra swards were always outyielded by the grass
leys is good evidence that increasing F. rubra composition was
contributing to the observed yield declines. Thus ecological
adjustment was occurring on blast furnace slag but not on the
zinc and copper wastes.

Loss of added organic-matter

Returning to the problem of regression on the zinc and copper
wastes, the re-appearance of metal toxicity suggests that the
initially applied organic materials were losing their ability to
complex metal ions in the soil solution. Determinations of the
residual organic matter in 1968 of earlier trials established in
1963 showed that only 2.4 per cent remained in zinc waste whereas
a 10 cm depth had been originally applied. The amount remaining
depended on whether fertilizers had been applied, the addition of
nutrients lowering the residual organic matter from 3.2 to 1.5
per cent. The situation was similar on copper waste, accounting
for the fact that annual fertilization appeared to hasten regression in some cases.

In the large-scale zinc waste trials it was found that by
1968 only 11.2 per cent of organic matter remained following initial treatment with sewage sludge, the figure being 4.2 per cent
for domestic refuse. This accounts for the superiority of sewage
sludge over domestic refuse in terms of yield production. Examination of the levels of plant-available zinc in the waste as
indicated by acetic acid extractions confirmed that sewage sludge
was more effective than domestic refuse in lowering the soil
concentration; the values were 6590 ppm for the sewage sludge
plots, 1370 ppm for the domestic refuse plots and 16750 ppm for
the unamended plots. Similar differences with respect to E.D.T.A-
extractable copper levels were found in the amended copper waste.

Although addition of pulverised fuel ash was designed to
provide a waste-free rooting medium for growth, analyses of the
surface layers of ash in 1968 showed that heavy metal contamination from the waste beneath had occurred. Whereas fresh pulver-

ised fuel ash contained only 80 ppm plant-available zinc, the surface ash from a 22.5 cm covering over zinc waste contained 4 950 ppm acetic acid extractable zinc in 1968 with considerably elevated copper and lead levels. Pulverised fuel ash overlying copper waste showed a similar but less dramatic increase with respect to E.D.T.A.-extractable copper, rising from 12 to 28 ppm. The appearance of localised bare patches and symptoms of metal toxicity on zinc waste amended with ash, but not on copper waste, indicates that the zinc waste poses greater metalliferous ion mobility problems. The evidence suggests that serious regression and dieback of the pulverised fuel ash swards may occur with time; it was observed in 1971 that complete dieback of the zinc waste swards did occur.

Although the dry matter production data presented gives a good indication of long-term yield performance, it does not show whether growth is acceptable visually or not. Observations indicated that regression first appeared as localised areas of leaf chlorosis and dieback. To obtain a quantitative estimate of sward uniformity the coefficients of variation of dry matter yield were calculated from the 1969 yield data. These show that on zinc waste (Table 11) growth was extremely variable except for swards established on 15.0 cm pulverised fuel ash which was the most successful amendment. On copper waste growth was generally more uniform and the blast furnace slag swards all exhibited less than 30 per cent variability. Field observations indicated that those swards exhibiting more than 40 per cent variability were visually unacceptable and regressing seriously because of toxicity. It should be pointed out here that regression caused only by nutrient deficiency, although depressing yields, does not result in dieback and therefore has little or no effect on variability of growth.

CONCLUSIONS

There are clearly two possible approaches to the reclamation of toxic metalliferous smelter wastes. On the one hand there is the use of standard and commercially available grass types planted on organically amended or covered wastes. The alternative is to sow metal-tolerant ecotypes on minimally amended wastes.

Steps must be taken to maintain a high level of organic matter by periodic additions of domestic refuse or sewage sludge or by allowing grass cuttings to rot in situ (they were always removed during the experiments to provide conditions most favourable for rapid regression). Otherwise, the use of standard grass seeds mixtures combined with amendments can achieve only a temporary cover on the non-ferrous wastes. Under the conditions of these experiments, fertilization and cutting treatments will

Waste	Treatment	Coeff. variation (%)	
		Festuca rubra	Ley
Zinc	5.0 cm domestic refuse	76	82
	5.0 cm sewage sludge	41	67
	7.5 cm pulverized fuel ash	-	40
	15.0 cm pulverized fuel ash	-	28
Copper	5.0 cm domestic refuse	43	24
	5.0 cm sewage sludge	31	29
	7.5 cm pulverized fuel ash	-	21
	15.0 cm pulverized fuel ash	-	16
Blast furnace slag	NPK only	28	23
	1.3 cm domestic refuse	27	24
	1.3 cm sewage sludge	26	21
	7.5 cm pulverized fuel ash	-	17
	15.0 cm pulverized fuel ash	-	17

Table 11. Coefficients of variation of dry matter yield $(g.m^{-2})$ in 1969 of grass swards established on wastes treated annually with fertilizer.

maintain acceptable growth for about five years on zinc waste and a few years longer on copper waste whereas growth on blast furnace slag is permanent if maintained by fertilization or the presence of nitrogen-fixing species. These methods will be suitable, however, if only a temporary grass cover is required to green-up the landscape and stimulate re-development.

If long-term grass cover is required, the non-ferrous wastes must be planted to metal-tolerant grass swards. If combined sowings of non-tolerant and metal-tolerant seeds are used, a seeds multiplication programme could be reduced to one year. Subsequent maintenance of the swards would involve standard fertilization and cutting treatments only.

ACKNOWLEDGEMENTS

We thank the Natural Environment Research Council who financed these investigations. Sewage sludge was provided by Llanelli and Neath Rural District Councils, domestic refuse by Swansea City Corporation, and pulverised fuel ash by the Central Electricity Generating Board. We are also grateful to Mr George Cohen (600 Group), Imperial Chemical Industries, and Swansea City Corporation for allowing trials to be conducted on their lands. The International Voluntary Service and Territorial Army provided valuable assistance with site preparation and maintenance.

REFERENCES

1. Goodman, G.T. & Bray, S.M., Ecological Aspects of the Reclamation of Derelict and Disturbed Land, Geo. Abstracts, Norwich, 1975.

2. Street, H.E. & Goodman, G.T., Techniques of revegetation in the Lower Swansea Valley, In The Lower Swansea Valley Project, (ed. K.J. Hilton), Longmans Green, London, 1967.

3. Pitcairn, C.E.R., An ecological study of the factors influencing revegetation of industrial waste heaps contaminated with heavy metals, Ph.D. thesis, Univ. Wales, 1969.

4. Lucas, R.E., Chemical and physical behaviour of copper in organic soils, Soil Sci., 66, 119, 1948.

5. Jensen, H.L. & Lamm, C.G., On the zinc content of Danish soils, Acta Agric. scand., 11, 63, 1961.

6. Goodman, G.T., Pitcairn, C.E.R. & Gemmell, R.P., Ecological factors affecting growth on sites contaminated with heavy metals, In Ecology and Reclamation of Devastated Land, (ed. R.J. Hutnik & G. Davis), Gordon & Breach, London, 1973.

7. Gadgil, R.L., Tolerance of heavy metals and the reclamation of industrial waste, J. appl. Ecol., 6, 247, 1969.

8. Gadgil, R.L., The plant ecology of the Lower Swansea Valley, University College of Swansea, Lower Swansea Valley Project, Study Rep. 9, 1965.

9. Bradshaw, A.D., McNeilly, T.S. & Gregory, R.P.G., Industrialisation, evolution, and the development of heavy metal tolerance in plants, In Ecology and the Industrial Society, (ed. G.T. Goodman, R.W. Edwards & J.M. Lambert), Blackwell, Oxford, 1965.

10. De Vries, D.M. & De Boer, T.A., Methods used in botanical grassland research in the Netherlands and their application, <u>Herb. Abstr.</u>, 29, 1, 1959

11. Kruijne, A.A. & De Vries, D.M., Data concerning herbage plants, <u>Inst. biol. & chem. Res. on Field Crops & Herbage</u>, Wageningen, 1963.

INDEX

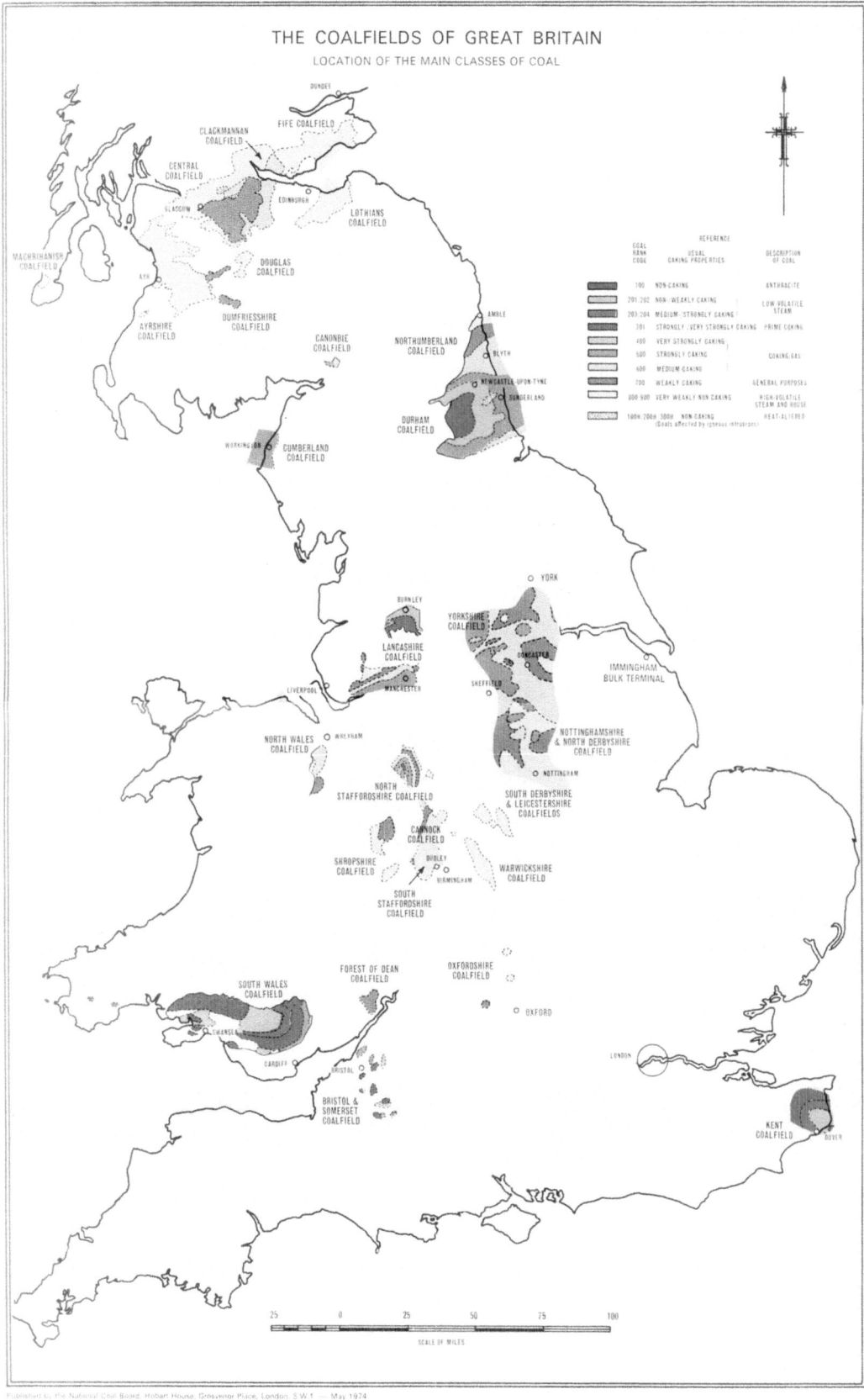

THE COALFIELDS OF GREAT BRITAIN
LOCATION OF THE MAIN CLASSES OF COAL

Published by The National Coal Board, Hobart House, Grosvenor Place, London, S.W.1 — May 1974